CAMBRIDGE STUDIES IN
ADVANCED MATHEMATICS 17

Groups acting on graphs

Already published

1 W.M.L. Holcombe *Algebraic automata theory*
2 Karl Petersen *Ergodic theory*
3 Peter T. Johnstone *Stone spaces*
4 W.H. Schikhof *Ultrametric calculus*
5 J-P. Kahane *Some random series of functions*
6 H. Cohn *Introduction to the construction of class fields*
7 J. Lambek and P.J. Scott *Introduction to higher order categorical logic*
8 H. Matsumura *Commutative ring theory*
9 C.B. Thomas *Characteristic classes and the cohomology of finite groups*
10 M. Aschbacher *Finite group theory*
11 J.L. Alperin *Local representation theory*
12 Paul Koosis *The logarithmic integral: 1*
13 A. Pietsch *Eigenvalues and s-numbers*
14 S.J. Patterson *An introduction to the theory of the Riemann zeta-function*
15 H-J. Baues *Algebraic homotopy*
17 Warren Dicks and M.J. Dunwoody *Groups acting on graphs*

Groups acting on graphs

WARREN DICKS
Professor in mathematics, Universitat Autònoma de Barcelona

M.J. DUNWOODY
Reader in mathematics, University of Sussex

CAMBRIDGE UNIVERSITY PRESS
Cambridge
New York New Rochelle Melbourne Sydney

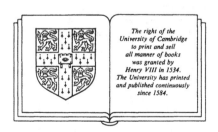

CAMBRIDGE UNIVERSITY PRESS
Cambridge, New York, Melbourne, Madrid, Cape Town, Singapore,
São Paulo, Delhi, Dubai, Tokyo, Mexico City

Cambridge University Press
The Edinburgh Building, Cambridge CB2 8RU, UK

Published in the United States of America by Cambridge University Press, New York

www.cambridge.org
Information on this title: www.cambridge.org/9780521180009

First published 1989
First paperback edition 2010

A catalogue record for this publication is available from the British Library

Library of Congress Cataloguing in Publication data

Dicks, Warren, 1947–
Groups acting on graphs/Warren Dicks and M.J. Dunwoody.
p. cm. – – (Cambridge studies in advanced mathematics; 17)
Bibliography: p.
Includes indexes.
1. Groups, Theory of. 2. Graph theory. I. Dunwoody, M.J.
II. Title. III. Series.
QA171.D47 1989 88–3661 CIP

ISBN 978-0-521-23033-9 Hardback
ISBN 978-0-521-18000-9 Paperback

To Alan, Elena and Xaro
To Jean, Luke and Megan

Contents

Preface

The simplest instance of the interplay between group theory and topology occurs where a group acts on a graph and information is obtained about the group or about the graph; it is these occurrences which form the theme of this book.

Chapter I offers a review of the Bass–Serre theory of groups acting on trees and graphs, with some typical combinatorial group theoretic applications. For the sake of novelty, we have included a very recent result from the literature on the fixed group of an automorphism of a free group. Although we have attempted to make the account self-contained, it is rather brusque for initiation purposes, and the reader should ideally already have some familiarity with group theory, group actions, presentations and combinatorial theory.

Chapters II, III and IV are essentially new.

Chapter II, using Boolean rings, associates to each connected graph and positive integer n, a tree which explains how the graph disconnects when any n edges are deleted. One application is the recent result from the literature characterizing infinite finite-valency distance-transitive graphs. This chapter is elementary in the sense that no background material is assumed.

Chapter III is devoted to drawing lines joining up functions in an equivariant way to get a previously unsuspected tree. The argument is technical and elementary. The result has some rather pleasing applications, which are collected together in Chapter IV. New results include the proof of a conjecture of Wall, and a characterization of arbitrary groups with more than one end; previously known results are the characterization of groups of cohomological dimension at most one over an arbitrary ring, and the characterization of groups which have a free subgroup of finite index. The reader is assumed to be familiar with module theory and exact sequences; a Sylow theorem is used in a remark; cohomology is introduced in a mild way, since the results can be phrased in terms of derivations to projective modules.

Chapters V and VI examine dimensions two and three, and consist of results which, having only recently appeared in the literature, are appearing in book form for the first time.

Chapter V is an algebraic account of the cohomological characterization of infinite surface groups as the groups which satisfy two-dimensional Poincaré duality. The cohomology and topology are more sophisticated than in the rest of the book. The reader is assumed to be familiar with the necessary homological algebra, which is quickly summarized without many proofs. The reader familiar with the topology of surfaces, or manifolds in general, will be able to appreciate the motivation behind the entire chapter; the reader without the background in topology will have to be sufficiently algebraically inclined to be motivated by the result in its own right.

Chapter VI examines groups acting on two-complexes and deduces that almost finitely presented groups are accessible in the sense of Wall. It concludes with a similar analysis of three-manifolds and deduces the equivariant loop and sphere theorems. Here the topological background is summarized without proofs.

There are no exercises, apart from four open conjectures and the occasional tedious argument left to the reader.

Each chapter concludes with some notes and comments citing our sources for results and ideas. The sources, which are listed in the absolutely minimal bibliography and author index, tend not to be primary, and our attributions should be taken lightly, especially by authors who have been omitted.

We are indebted to the many mathematicians who have made helpful comments and devoted much time and effort to helping us understand the literature; their sole reward is the knowledge that the book would have been even worse without their help.

We thank Ed Formanek and Peter Linnell for generously contributing unpublished results and arguments.

The first-named author thanks the Mathematics Department of Pennsylvania State University for providing a graduate course forum to air and develop some of the results, and the CRM in Barcelona for support and gracious hospitality during the summer of 1985. The second-named author thanks the Royal Society and the IHES for support and gracious hospitality during the first six months of 1983.

Warren Dicks
Barcelona, Spain
M.J. Dunwoody
Sussex, England

WARREN DICKS AND M. J. DUNWOODY,

GROUPS ACTING ON GRAPHS, CAMBRIDGE STUDIES IN ADVANCED MATHEMATICS 17, CAMBRIDGE UNIVERSITY PRESS, 1989.

ERRATA

(JANUARY 31, 2008)

PLACE	**CHANGE**
5^4	Change "(e)" to "(v)" twice.
6^7	Interchange labels "$srse$" and "$srsre$".
9^{1-2}	Change to "repetitions of vertices and no repetitions of edges. Clearly such a path is reduced.".
11_8	Change "EX" to "VX".
11_5	Change boldface subscript "ι" to ordinary subscript "ι".
11_5	Change "*and*" to "and".
12^6	Change "elements 1,1" to "element 1".
13^{11}	Change "$t_e(G(e)$" to "$G(e)^{t_e}$".
17^{13}	Delete "so $G/N \approx \pi(G\backslash T)$".
18_{11}	Change "$EX^{\pm 1}$" to "$ET^{\pm 1}$".
27^1	Interchange "$G_{\tau e} = G_{\{p,p'\}}$" and "$G_e$".
29_5	Interchange "V" and "E".
32_{12}	Change "u_1" to "g_1".
33^{13}	Change "$\alpha(e)(\alpha_{\tau e}(g^{t_e})u)$" to "$\alpha(e)(\alpha_{\bar{\tau} e}(g^{t_e})(u))$".
35_9	Change "$\underset{v\in V}{*} G(v)$" to "$G = \underset{v \in V}{*} G(v)$".
39^{10}	Change "69" to "71".
39_{10}	Change "$e = EY_0$" to "$e \in EY_0$".
40^7	Delete one "meet any".
41_{17}	Change "thdn" to "then".
41_{19}	Change "$g \cdots$" to "$g_2 \cdots$".
44	Stallings (1991) uses similar techniques to prove many other results.
45^1	Change "G" to "$*_{i\in I} G_i$".
45^7	Insert Corollary: If $\alpha : F \to G$ is a homomorphism of free groups then there exist subgroups F_1 and F_2 of F such that $F = F_1 * F_2$, and α is injective on F_1, and α is trivial on F_2. (Herbert Federer and Bjarni Jónsson, Some properties of free groups, Trans. Amer. Math. Soc., 63 (1950), 1-27.)
45_{10}	Change "take" to "takes".
45_3	Change "Gerardin" to "Gérardin".
$46_{6,5}$	Delete "from ... Brown".
50_{10}	Change "(v, v')" to "(v', v)".
53^5	Change "$= v(e^*)$" to "$= v(e)^*$".

PLACE	CHANGE
54_{18}	Change “$s(v) = s(w)$ then $e(v) = e(w)$ for all $e \in E$” to “$e(v) = e(w)$ for all $e \in E$ then $s(v) = s(w)$”.
54_{3}	After “$\delta(ss') \subseteq \delta s$” insert “ $\cup\, \delta s'$”.
55^{2}	Change “$s'(v)$” to “$s(v')$”.
55_{11}	Add “and not containing 0,1” after “E_{n-1}”.
$57_{13,3}$	Change “$E' \cup E$” to “$E \cup E'$”.
$61^{17,23}$	Change “$\iota s, \tau s = \tau s{*}, \iota s{*}$” to “$\iota s = \iota s^{*}, \tau s, \tau s^{*}$” twice.
71_{10}	Change“Fredenthal” to “Freudenthal”.
77^{18}	Change“*and*” to “and”.
83^{10}	Change “S'” to“E'”.
83^{11}	Change “{e}” to “$\{e\}$”.
92^{16}	Delete “$\tilde{T}_1$ to a path of”.
99_{7}	Change “succesor” to “successor”.
100_{11}	Change “*H*-*H*” to “*G*-*H*”.
101_{3}	Change “amost” to “almost”.
102_{5}	Change “$<$” to “$\leq$”.
103^{9-13}	Change to “If $(p, q, r) = (2, 3, 6)$ then *acb*, *bac* generate a free abelian subgroup of rank two and index 6; see Magnus (1974),p.69.”
105^{19}	Bass (1993) gives a (short) proof that G has a free subgroup of index n.
107_{9}	Change “G indexed by A” to “A indexed by G”.
107_{5}	Change“of AG” to “of (G, A)”.
115^{14}	Change “Theorem 3.1” to “Theorem 3.13”.
134_{9}	Change “(1968)” to “(1971)”.
132_{14}	Change “B” to“R”.
134_{5}	Change to “Theorem 6.12 is due to Hopf for G finitely generated, and the general case is new”.
134_{3}	Change “Theorem 4.12” to “Theorem 4.11”.
136^{9}	Change the first “P” to “G”.
136^{10}	Change “H^{*}” to “H_{*}” and “H^{i}” to “H_{i}”.
134^{2}	After “Brown(1982)” add “and Zimmerman (1981)”.
136^{12}	Change “H_{*}” to “H^{*}” and “H_{i}” to “H^{i}”.
140_{2}	Change “he” to “the”.
142_{13}	Change “P” to “Q” and “P'” to “Q'”.
142_{12}	Change “P” to “Q”.
142_{11}	Change “$P \to P'$” to “$Q \to Q'$”.
142_{10}	Change “$P \to P'$” to “$Q \to Q'$”.
142_{7}	Change “Q” to “B” twice.
143^{11-13}	Change “P” to “Q”, “P'” to “Q'”,“P''” to “Q''”, twice each.
143_{3}	Change “moduls” to “modules”.
$144_{3,2}$	Change the seven occurrences of “P” to “Q”.
$146^{2,3}$	Change “$\partial(p \otimes q) = \partial_P p \otimes q + (-1)^{\deg q} p \otimes \partial_q q$” to “$\partial(p \otimes q) = (-1)^{\deg q} \partial_P p \otimes q + p \otimes \partial_q q$”.

PLACE	CHANGE
146_{10-4}	Should read "$(\partial_{P\otimes Q}x)\cap\phi$ $= [(-1)^{\deg\, q}\partial_P p\otimes q + p\otimes\partial_Q q]\cap\phi$ $= (-1)^{\deg\, q}\partial_P p\otimes\phi q + p\otimes\phi\partial_Q q$ $= (-1)^{\deg\, q}\partial_{P\otimes C}(p\otimes\phi q) + [(p\otimes q)\cap\phi\partial_Q]$ $= (-1)^{\deg\, q}\partial_{P\otimes C}[(p\otimes q)\cap\phi] + [(p\otimes q)\cap\partial_{\mathrm{Hom}(Q,C)}\phi]$ $= (-1)^{\deg\, q}\partial_{P\otimes C}(x\cap\phi) + (x\cap\partial_{\mathrm{Hom}(Q,C)}\phi)$ If, further, ϕ is homogeneous, then either $\partial_{P\otimes C}(x\cap\phi) = 0$ or $\deg\phi = -\deg q$, and in both cases we can write (4) $\quad(\partial_{P\otimes Q}x)\cap\phi = ((-1)^{\deg\,\phi}\partial_{P\otimes C}(x\cap\phi)) + x\cap\partial_{\mathrm{Hom}(Q,C)}\phi$".
148^{13}	In 2.16 Proposition, in the display change "$\mathrm{Ext}_R(B,C)$" to "$\mathrm{Ext}_R^n(B,C)$" twice, and after the display change "commutes" to "commutes with sign $(-1)^n$".
148_9	In 2.17 Proposition, in the display change "$\mathrm{Ext}_R^n(B',C)$" to "$\mathrm{Ext}_R(B',C)$", and change"$\mathrm{Ext}_R^{n+1}(B'',C)$" to "$\mathrm{Ext}_R(B'',C)$", and after the display delete "with sign $(-1)^{n+1}$".
148_5	Interchange "η" and "ξ".
149^{10}	Change "$\partial_{P\otimes C}(x\cap\phi) - (-1)^{\deg\phi}(x\cap\partial_{\mathrm{Hom}(Q,C)}\phi$" to "$((-1)^{\deg\phi}\partial_{P\otimes C}(x\cap\phi)) + x\cap\partial_{\mathrm{Hom}(Q,C)}\phi$".
$149^{11,12}$	Delete "with sign $-(-1)^{\deg\phi} = (-1)^{n+1}$".
149^{15}	In 2.18 Proposition, in the display change "Ext_R^{n-1}" to "Ext_R^{n+1}".
149^{17}	In 2.18 Proposition, after the display change "commutes with sign $(-1)^n$" to "commutes with sign $(-1)^{n+1}$".
154^{16}	Change the first "is" to "in".
155^1	In top display change "$(-1)^{n+1}\xi\cap -$" to "$\xi\cap -$".
155_1	Change "exact at R" to "exact at RG".
156^5	Change "Theorem I.9.2" to "Corollary I.9.4".
158_4	Change "contractible n-manifold X" to "K-orientable K-acyclic K-homology n-manifold X, as defined in Section 3 of Dicks-Leary (1995)".
$163^{5,7}$	Change "$[1,R_p]$" to "$[2,R_p]$" twice.
163^{13}	Add "and $m_{p,1}$ denotes 1".
170_{13}	Change "Thus H is FP_∞" to "Thus G is FP_∞".
171^{17-19}	Change " notice ... ." to "notice that the G-action arises by embedding G in $H\wr\mathrm{Sym}_n$ and defining actions of H^n and Sym_n separately.".
173^9	Delete "G-finite" after "locally finite".

PLACE **CHANGE**

173_{20-4} Replace with

"Fix a vertex v_0 of Y_0.

Consider any $w \in V_0$, and recursively construct an infinite reduced path P_w as follows. Start with the vertex w, thought of as a base vertex, and take the neighbours of w to be the base vertices of their respective components in the forest $Y_0 - \text{star}(w)$. Since $Y_0 - \text{star}(w)$ is infinite and has only finitely many components, one of the components is infinite. Choose one of these infinite components, and if there are more than one, choose one which does not contain v_0. This choice of infinite subtree corresponds to choosing an edge incident to w to be the first edge in our infinite path. We now repeat the same procedure with our chosen infinite subtree with base vertex. In this way, we recursively construct an infinite reduced path P_w which starts at w, and does not contain any edge f such that w lies in an infinite component of $Y_0 - \{f\}$ not containing v_0.

Let e be an edge of Y_0. Let $Y_0(v_0, e)$ denote the Y_0-geodesic from v_0 to a vertex of e, but not passing through e. Let $\delta Y_0(v_0, e)$ denote the finite set of edges of Y_0 which have one vertex in $\delta Y_0(v_0, e)$ and the other vertex not in $Y_0(v_0, e)$. Thus e lies in $\delta Y_0(v_0, e)$, and $Y_0(v_0, e)$ forms one of the finite components of $Y_0 - \delta Y_0(v_0, e)$. Let Y_e denote the the subtree of Y_0 generated by the finitely many finite components of $Y_0 - \delta Y_0(v_0, e)$, so Y_e is finite.

For $e \in EY_0$, $w \in VY_0$, we claim that if $e \in P_w$ then $w \in Y_e$. Suppose that w does not lie in Y_e, so w lies in an infinite component Y_1 of $Y_0 - \delta Y_0(v_0, e)$. Let f denote the element of $\delta Y_0(v_0, e)$ incident to Y_1. Then f lies between w and $Y_0(v_0, e)$. Hence f lies between w and v_0. Also, Y_1 is an infinite component of $Y_0 - \{f\}$ containing w, so by its construction, P_w stays in Y_1 and does not cross f. Hence P_w does not meet $Y_0(v_0, e)$, so does not contain e. This proves the claim.

For any v in V, there is a unique element g of G such that $gv \in V_0$, because G acts freely on V, and we define $P_v = g^{-1} P_{gv}$. Thus P_v is an infinite reduced path in T which begins at v.

Consider any edge e of T. We claim that there are only finitely many $v \in V$ such that e belongs to P_v. Suppose then that $v \in V$ such that $e \in P_v$. There is a unique g in G such that $gv \in Y_0$, and then $ge \in gP_v = P_{gv}$. Hence ge lies in Y_0, and gv lies in the finite subtree Y_{ge} of Y_0. Here g is the unique element of G such that $ge \in EY_0$, and we have $v \in g^{-1} Y_{ge}$, so there are only finitely many possibilities for v, as desired.".

174_3 Change "Thus we may assume that $n \geq 1$." to "If $n = 1$ then G has an infinite cyclic subgroup of finite index by Theorem 4.4, and this case is easy. Thus we may assume that $n \geq 2$.".

176_5 Change "K^k" to "H^k.

176_2 Change "Thus in" to "Now let (G, W) be a PD^n pair, so, by ".

177_{11} Change whole line to"K-orientable K-acyclic K-manifold X of dimension n,

PLACE	CHANGE
	whose boundary components are K-acyclic".
177_6	Change "ξ°" to "ε°".
178^{20}	Change "Definitions" to "Definition".
183_9	There is a vertical arrow missing on the left of the diagram (1).
185_{12}	In the display that comes two before (7), in the top row, change "$K\omega KE$" to "ωKE".
185_{11}	In the display that comes two before (7), in the label on the rightmost vertical arrow, delete "$(-1)^n$".
185_7	In the display that comes before (7), in the label on the rightmost vertical arrow, delete "$(-1)^n$".
186^{15}	In the display in mid-page, change the two rightmost "$\xi \cap -$" to "$-\xi \cap -$".
186_{12}	In (8) change "$\xi \cap \eta_e$" to "$-\xi \cap \eta_e$".
198^7	Change "$W - Gw$" to "$W - Gw_0$".
198_{15}	Insert "$ET = Ge$" after "G_e is finite".
199^{14}	Insert "$ET = Ge$" after "G_e is finite".
202_{10}	Change "$\sum\limits_{j\in[1,N]}$" to "$\sum\limits_{i,j\in[1,N]}$".
203^{10}	Change "$\lVert a\rVert\,\lVert b\rVert \geq \lvert\mathrm{tr}\,(\bar{b})\rvert$" to "$\lVert a\rVert\,\lVert b\rVert \geq \lvert\mathrm{tr}\,(a\bar{b})\rvert$".
202_4	Change "then – induces" to "then $\overline{}$ induces".
203_8	Change to "(1) $\lVert a_n e\rVert^2 = \mathrm{tr}(a_n e\overline{a_n e}) = \mathrm{tr}(a_n e\bar{e}\,\overline{a_n}) = \mathrm{tr}(a_n c\overline{a_n}) = \mathrm{tr}(a_n \bar{c}\,\overline{a_n})$.".
$203_{7,6}$	Change "$\mathrm{tr}(a_n\overline{ca_n})$" to "$\mathrm{tr}(a_n\bar{c}\,\overline{a_n})$" twice.
205^8	Change "$P \mapsto KG \otimes_K P$" to "$P \to KG \otimes_K P$".
205^{14}	Change "for all $j \in [1, m]$" to "for all $i \in [1, m]$".
205_{12}	Change "$\sum\limits_{i,g\in G}$" to "$\sum\limits_{i,g}$".
$206_{12,13}$	Change "$[w_1 \cdots w_q] = [w_2 \cdots w_q w_1]$" to "$\mathrm{Tr}(w_1 \cdots w_q) = \mathrm{Tr}(w_2 \cdots w_q w_1)$".
207^4	Before "invertible" insert "is".
208_{14}	Change "$A *_C X_0$" to "$A *_C x_0$".
209_2	Change "V-term" to "E-term".
211_1	Change "$\to {}_0$" to "$\to P_0$".
212_{22-21}	Change "$= K$ is right annihilated by ωKG" to "$= K = KG/\omega KG$".
212_{20}	Change "$\alpha^*(P^*) \subseteq \omega KG$" to "$\alpha^*(P^*) = \omega KG$".
219^5	After "if" insert "and only if".
219^{15}	Change "$Z_0(K, G)$" to "$\mathrm{Hom}(C_0(K), G)$".
220_4	Change "$\mathbb{Z} \otimes \mathbb{Z}$" to "$\mathbb{Z} \times \mathbb{Z}$".
222_{15}	Change "G" to "K".
224^5	Change "s" to "σ".
224_{14}	One can change "$P \cap \lvert K\rvert$" to "P", since $P \subseteq \lvert K\rvert$.
224_7	Change "$j(\gamma_i)$" to "$j_P(\gamma_i)$".
225^{11}	Change "h^1" to "h_1".
225_2	After "colouring" insert "with two colours".
229_3	Change "X" to "S".
231^5	Change the second "v_1" to "v_2".

PLACE	CHANGE
232^{15}	Change "ET" to "VT".
232^{16}	Change the second "ET" to "T".
236^{9}	At the end of the line add "Moreover it follows from the thinness of b_2^* or b_1^* that $\nu = \delta$."
240_{13-12}	Change "G, the automorphism group of K, is generated" to "G is the group of automorphisms of K generated".
245^{1-3}	Delete "$H^1(K, \mathbb{Z}_2)$ ... that".
245_{3}	Change "$H^1(K, \mathbb{Z}_2) = 0$" to "every scc separates M".
272^{7}	Insert in left hand column: "Bass, H. {45, 46, 71} 1993. Covering theory for graphs of groups, *J. Pure and Appl. Algebra* **89**, 3–47. {105≈} ."
272^{7}	In right hand column change the Burns entry to "Burns, R.G. 1971. On the intersection of finitely generated subgroups of a free group, *Math. Z.* **119**, 121–130. {39} "
273_{6}	In right hand column change "Gerardin" to "Gérardin".
273_{5}	In left hand column change **"15" to "25".**
273^{16-18}	In right hand column delete the entry.
273^{25}	In right hand column change "Normal Flächen" to "Normalflächen".
273_{4}	In right hand column change "Raüme" to "Räume".
274_{22}	Delete from left hand column "134,".
274^{20-45}	In the right hand column, interchange lines 20-32 with lines 33-45, to obtain alphabetic order.
274^{27}	In left hand column change "dreidemensionalen" to "dreidimensionalen".
274^{25}	In right hand column change "isomorphismen" to "Isomorphismen".
275^{11}	Insert in right hand column 1991. Foldings of G-trees, pp. 355-368 in *Arboreal Group Theory* (Roger C. Alperin, Editor), MSRI Publications 19, Springer-Verlag, Berlin, 1991. {44≈}
275_{7}	Change "Räume" to "Räumen".
276^{3}	Insert in left hand column "(-)′".
274_{7}	Insert in left hand column: "Magnus, W. *Noneuclidean Tesselations and their Groups*, Academic Press, New York, 1974. {103}"
275^{10}	In the right hand column change " {71, 100} " to " {71, 100, 134}".
275_{1}	In the right hand column add "Zimmerman, B., 1981. Über Homeömorphismen n-dimensionaler Henkelkörper und endliche Erweiterungen von Schottky-Gruppen, *Comm. Math. Helv.* **56**, 474–481. {134}"
276_{12}	The triangle in the right hand column should be unshaded.

Conventions

G denotes a group, fixed throughout the book.

$\varnothing$ denotes the empty set.

Sets are indicated by $\{x | x \cdots\}$ or sometimes $\{x : x \cdots\}$ for typographical reasons.

$B \subseteq A$ means B is a subset of A.

$B \subset A$ means B is a *proper* subset of A, that is, distinct from A.

If $B \subseteq A$ then $A - B$ denotes the complement of B in A.

$A \cup B$, $A \vee B$, $A \cap B$, $A \times B$, respectively, denote the union, the disjoint union, the intersection and the Cartesian product of two sets, A, B.

$\bigcup_{i \in I} A_i$, $\bigvee_{i \in I} A_i$, $\bigcap_{i \in I} A_i$, $\prod_{i \in I} A_i$, respectively, denote the union, the disjoint union, the intersection and the Cartesian product of a family of sets A_i indexed by the elements i of a set I.

A^n denotes the Cartesian product of copies of a set A indexed by a non-negative integer n, and the elements are written as n-tuples $(a_1, \ldots, a_n)$.

$|A|$ denotes the cardinal of a set A.

If m, n are integers then $[m, n]$ denotes the set of integers i such that $m \leqslant i \leqslant n$.

If α, γ are ordinals then $[\alpha, \gamma]$ and $[\alpha, \gamma)$ respectively denote the set of ordinals β with $\alpha \leqslant \beta \leqslant \gamma$ and $\alpha \leqslant \beta < \gamma$.

If I is a set and m_i, $i \in I$, are cardinals and m is a cardinal then $m = \mathrm{HCF}_{i \in I} m_i$ means that m is the largest cardinal which divides all the m_i. In practice, m is an integer, or equivalently some m_i is an integer.

$\mathbb{N}, \mathbb{Z}^+, \mathbb{Z}, \mathbb{Q}, \mathbb{R}, \mathbb{C}, \mathbb{R}^n$, respectively, denote the positive integers, the non-negative integers, the integers, the rationals, the reals, the complex numbers, and Euclidean n-space.

$\mathbb{Z}_2$ denotes the set consisting of two elements 0 and 1; it performs as a set, a group, a ring, a Boolean ring, a field, and a discrete topological space.

Except where otherwise indicated, functions will be written on the left of the argument, and composed accordingly. We write $\alpha : X \to Y$ or $X \xrightarrow{\alpha} Y$ to denote a function, and $x \mapsto \alpha x$ to denote its action on elements. Here $\alpha^{-1}(y) = \{x \in X \mid \alpha x = y\}$ for any $y \in Y$.

Except where otherwise specified, groups will be written multiplicatively, and abelian groups will be written additively. If $x, y \in G$ then $[x, y]$ denotes the *commutator* $x^{-1}y^{-1}xy$; this should not be confused with the above interval notation.

$H \leqslant G$ means that H is a subgroup of G.

Rings are associative and have a 1; in all situations of interest, the 0 and 1 are distinct.

(Left or right) module actions respect the 1 of the ring.

$M \oplus N$ denotes the direct sum of two modules, and $\bigoplus_{i \in I} M_i$ denotes the direct sum of a family of modules M_i indexed by the elements i of a set I.

If R is a ring, M a right R-module, and N a left R-module, then $M \otimes_R N$ denotes the tensor product, viewed as an abelian group.

The numbering treats theorems, definitions, examples, remarks, etc. as subsections, labelled, for example, as 2.9 Remarks, in Section IV.2 in Chapter IV, and referred to as Remark 2.9 within Chapter IV, and as Remark IV.2.9 within all other chapters. The end of such a subsection is indicated by ▮.

References to the bibliography are by the author–date system, with primes to distinguish publications by the same author in the same year.

I

Groups and graphs

Sections 1 and 2 collect together the basic definitions on group actions and graphs, and Section 3 introduces the concept of a graph of groups. Section 4 then describes the structure of a group acting on a tree in terms of the fundamental group of a graph of groups. Section 5 lists some examples of trees arising in nature. Section 6 motivates the main argument of Section 7, which shows the converse of the structure theorem for groups acting on trees, that is, the fundamental group of a graph of groups acts on a tree; some applications in combinatorial group theory are then given. This is continued in Sections 8 and 10, where some important theorems on free groups and free products are proved, while Section 9 gives the structure theorem for groups acting on connected graphs.

1 Groups

The purpose of this section is to recall a list of basic definitions which will be needed throughout.

1.1 Definitions. Let S be a set.

We write $S^{\pm 1}$ for $S \times \{1, -1\}$, and denote an element (s, ε) by s^{ε}.

By a *word* in $S^{\pm 1}$ we mean a finite sequence in $S^{\pm 1}$, possibly empty. The word $(s_1^{\varepsilon_1}, \ldots, s_n^{\varepsilon_n})$ will usually be abbreviated $s_1^{\varepsilon_1} \cdots s_n^{\varepsilon_n}$.

Let $W(S)$ be the set of all words in $S^{\pm 1}$. There is a binary operation $W(S) \times W(S) \to W(S), (w, w') \mapsto ww'$, given by concatenation, and a unary operation $W(S) \to W(S)$, $w \mapsto w^{-1}$, given by $(s_1^{\varepsilon_1} \cdots s_n^{\varepsilon_n})^{-1} = s_n^{-\varepsilon_n} \cdots s_1^{-\varepsilon_1}$.

For any function $\alpha: S \to G, s \mapsto \alpha s$, there is induced a function $\alpha: W(S) \to G$, $s_1^{\varepsilon_1} \cdots s_n^{\varepsilon_n} \mapsto \alpha(s_1)^{\varepsilon_1} \cdots \alpha(s_n)^{\varepsilon_n}$.

Let R be a subset of $W(S)$. We say G has a *presentation* with *generating set* S and *relation set* R, and write $G = \langle S \mid R \rangle$, if the following holds: there is specified a function $\alpha: S \to G$ such that $\alpha(w) = 1$ for all $w \in R$, having the property that for any group H and function $\beta: S \to H$ such that $\beta(w) = 1$ for all $w \in R$, there exists a unique group homomorphism $\phi: G \to H$ such that $\beta = \phi\alpha$. Even though α need not be injective, we usually suppress α and use the same symbol to denote an element of S and its image in G, hoping that the meaning is clear from the context. In essence, S can be thought of as a family of elements of G, possibly having repetitions.

Variations of the prose are: $\langle S \mid R \rangle$ presents G; G has a presentation with generators $s \in S$ and relators $r \in R$, or relations $r = 1$, $r \in R$. In the latter formulation it is often convenient to write a relation of the form $w_1 w_2 = 1$ as $w_1 = w_2^{-1}$.

Given any subset R of $W(S)$ there exists a group presented by $\langle S \mid R \rangle$; to prove this, one considers the intersection of all equivalence relations induced on $W(S)$ by the various possible β's, and takes as G the set of equivalence classes, with multiplication induced by concatenation.

Any two groups presented by $\langle S \mid R \rangle$ are isomorphic, and the isomorphism is unique if the family S is respected.

Conversely, G always has some presentation, for example $\langle G \mid R \rangle$ where $R = \{((a, 1), (b, 1), (ab, -1)) \in W(G) \mid a, b \in G\}$; we refer to the elements of the latter set as *the relations for* G.

In specific cases, it is usual to list the elements of S and R, casually omitting the set brackets. We also use exponents to indicate repetition. For example, for any $n \geqslant 1$, $\langle s \mid s^n \rangle$ presents the cyclic group C_n of order n, and $\langle r, s \mid r^2, s^2, (rs)^n \rangle$ presents the dihedral group D_n of order $2n$. This extends by analogy to $n = \infty$, with $C_\infty = \langle s \mid \varnothing \rangle$, $D_\infty = \langle r, s \mid r^2, s^2 \rangle$.

The *rank* of G, denoted $\operatorname{rank}(G)$, is the minimum number of generators of G; that is, the least cardinal n such that there exists a presentation $\langle S \mid R \rangle$ of G with $|S| = n$.

For example, the only group of rank zero is the *trivial* group $G = 1$.

For another example, for any set S, if $R = \{w^2 \mid w \in W(S)\}$, then $\langle S \mid R \rangle$ has the structure of a vector space of dimension $|S|$ over the field of two elements; as this cannot be generated by fewer than $|S|$ elements, its rank is $|S|$.

We say that G is a *free group* if it has a presentation of the form $\langle S \mid \varnothing \rangle$. In this event, G is said to be *freely generated* by S, and that S is a *free generating set* of G. The previous example shows that $|S| = \operatorname{rank}(G)$. For any cardinal n, we write F_n for the free group of rank n.

If S is a subset of G, we write $\langle S \rangle$ for the subgroup of G *generated by* S, that is the smallest subgroup of G containing S. ∎

1.2 Definitions. By a *G-set* X we mean a set given with a function $G \times X \to X$, $(g, x) \mapsto gx$, such that $1x = x$ for all $x \in X$, and $g(g'x) = (gg')x$ for all $g, g' \in G$, $x \in X$. This is equivalent to specifying a group homomorphism from G to $\operatorname{Sym} X$, the group of all permutations of X, written on the left. We say also that G *acts* on X, and that there is a *G-action* on X.

For example, G is a G-set under left multiplication; more generally, if H is any subgroup of G then the set of right cosets, $G/H = \{xH \mid x \in G\}$, is a G-set with G-action given by $g(xH) = (gx)H$. We denote the cardinal of this set by $(G:H)$, called the *index* of H in G.

For another example, G is a G-set under *left conjugation*, given by ${}^g x = gxg^{-1}$.

If $X_i, i \in I$, is a family of G-sets then the disjoint union $\bigvee_{i \in I} X_i$ is a G-set, as is the Cartesian product $\prod_{i \in I} X_i$, where G is said to act *diagonally*.

A function $\alpha: X_1 \to X_2$ between G-sets is said to be a *G-map* if $\alpha(gx) = g(\alpha x)$ for all $g \in G, x \in X_1$. We say X_1, X_2 are *G-isomorphic*, denoted $X_1 \approx X_2$, if there exists a bijective G-map from one to the other.

By a *right G-set* X we mean a set given with a function $X \times G \to X$, $(x, g) \mapsto xg$, such that $x1 = x$ for all $x \in X$, and $(xg)g' = x(gg')$ for all $g, g' \in G$, $x \in X$. This is equivalent to X being a G-set with G-action $gx = xg^{-1}$. For example, we have *right conjugation* $x^g = g^{-1}xg$. ∎

1.3 Definitions. Let X be a G-set.

Let $x \in X$. By the *G-stabilizer* of x we mean the subgroup $G_x = \{g \in G \mid gx = x\}$ of G; if P is any subset or element of G_x we say that x is *stabilized by P*, or is *P-stable*. If $g \in G$, then $G_{gx} = {}^g G_x$, where for a subgroup H of G, we write ${}^g H$ and H^g for the *left conjugate* and *right conjugate* $gHg^{-1}, g^{-1}Hg$, respectively.

We say that G acts *trivially* if $gx = x$ for all $g \in G, x \in X$.

We say that X is a *G-free* G-set if $G_x = 1$ for all $x \in X$. For example, if S is a set with trivial G-action then $G \times S$ is G-free.

Since G acts on the set of subsets of X with $gX' = \{gx \mid x \in X'\}$ for $g \in G, X' \subseteq X$, this terminology extends to subsets of X. If X' is G-stable then we say that X' is a *G-subset* of X.

Similarly, G acts on the set of finite sequences $x_1, \ldots, x_n$ in X, so the notation applies here, and $G_{x_1, \ldots, x_n} = G_{x_1} \cap \cdots \cap G_{x_n}$.

For $x \in X$, the *G-orbit* of x is $Gx = \{gx \mid g \in G\}$, a G-subset of X which is G-isomorphic to G/G_x with $gx \in Gx$ corresponding to $gG_x \in G/G_x$.

By the *quotient set* for the G-set X, we mean $G\backslash X = \{Gx \mid x \in X\}$, the set of G-orbits; there is a natural map $X \to G\backslash X, x \mapsto Gx$. If $G\backslash X$ is finite we say that X is *G-finite.*

By a *G-transversal* in X we mean a subset S of X which meets each G-orbit exactly once, so the composite $S \subseteq X \to G\backslash X$ is bijective. Then X is G-isomorphic to $\bigvee_{s\in S} G/G_s$ with $gG_s \in \bigvee_{s\in S} G/G_s$ corresponding to $gs \in X$, for all $g \in G$, $s \in S$. Hence X is the G-set presented on the generating set S with relations saying that s is G_s-stable for each $s \in S$. ∎

1.4 Remarks. (i) Notice we have a structure theorem for G-sets, which says that a G-set is specified up to G-isomorphism by a G-transversal and the G-stabilizers of the elements of the G-transversal.

For example, a G-set is G-free if and only if it is a disjoint union of copies of G, or equivalently, of the form $G \times S$.

(ii) If $\alpha: X \to Y$ is a map of G-sets then $G_x \subseteq G_{\alpha x}$ for all $x \in X$, and if α is injective then $G_x = G_{\alpha x}$ for all $x \in X$. For example, the only G-sets which have G-maps to free G-sets are the free G-sets.

(iii) Conversely, suppose X, Y, are G-sets, and for each $x \in X$, G_x stabilizes an element of Y. Then we can choose any G-transversal S in X and construct a function $\alpha: S \to Y$ such that $G_s \subseteq G_{\alpha s}$ for all $s \in S$. Now α extends to a well-defined G-map $X \to Y, gs \mapsto g\alpha(s)$. ∎

2 Graphs

We now come to another list of basic concepts, this time somewhat less standard.

2.1 Definitions. By a *G-graph* (X, V, E, ι, τ) we mean a nonempty G-set X with a specified nonempty G-subset V, its complement $E = X - V$, and two G-maps $\iota, \tau: E \to V$. In this event we say simply that X is a G-graph.

For any G-subset Y of X we write $VY = V \cap Y, EY = E \cap Y$. If Y is nonempty, and for each $e \in EY$ both ιe and τe belong to VY, then Y is said to be a *G-subgraph* of X.

In particular, $VX = V, EX = E$. We call V and E the *vertex set* and *edge set* of X, and the elements *vertices* and *edges* of X, respectively. The functions $\iota, \tau: E \to V$ are the *incidence functions* of X.

If e is any edge then ιe and τe are the vertices *incident* to e, and are called the *initial* and *terminal* vertices of e, respectively. The definition allows the possibility that ιe and τe may be equal, in which case e is called a *loop.* In almost all our examples the G-map $(\iota, \tau): E \to V \times V$ will be injective, and here $G_e = G_{\iota e, \tau e} = G_{\iota e} \cap G_{\tau e}$ for all $e \in E$.

For $v \in V$, we define $\text{star}(v) = \iota^{-1}(v) \vee \tau^{-1}(v)$, sometimes called the *neighbourhood* of v. The number of elements in star (v) is called the *valency* of v; the elements of star (v) are the edges *incident to* v, either *going into* v or *going out of* v, depending on whether they belong to $\tau^{-1}(e)$ or $\iota^{-1}(e)$, respectively, possibly both. The vertices joined to v by an edge are called the *neighbours* of v.

If every vertex of X has finite valency then X is said to be *locally finite.*

By a *geometric realization of* X we mean an oriented one-dimensional CW-complex with V the set of zero-cells and E the set of one-cells with each edge e starting at ιe and finishing at τe.

For G-graphs X, Y, a *G-graph map* $\alpha: X \to Y$ is a G-map such that $\alpha(VX) \subseteq VY, \alpha(EX) \subseteq EY$, and for each $e \in EX$, $\alpha(\iota e) = \iota(\alpha e)$, $\alpha(\tau e) = \tau(\alpha e)$.

The terms *G-graph isomorphism* and *G-graph automorphism* are then defined in the natural way.

In all the above phrases, if G is omitted we understand that $G = 1$; in this way we recover the concepts of *graph, subgraph* and *graph map.* Thus a G-graph may be viewed as a graph given with a homomorphism from G to its automorphism group.

By the quotient graph $G\backslash X$ we mean the graph $(G\backslash X, G\backslash V, G\backslash E, \bar{\iota}, \bar{\tau})$ where $\bar{\iota}(Ge) = G\iota e$, $\bar{\tau}(Ge) = G\tau e$ for all $Ge \in G\backslash E$; it is straightforward to see that $\bar{\iota}, \bar{\tau}$ are well-defined. There is then a graph map $X \to G\backslash X, x \mapsto Gx$.

The *Cayley graph* of G with respect to a subset S of G, denoted $X(G, S)$, is the G-graph with vertex set G, edge set $G \times S$, and incidence functions $\iota(g, s) = g, \tau(g, s) = gs$ for all $(g, s) \in G \times S$. This is a G-free G-graph. ∎

2.2 Examples. (i) If $G = \langle s \mid s^4 \rangle = C_4$, $S = \{s\}$, then

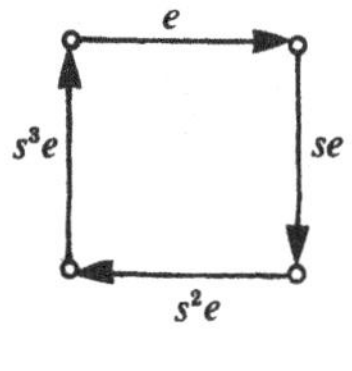

is a geometric realization of $X = X(G, S)$, where $e = (1, s) \in G \times S = EX$. The quotient graph is

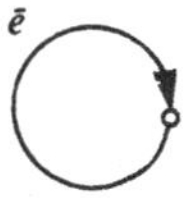

which lifts back to a G-transversal

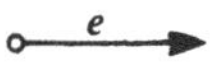

Notice this is not a subgraph, since the terminal vertex is absent.

(ii) If $G = \langle s \mid \varnothing \rangle = F_1 = C_\infty$, and $S = \{s\}$, then

indicates a geometric realization of $X = X(G, S)$ homeomorphic to $\mathbb{R}$, with $e = (1, s) \in EX$. The quotient graph and G-transversal are as in (i).

(iii) If $G = \langle r, s \mid r^2, s^2, (rs)^4 \rangle = D_4$, and $S = \{r, s\}$ then

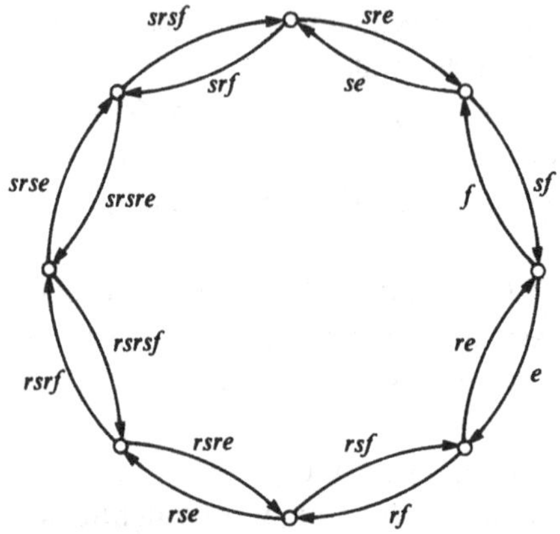

is a geometric realization of $X = X(G, S)$ where $e = (1, r)$, $f = (1, s) \in G \times S = EX$. The quotient graph is

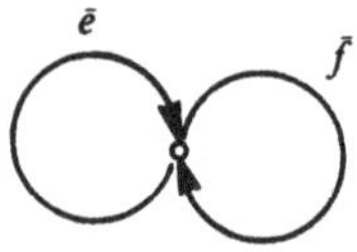

which lifts back to a G-transversal

(iv) If $G = \langle s, r \mid \varnothing \rangle = F_2$, and $S = \{s, r\}$, then Fig. I.1 indicates a geometric realization of $X = X(G, S)$ omitting the arrows. The quotient graph and G-transversal are essentially as in (iii).

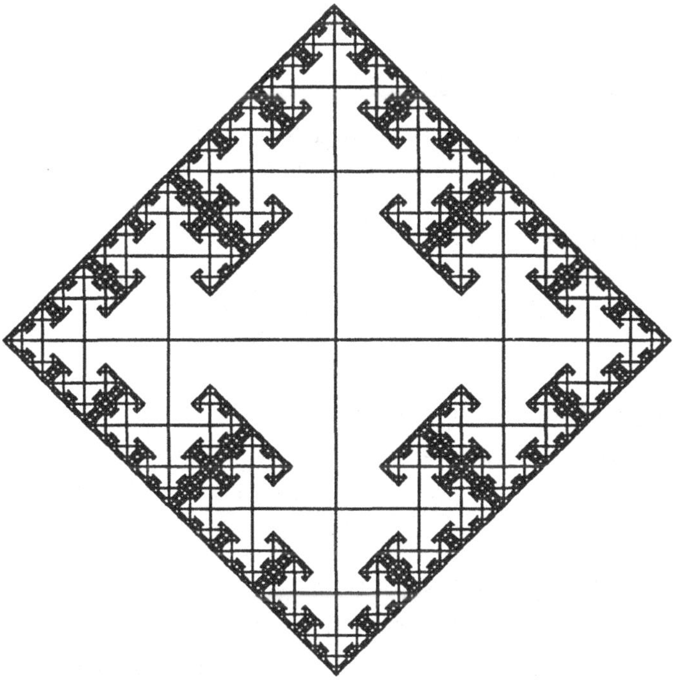

Fig. I.1

(v) If $G = \langle r, s \mid r^2, s^2, (rs)^4 \rangle = D_4$, then

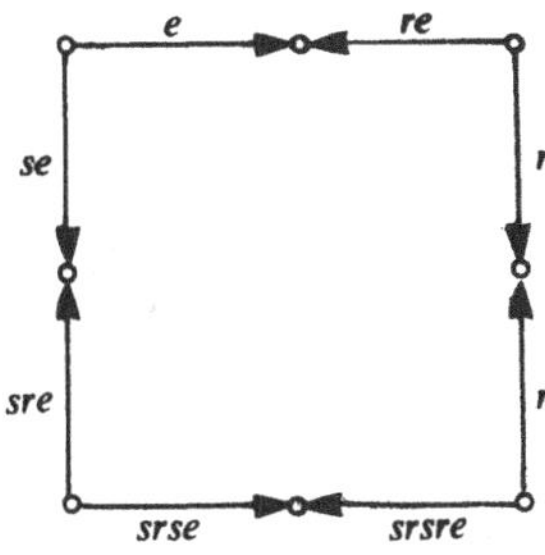

is a geometric realization of a G-graph. The quotient graph is

which lifts back to a G-transversal

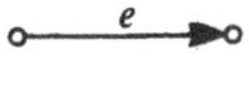

(vi) If $G = \langle r, s | r^2, s^2 \rangle = D_\infty$, then

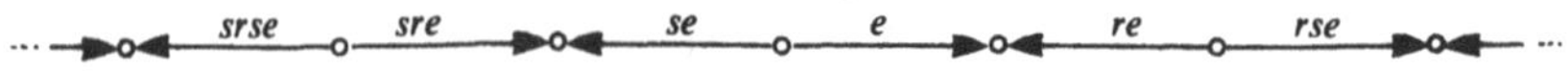

indicates a geometric realization of a G-graph homeomorphic to $\mathbb{R}$; the quotient graph and G-transversal are as in (v). ∎

2.3 Definitions. Let X be a graph.

More incidence functions, again denoted ι, τ, are defined on $EX^{\pm 1}$ by setting $\iota e^1 = \iota e, \tau e^1 = \tau e$, and $\iota e^{-1} = \tau e, \tau e^{-1} = \iota e$ for all $e \in EX$. We think of e^1, e^{-1} as travelling along e the right way and the wrong way, respectively.

A *path* p in X is a finite sequence

(1) $\qquad v_0, e_1^{\varepsilon_1}, v_1, \ldots, v_{n-1}, e_n^{\varepsilon_n}, v_n,$

where

$$n \geqslant 0,$$
$$v_i \in VX \quad \text{for each } i \in [0, n],$$
$$e_i^{\varepsilon_i} \in EX^{\pm 1}, \iota e_i^{\varepsilon_i} = v_{i-1}, \tau e_i^{\varepsilon_i} = v_i \quad \text{for each } i \in [1, n].$$

Incidence functions, still denoted ι, τ, are defined on the set of paths in X by setting $\iota p = v_0, \tau p = v_n$; p is said to be a path of *length* n *from* v_0 *to* v_n, and $v_0, \ldots, v_n, e_1, \ldots, e_n, e_1^{\varepsilon_1}, \ldots, e_n^{\varepsilon_n}$ are said to *occur* in p.

It is customary to abbreviate p to $e_1^{\varepsilon_1}, \ldots, e_n^{\varepsilon_n}$. If $n = 0$ then p is said to be *empty*, and here we must specify v_0; if $n \geqslant 1$ the vertices can be recovered from the abbreviated data.

The *inverse* of p, denoted p^{-1}, is the path $v_n, e_n^{-\varepsilon_n}, v_{n-1}, \ldots, v_1, e_1^{-\varepsilon_1}, v_0$. If q is a path with $\iota q = \tau p$ then in an obvious way we can form a path by *concatenation*, denoted p, q.

If for each $i \in [1, n-1]$, $e_{i+1}^{\varepsilon_{i+1}} \neq e_i^{-\varepsilon_i}$ then p is said to be *reduced*. Notice that if $e_{i+1}^{\varepsilon_{i+1}} = e_i^{-\varepsilon_i}$ for some $i \in [1, n-1]$ then $e_1^{\varepsilon_1}, \ldots, e_{i-1}^{\varepsilon_{i-1}}, e_{i+2}^{\varepsilon_{i+2}}, \ldots, e_n^{\varepsilon_n}$ is a path of length $n-2$ from v_0 to v_n.

We say X is a *tree* if for any vertices v, w of X there is a unique reduced path from v to w; this path is then called the X-*geodesic* from v to w. The length of the geodesic is called the *distance* between v and w. For any subset W of V, by the subtree of X *generated by* W we mean the subgraph of X consisting of all edges and vertices which occur in the X-geodesics between the pairs of elements of W.

A subgraph of X which is a tree is called a *subtree* of X.

A path p is said to be a *closed* path *at* a vertex v if $\iota p = \tau p = v$, and is said to be a *simple closed* path if it is nonempty and there are no other repetitions of vertices. Clearly such a path is reduced, and conversely, any reduced closed path is a (possibly empty) sequence of simple closed paths. A graph with no simple closed paths is called a *forest*; equivalently, the only reduced closed paths are the empty ones.

Two elements of X are said to be *connected* in X if there exists a path in X in which they both occur; in this event there is a reduced path in which they both occur. It is straightforward to show that being connected in X is an equivalence relation. The equivalence classes of this relation are called the *components* of X, and they are subgraphs of X. A graph with only one component is said to be *connected*. On VX the relation of being connected in X is the equivalence relation generated by $\{\{\iota e, \tau e\} \mid e \in EX\}$.

Let E' be a set of edges of X. Write $\bar{E}$ for $EX - E'$ and $\bar{V}$ for the set of components of the graph $X - \bar{E}$ obtained from X by removing $\bar{E}$. There is a natural map $V \to \bar{V}, v \mapsto \bar{v}$, and one can think of $\bar{v}$ as the equivalence class of v relative to the equivalence relation on V generated by $\{(\iota e, \tau e) \mid e \in E'\}$. Let $\bar{X}$ be the graph with vertex set $\bar{V}$, edge set $\bar{E}$ and incidence functions $\bar{\iota}, \bar{\tau}$ with $\bar{\iota} e = \overline{\iota e}, \bar{\tau} e = \overline{\tau e}$ for all $e \in \bar{E}$. There is a map $X \to \bar{X}, x \mapsto \bar{x}$, which on V is as above, on $\bar{E}$ is the identity, and on E' sends e to $\overline{\iota e} = \overline{\tau e}$; this is not a graph map unless E' is empty. We say $\bar{X}$ is the graph obtained from X by *contracting* all the edges in E', and call $X \to \bar{X}$ the *contracting map*.

For example, if $E' = EX$ then $\bar{E} = \varnothing$ and $\bar{X}$ is a graph with no edges, and vertex set the set of components of X. This provides terminology which is frequently useful for seeing that a graph is connected.

If X is a G-graph and E' is a G-subset of EX then $\bar{X}$ is a G-graph and $X \to \bar{X}$ is a G-map. ∎

2.4 Example. Let S be a subset of G and $X = X(G, S)$.

Let $\bar{X}$ be the graph obtained by contracting all the edges of X, and let $X \to \bar{X}, x \mapsto \bar{x}$, be the contracting map. Then $\bar{X}$ is the G-set with one generator $v = \bar{1}$ and relations $gv = gsv$ for all $(g, s) \in G \times S$, that is, $sv = v$ for all $s \in S$. Hence, G_v is the subgroup of G generated by S, and the components of X correspond to cosets $gG_v \in G/G_v$. Thus X is connected if and only if S generates G.

It will be shown in Theorem 8.2 that X is a tree if and only if S freely generates G; see Examples 2.2(ii), (iv). ∎

2.5 Proposition. *A graph is a tree if and only if it is a connected forest.*

Proof. Let X be a tree. Clearly X is connected. Suppose X has a simple closed path p at some vertex v. Then p and the empty path at v are distinct reduced paths in X from v to itself, which contradicts uniqueness. Hence X is a forest.

Conversely, suppose that X is a connected forest. Let v, w be vertices of X. Since X is connected there is a reduced path from v to w, and it remains to show uniqueness. Suppose that $p = e_1^{\varepsilon_1}, \ldots, e_n^{\varepsilon_n}$ and $q = f_1^{\eta_1}, \ldots, f_m^{\eta_m}$ are reduced paths from v to w. Then $p, q^{-1} = e_1^{\varepsilon_1}, \ldots, e_n^{\varepsilon_n}, f_m^{-\eta_m}, \ldots, f_1^{-\eta_1}$ is a closed path at v. If p, q^{-1} is reduced then it must be empty so clearly $p = q$. If p, q^{-1} is not reduced then $n \geqslant 1, m \geqslant 1$ and $e_n^{\varepsilon_n} = f_m^{\eta_m}$. Here $e_1^{\varepsilon_1}, \ldots, e_{n-1}^{\varepsilon_{n-1}}$ and $f_1^{\eta_1}, \ldots, f_{m-1}^{\eta_{m-1}}$ are reduced paths from v to $\tau e_{n-1}^{\varepsilon_{n-1}}$; by induction on n, these paths are equal. Thus $p = q$ as desired. ∎

We now verify the existence of a very important type of transversal already illustrated in Example 2.2.

2.6 Proposition. *If X is a G-graph and $G \backslash X$ is connected then there exist subsets $Y_0 \subseteq Y \subseteq X$ such that Y is a G-transversal in X, Y_0 is a subtree of X, $VY = VY_0$ and for each $e \in EY$, $\iota e \in VY = VY_0$.*

We say Y is a *fundamental G-transversal in X, with subtree Y_0.*

Proof. Write $\bar{X} = G \backslash X$ and $\bar{x} = Gx$ for all $x \in X$.

Choose a vertex v_0 of X. By Zorn's Lemma we can choose a maximal subtree Y_0 of X containing v_0 such that the composite $Y_0 \subseteq X \to \bar{X}$ is injective. Let $\bar{Y}_0$ denote the image of Y_0. We claim that $V\bar{Y}_0 = V\bar{X}$. If not, since $\bar{X}$ is connected, any vertex in $\bar{Y}_0$ is connected to any vertex in $\bar{X} - \bar{Y}_0$ by a path in $\bar{X}$, so some edge $\bar{e}$ of $\bar{X}$ has one vertex $\bar{v}$ in $\bar{Y}_0$ and one vertex in $\bar{X} - \bar{Y}_0$. Here $\bar{v}$ comes from an element v of VY_0 and $\bar{e}$ from an edge e of X; since v lies in the same orbit as a vertex of e, it is a vertex of ge for some $g \in G$, and by replacing e with ge we may further assume that v is a vertex of e. Let w be the other vertex of e. Notice e, w do not lie in Y_0, since their images do not lie in $\bar{Y}_0$. But $Y_0 \cup \{e, w\}$ contradicts the maximality of Y_0. This proves the claim that $V\bar{Y}_0 = V\bar{X}$.

For each edge $\bar{e}$ in $E\bar{X} - E\bar{Y}_0$, $\iota\bar{e}$ comes from a unique vertex of Y_0, and as before we can assume $\iota e \in Y_0$. Adjoining the resulting edges to Y_0 gives a subset Y of X such that the composite $Y \subseteq X \to \bar{X}$ is bijective and if $e \in EY$ then $\iota e \in Y$. ∎

For example, if X is the Cayley graph of G with respect to a generating set S then the subset consisting of 1 and the edges $(1, s)$, $s \in S$, is a fundamental G-transversal. See Example 2.2.

For $G = 1$ the above argument shows the following.

2.7 Corollary. *If X is a connected graph then X has a maximal subtree. Any maximal subtree of X has vertex set all of VX.* ∎

3 Graphs of groups

We now introduce the main object of study for this chapter.

3.1 Definitions. By a *graph of groups* $(G(-), Y)$ we mean a connected graph $(Y, V, E, \bar{\iota}, \bar{\tau})$ together with a function $G(-)$ which assigns to each $v \in V$ a group $G(v)$, and to each edge $e \in E$ a distinguished subgroup $G(e)$ of $G(\bar{\iota}e)$ and an injective group homomorphism $t_e: G(e) \to G(\bar{\tau}e), g \mapsto g^{t_e}$. We call the $G(v), v \in V$, the *vertex groups*, the $G(e), e \in E$, the *edge groups*, and the t_e the *edge functions.*

In examples we sometimes depict this by labelling the vertices and edges of Y with the corresponding groups, see Example 3.3. ∎

3.2 General example. Let X be a G-graph such that $G \backslash X$ is connected and choose a fundamental G-transversal Y for X with subtree Y_0.

Since each element of X lies in the same G-orbit as a unique element of Y, for each $e \in EY$ there are unique $\bar{\iota}e, \bar{\tau}e \in VY$ which lie in the same G-orbits as $\iota e, \tau e$, respectively, and in fact $\bar{\iota}e = \iota e$. Using the incidence functions $\bar{\iota}, \bar{\tau}: EY \to VY$ we make Y into a graph, and clearly $Y \approx G \backslash X$. Notice Y_0 is simultaneously a maximal subtree of Y and a subtree of X. Observe that Y is not a subgraph of X unless $\bar{\tau}$ agrees with τ; in particular, an arbitrary maximal subtree of Y need not be a subgraph of X.

For each $e \in EY, \tau e$ and $\bar{\tau}e$ lie in the same G-orbit in EX, so we can choose an element t_e of G such that $t_e \bar{\tau}e = \tau e$; if $e \in EY_0$ then $\bar{\tau}e = \tau e$ and we take $t_e = 1$. We then call $(t_e | e \in EY)$ a *family of connecting elements.*

Now $G_e \subseteq G_{\iota e}$ *and* $G_e \subseteq G_{\tau e} = t_e G_{\bar{\tau}e} t_e^{-1}$ so there is an embedding $t_e: G_e \to G_{\bar{\tau}e}, g \mapsto g^{t_e} = t_e^{-1} g t_e$.

This data gives the graph of groups *associated to* X with respect to the fundamental G-transversal Y, the maximal subtree Y_0, and the family of connecting elements t_e. ∎

3.3 Specific examples. (i) In Example 2.2(i),(ii), the illustrated fundamental G-transversal, with connecting element s, gives the graph of groups

(ii) In Example 2.2(iii),(iv), the illustrated fundamental G-transversal,
with connecting elements r, s gives the graph of groups

(iii) In Example 2.2(v),(vi), the illustrated fundamental G-transversal,
with connecting elements $1, 1$, gives the graph of groups

$$\{1,s\} \quad 1 \quad \{1,r\}$$
$$\circ\!\!\longrightarrow\!\!\bullet\!\circ$$

■

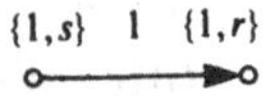

Thus a group acting on a graph with connected quotient graph determines
a graph of groups. Conversely we now show how a graph of groups
determines a group acting on a graph with connected quotient graph.

3.4 Definitions. Let $(G(-), Y)$ be a graph of groups as in Definition 3.1.
Choose a maximal subtree Y_0 of Y, so $VY_0 = VY$ by Corollary 2.7. We
define the associated *fundamental group* $\pi(G(-), Y, Y_0)$ to be the group
presented with

  generating set: $\{t_e | e \in E\} \vee \underset{v \in V}{\vee} G(v)$

  relations:

    the relations for $G(v)$, for each $v \in V$,
    $t_e^{-1} g t_e = g^{t_e}$ for all $e \in E, g \in G(e) \subseteq G(\bar{\iota}e)$, so $g^{t_e} \in G(\bar{\iota}e)$,
    $t_e = 1$, for all $e \in EY_0$.

Let $G = \pi(G(-), Y, Y_0)$.

  In Corollary 7.5 we shall see that the $G(v)$ embed in G, and can be
treated as subgroups.

  To reduce the risk of confusion about the symbol t_e being used to denote
an element of (the generating set of) G and a homomorphism, we make
the convention that $(-)^{t_e}$ always refers to the homomorphism in this
context.

Let T be the G-set presented with generating set Y, and relations saying that each $y \in Y$ is $G(y)$-stable. Then T has G-subsets $VT = GV$, $ET = GE$ and here $ET = T - VT$. It is straightforward to verify that there are well-defined G-maps $\iota, \tau: ET \to VT$ such that $\iota(ge) = g\bar{\iota}e$, $\tau(ge) = gt_e\bar{\tau}e$ for all $g \in G, e \in E$. This data then gives a G-graph T, which is in fact a tree, as will be seen in Theorem 7.6. We call T the associated *standard graph* or *standard tree*, denoted $T(G(-), Y, Y_0)$. It is straightforward to show that Y is a fundamental G-transversal in T, and for each $v \in V$,

$$\iota^{-1}(v) = \bigcup_{e \in \bar{\iota}^{-1}(v)} G(v)e \approx \bigvee_{e \in \bar{\iota}^{-1}(v)} G(v)/G(e)$$

$$\tau^{-1}(v) = \bigcup_{e \in \bar{\tau}^{-1}(v)} G(v)t_e^{-1}e \approx \bigvee_{e \in \bar{\tau}^{-1}(v)} G(v)/G(e)^{t_e}.$$

Thus in T, v has $\sum_{e \in \bar{\iota}^{-1}(v)} (G(v):G(e))$ edges going out and $\sum_{e \in \bar{\tau}^{-1}(v)} (G(v):t_e(G(e))$ edges going in. ∎

Notice that once one knows that the vertex groups embed in the fundamental group, it is a simple exercise to verify that a graph of groups can be recovered from the fundamental group acting on the standard graph.

In the next section we shall see that, conversely, starting from a group acting on a tree and forming the graph of groups we recover the group acting on the tree.

Hence the graphs of groups occurring in Example 3.3(i), (ii), (iii) have associated fundamental group and standard tree given in Example 2.2(ii), (iv), (vi), respectively.

We conclude this section with examples of special graphs of groups which occur throughout the sequel.

3.5 Examples. Let $(G(-), Y)$ be a graph of groups and Y_0 a maximal subtree of Y.

(i) Suppose $G(y) = 1$ for all $y \in Y$. Here $\pi(G(-), Y, Y_0)$ is denoted $\pi(Y, Y_0)$. It is the group presented on generators $\{t_e \mid e \in E\}$ with relations $t_e = 1$ for all $e \in EY_0$, so is a free group of rank $|EY - EY_0|$.

We have seen this situation in Example 3.3(i), (ii) and Example 2.2(ii), (iv) with Y consisting of one, two loops, respectively.

Since its isomorphism type is independent of the choice of Y_0, the group $\pi(Y, Y_0)$ is usually denoted $\pi(Y)$ called the *fundamental group* of Y. This will be discussed further in Definition 8.1.

(ii) Suppose Y has one edge e and two vertices $\iota e, \tau e$. Let $A = G(\iota e)$,

$B = G(\tau e)$, $C = G(e)$, so C is a subgroup of A and there is specified an embedding $C \to B, c \mapsto c^t$. The graph of groups is then depicted

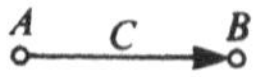

Here $Y_0 = Y$, and the fundamental group is called the *free product of A and B amalgamating C*, denoted $A \underset{C}{*} B$; it is presented on generating set $A \vee B$ with relations saying $c = c^t$ for all $c \in C$, together with the relations of A and B. In the tree T, the vertices are of two sorts, with either $(A:C)$ edges going out, or $(B:C^t)$ edges going in.

In the case $C = 1$, we write simply $A * B$, called the *free product* of A and B.

This is the situation in Example 3.3(iii) and Example 2.2(vi), and we can write $D_\infty = C_2 * C_2$.

(iii) Suppose Y is a tree and $G(e) = 1$ for all $e \in EY$. Then the fundamental group is the *free product* of the vertex groups $G(v)$, $v \in VY$, denoted $\underset{v \in VY}{*} G(v)$.

(iv) Suppose $G(e) = 1$ for all $e \in EY$. Then the fundamental group is $\pi(Y, Y_0) * \underset{v \in VY}{*} G(v)$.

(v) Suppose Y has one edge e and one vertex $v = \iota e = \tau e$. Let $A = G(v)$, $C = G(e)$, so C is a subgroup of A and there is specified an embedding $C \to A, c \mapsto c^t$.

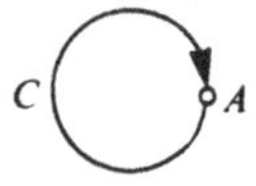

Here Y_0 consists of the single vertex, and the fundamental group is called the *HNN extension of A by* $t: C \to A$, denoted $A \underset{C}{*} t$; it is formed by adjoining to A an indeterminate t satisfying relations $t^{-1} c t = c^t$ for all $c \in C$. Every vertex of the tree T has $(A:C^t)$ edges going in, and $(A:C)$ edges going out.

If $C = 1$ we write $A * t$, so $A * t \approx A * C_\infty$.

We have seen the case $A = C = 1$ in Examples 3.3(i) and 2.2(ii).

A more complicated example arises by taking $A = \langle s \mid \varnothing \rangle, C = \langle s^3 \rangle$, and $t: C \to A$ with $(s^3)^t = s^2$. Here $G = \langle s, t \mid t^{-1} s^3 t = s^2 \rangle$ and T is a tree in which every vertex has two edges going in and three edges going out.

(vi) The fundamental group of any graph of groups can be obtained by successively performing one free product with amalgamation for each edge in the maximal subtree and then one HNN extension for each edge not in the maximal subtree. ∎

4 Groups acting on trees

Throughout this section let T be a G-tree.

4.1 Structure Theorem for groups acting on trees. *In the G-tree T choose a fundamental G-transversal Y with subtree Y_0 and denote the incidence functions by $\bar{\iota}, \bar{\tau}$; choose, for each $e \in EY$, $t_e \in G$ such that $t_e \bar{\tau} e = \tau e$, with $t_e = 1$ if $e \in EY_0$; and form the resulting graph of groups $(G(-), Y)$. Then G is naturally isomorphic to $\pi(G(-), Y, Y_0)$.*

Explicitly G has as a presentation:

(1) generating set: $\{t_e | e \in EY\} \vee \bigvee_{v \in VY} G_v$.

(2) relations:

the relations for G_v, for each $v \in VY$;

$t_e^{-1} g t_e = g^{t_e}$, for all $e \in EY, g \in G_e \subseteq G_{\bar{\iota} e}$ so $g^{t_e} \in G_{\bar{\tau} e}$;

$t_e = 1$, for all $e \in EY_0$.

Proof. Let us consider any $v \in VY$, and analyze the neighbours of v in T.

Consider any edge of T incident to v and express it in the form ge with $g \in G$, $e \in EY$.

If $\iota ge = v$ then $v = \iota ge = g\iota e = g\bar{\iota} e$, and, as Y is a G-transversal, $\bar{\iota} e = v$, $g \in G_v$ and $\tau ge = g\tau e = gt_e \bar{\tau} e$.

If $\tau ge = v$ then $v = \tau ge = g\tau e = gt_e \bar{\tau} e$, and, as Y is a G-tranversal, $\bar{\tau} e = v$ and $gt_e \in G_v$, so we can write $g = ht_e^{-1}$ with $h \in G_v$, and $\iota ge = g\bar{\iota} e = ht_e^{-1} \bar{\iota} e$.

Conversely, all edges of T constructed in this way are incident to v.

We can summarize this by saying that the paths of length 1 in T starting at v are the sequences of the form $v, gt_e^{\frac{1}{2}(\varepsilon - 1)} e^{\varepsilon}, gt_e^{\varepsilon} w$ where v, e^{ε}, w is a path in Y, and $g \in G_v$.

Let $P = \pi(G(-), Y, Y_0)$, so P has presentation (1), (2). Since (1) is a subfamily of G and all the relations (2) hold in G, there is a natural homomorphism $P \to G$, and we wish to show it is an isomorphism.

Since G acts on T, the map $P \to G$ induces a P-action on T. Consider the subset PY of T. By the preceding paragraph, all edges of T incident to Y lie in PY, and so do their vertices; hence all edges of T incident to PY lie in PY and so do their vertices. Since T is connected it follows that $PY = T$. Choose any vertex v_0 of Y. For any $g \in G$, we have $gv_0 \in T = PY$ so $gv_0 = pv$ for some $p \in P$, $v \in Y$. But Y is a G-transversal so $v = v_0$. Since G_{v_0} is in the image of P, we see $P \to G$ is surjective.

Consider any $p \in P$. We claim we can choose a path $v_0, e_1^{\varepsilon_1}, v_1, e_2^{\varepsilon_2}$,

$v_2, \ldots, v_{n-1}, e_n^{\varepsilon_n}, v_n = v_0$ in Y, and elements $g_i \in G_{v_i}$, $i \in [0, n]$, such that

(3) $$p = g_0 t_{e_1}^{\varepsilon_1} g_1 t_{e_2}^{\varepsilon_2} g_2 \cdots g_{n-1} t_{e_n}^{\varepsilon_n} g_n.$$

This is achieved by expressing p as a product of the given generators and their inverses, then using the relations for the G_v to collect together generators from the same G_v into single expressions, and finally inserting 1's as dictated by paths in the maximal subtree Y_0 to obtain an expression as in (3).

It is straightforward to check that

(4) $$\begin{aligned} &v_0, g_0 t_{e_1}^{\frac{1}{2}(\varepsilon_1 - 1)} e_1^{\varepsilon_1}, g_0 t_{e_1}^{\varepsilon_1} v_1, g_0 t_{e_1}^{\varepsilon_1} g_1 t_{e_2}^{\frac{1}{2}(\varepsilon_2 - 1)} e_2^{\varepsilon_2}, \\ &g_0 t_1^{\varepsilon_1} g_1 t_{e_2}^{\varepsilon_2} v_2, \ldots, g_0 t_{e_1}^{\varepsilon_1} g_1 t_{e_2}^{\varepsilon_2} \cdots g_{n-1} t_{e_n}^{\frac{1}{2}(\varepsilon_n - 1)} e_n^{\varepsilon_n}, \\ &g_0 t_{e_1}^{\varepsilon_1} g_1 t_{e_2}^{\varepsilon_2} \cdots g_{n-1} t_{e_n}^{\varepsilon_n} v_n = p v_n \end{aligned}$$

is then a path in T. Notice that, as P acts via the map $P \to G$, it is irrelevant whether the expressions are considered as representing elements of P or G.

We shall show by induction on n that if p is mapped to 1 in G then $p = 1$ in P. Since the composite $G_{v_0} \to P \to G$ is the inclusion map, we may assume $n \geqslant 1$. Since T is a tree and $n \geqslant 1$, the path (4) is not reduced, and for some $i \in [1, n-1]$ the ith edge and the $(i+1)$th edge are inverse to each other. It follows that $\varepsilon_{i+1} = -\varepsilon_i$ and $t_{e_i}^{\frac{1}{2}(\varepsilon_i - 1)} e_i = t_{e_i}^{\varepsilon_i} g_i t_{e_{i+1}}^{\frac{1}{2}(\varepsilon_{i+1} - 1)} e_{i+1}$. Since Y is a G-transversal in T, $e_{i+1} = e_i$ and $t_{e_i}^{\frac{1}{2}(\varepsilon_i + 1)} g_i t_{e_i}^{-\frac{1}{2}(\varepsilon_i + 1)} e_i = e_i$. Thus we have two generators $h = t_{e_i}^{\frac{1}{2}(\varepsilon_i + 1)} g_i t_{e_i}^{-\frac{1}{2}(\varepsilon_i + 1)} \in G_{e_i} \subseteq G_{\iota e_i}$ and $h^{t_{e_i}} = t_{e_i}^{\frac{1}{2}(\varepsilon_i - 1)} g_i t_{e_i}^{-\frac{1}{2}(\varepsilon_i - 1)} \in G_{\tau e_i}$, and, by (2), $t_{e_i}^{-1} h t_{e_i} = h^{t_{e_i}}$ in P.

If $\varepsilon_i = 1$ then $\iota e_i = v_{i-1} = v_{i+1}$, $\tau e_i = v_i$ so $h \in G_{\iota e_i} = G_{v_{i-1}}$, and $h^{t_{e_i}} = g_i$ so $t_{e_i}^{-1} h t_{e_i} = g_i$ in P. Hence in (3) we can replace $g_{i-1} t_{e_i} g_i t_{e_i}^{-1} g_{i+1}$ with the single generator $g_{i-1} h g_{i+1} \in G_{v_{i+1}}$ and omit e_i, v_i, e_i^{-1} from the path in Y, and so reduce n by 2.

Similarly, if $\varepsilon_i = -1$ then $\iota e_i = v_i$, $\tau e_i = v_{i-1} = v_{i+1}$, so $h^{t_{e_i}} \in G_{\tau e_i} = G_{v_{i-1}}$, and $h = g_i$ so $h^{t_{e_i}} = t_{e_i}^{-1} g_i t_{e_i}$ in P. Hence in (3) we can replace $g_{i-1} t_{e_i}^{-1} g_i t_{e_i} g_{i+1}$ with the single generator $g_{i-1} h^{t_{e_i}} g_{i+1} \in G_{v_{i-1}}$ and omit e_i^{-1}, v_i, e_i from the path in Y, and so reduce n by 2.

It follows by induction on n that $p = 1$, so $P \to G$ is injective, and G has the desired presentation. ∎

The case of a free action is particularly interesting.

4.2 Corollary. *If G acts freely on T then G is a free group, in fact, $G \approx \pi(G \backslash T)$.* ∎

Since trivial vertex groups correspond to free groups, there is a type of duality between vertex groups and free groups. We now state some results at the two extremes, for which we require the following observation.

4.3 Lemma. *If N is a normal subgroup of G then $N\backslash T$ is a connected G/N-graph and each $Nt \in N\backslash T$ has stabilizer $(G/N)_{Nt} = NG_t/N$.* ∎

4.4 Proposition. *If N is the subgroup of G generated by the G_v, $v \in VT$, then N is normal and G/N is free. Moreover, $N\backslash T$ is a G/N-free G/N-tree and $G/N \approx \pi(G\backslash T)$.*

Proof. It follows easily from the Structure Theorem 4.1 that $G/N \approx \pi(G\backslash T)$.

We now apply this with N in place of G. Since N is generated by the vertex stabilizers $N_v = G_v$, $v \in VT$, we see that $\pi(N\backslash T) \approx N/N = 1$. Hence $N\backslash T$ is a tree. By Lemma 4.3, G/N acts freely on $N\backslash VT$, and hence on $N\backslash T$, so $G/N \approx \pi(G\backslash T)$. ∎

4.5 Proposition. *If a subgroup H of G does not meet any vertex stabilizer then H acts freely on T, so H is free. For example, if H is torsion-free and the vertex stabilizers are torsion groups then H is free.* ∎

4.6 Proposition. *If $G \to A$ is a homomorphism of groups which is injective on each vertex stabilizer then the kernel N is free. In fact $N \approx \pi(X)$, where X is the connected G/N-graph $N\backslash T$.* ∎

If the homomorphism $G \to A$ is surjective then X is a connected A-graph. In Theorem 9.2 we shall see that all group actions on connected graphs can be realized in this way.

The Structure Theorem 4.1 suggests that there are only limited possibilities for a group to act on a tree if the group has certain special properties such as being finite, cyclic, soluble, free, etc. The finite case will be of great importance in Chapters 3 and 4, and it will be useful to know the following.

4.7 Proposition. *Let v be a vertex of T. Then G stabilizes a vertex of T if and only if there is an integer N such that the distance from v to each element of Gv is at most N.*

Proof. If G stabilizes a vertex v_0 of T, and the T-geodesic p from v to v_0 has length n, then for each $g \in G$ we have a path p, gp^{-1} of length $2n$ from v to gv.

Conversely, suppose there is an integer N such that for each $g \in G$ the T-geodesic from v to gv has length at most N. Let T' be the subtree generated by Gv.

It is easy to see that T' is a G-subtree of T and no reduced path has length greater than $2N$.

If T' has at most one edge then every element of T' is G-stable, and we have the desired G-stable vertex. Thus we may assume that T' has at least two edges, so some vertex of T' has valency at least two. Now delete from T' all vertices of valency one, and their incident edges. This leaves a G-subtree T'' in which no reduced path has length greater than $2N-2$. By induction, G stabilizes a vertex. ∎

Let us note the cases where Gv is finite, and where G is finite.

4.8 Corollary. *If there is a finite G-orbit in VT then G stabilizes a vertex of T.* ∎

4.9 Corollary. *A finite group acting on a tree must stabilize a vertex.* ∎

At this stage it is convenient to introduce some very useful terminology.

4.10 Definitions. Let e, f be edges of T and v a vertex of T.

We say that e *points to* v if e is the first edge in the T-geodesic from ιe to v, or equivalently, τe and v lie in the same component of $T - \{e\}$. Otherwise, e *points away from* v.

Consider a reduced path $e_1^{\varepsilon_1}, \ldots, e_n^{\varepsilon_n}$ in T with $e_1 = e, e_n = f$; it is unique unless $e = f$ and $n = 1$. If $\varepsilon_1 = \varepsilon_n$ we say that e and f *point in the same direction*; otherwise e and f *point in opposite directions.* We define a partial order $\geqslant$ on $EX^{\pm 1}$ by setting $e_1^{\varepsilon_1} \geqslant e_n^{\varepsilon_n}$.

An element g of G is said to *translate* e if ge and e are distinct and point in the same direction, that is, $ge > e$ or $e > ge$.

By an *infinite path* in X we mean a sequence $v_0, e_1^{\varepsilon_1}, v_1, \ldots, v_{n-1}, e_n^{\varepsilon_n}, v_n, \ldots$ where for each $n \geqslant 0$, $v_n \in VX$, and for each $n \geqslant 1$, $e_n^{\varepsilon_n} \in EX^{\pm 1}, \iota e_n^{\varepsilon_n} = v_{n-1}$, $\tau e_n^{\varepsilon_n} = v_n$.

By a *doubly infinite path* in X we mean a sequence $\ldots, v_{n-1}, e_n^{\varepsilon_n}, v_n, \ldots$ where for each $n \in \mathbb{Z}$, $v_n \in VX$, $e_n^{\varepsilon_n} \in EX^{\pm 1}$, $\iota e_n^{\varepsilon_n} = v_{n-1}$, $\tau e_n^{\varepsilon_n} = v_n$. ∎

4.11 Proposition. *For any $g \in G$ the following are equivalent:*

(*a*) *g does not stabilize a vertex of T.*

(*b*) *g translates an edge of T.*

(c) *g acts by translation on a subtree of T homeomorphic to $\mathbb{R}$. In this event g has infinite order.*

Proof. $(a)\Rightarrow(c)$. Choose a vertex v of T in such a way that the T-geodesic $p = e_1^{\varepsilon_1}, \ldots, e_n^{\varepsilon_n}$ from v to gv is as short as possible. Assume (a) holds so $n \geqslant 1$. Notice that $e_n^{-\varepsilon_n} \neq g e_1^{\varepsilon_1}$, for if $e_n^{-\varepsilon_n} = g e_1^{\varepsilon_1}$ then $n \geqslant 2$ and $e_2^{\varepsilon_2}, \ldots, e_{n-1}^{\varepsilon_{n-1}}$ is the T-geodesic from $\iota e_2^{\varepsilon_2}$ to $\tau e_{n-1}^{\varepsilon_{n-1}} = \iota e_n^{\varepsilon_n} = g\tau e_1^{\varepsilon_1} = g\iota e_2^{\varepsilon_2}$ which contradicts the minimality of n. Hence by concatenating the $g^m p$, $m \in \mathbb{Z}$, we can construct a reduced doubly infinite path which gives a subtree homeomorphic to $\mathbb{R}$, on which g acts by translation by n. It is not difficult to see that g acts by translating a fundamental G-transversal containing p.

$(c)\Rightarrow(b)\Rightarrow(a)$ is clear. ∎

4.12 Theorem. *Exactly one of the following holds:*

(a) *G stabilizes a vertex of T.*

(b) *There is a reduced infinite path $v_0, e_1^{\varepsilon_1}, v_1, e_2^{\varepsilon_2}, \ldots$ in T such that $G_{v_0} \subseteq G_{v_1} \subseteq \cdots$, $G = \bigcup_{n \geqslant 0} G_{v_n} = \bigcup_{n \geqslant 1} G_{e_n}$ and for all $n \geqslant 1$, $G \neq G_{e_n}$.*

(c) *Some element of G translates some edge e of T, and then for $C = G_e$, either $G = B \underset{C}{*} D$ with $B \neq C \neq D$ or $G = B \underset{C}{*} x$.*

Proof. It is an easy matter to verify that $(a), (b), (c)$ are pairwise incompatible, and we wish to show that at least one of them holds.

Suppose that (a) and (c) fail and consider any vertex v.

It follows from the failure of (a) that $G \neq G_v$. Consider any $g \in G - G_v$. Since (c) fails, it follows from Proposition 4.11 $(a)\Rightarrow(b)$ that g stabilizes some vertex $w \neq v$ of T, and we may choose w as close as possible to v. Let e^ε be the first edge in the T-geodesic p from v to w. Then p, gp^{-1} is reduced, and hence $e^\varepsilon > ge^{-\varepsilon}$.

We claim that e^ε is independent of the choice of g. Suppose that $e'^{\varepsilon'}$ is the first edge in the geodesic from v to a vertex w' stabilized by $g' \in G - G_v$. Then $e'^{\varepsilon'} > g'e'^{-\varepsilon'}$. If $e^\varepsilon \neq e'^{\varepsilon'}$ then $e^{-\varepsilon}, e'^{\varepsilon'}$ is a reduced path so $e^{-\varepsilon} > e'^{\varepsilon'}$, and thus $g'e^{-\varepsilon} > g'e'^{\varepsilon'} > e'^{-\varepsilon'} > e^\varepsilon > ge^{-\varepsilon}$, which means that gg'^{-1} translates $g'e^{-\varepsilon}$, contradicting the failure of (c). Hence, $e^\varepsilon = e'^{\varepsilon'}$ as desired.

Hence, τe^ε is uniquely determined by v, and we denote it by $\phi(v)$. This gives a well-defined map $\phi: VT \to VT$.

It is clear that ϕ is a G-map, so $G_v \subseteq G_{\phi(v)}$.

By choosing $g \in G - G_{\phi(v)}$, we see $\phi^2(v) \neq v$; hence $v, \phi(v), \phi^2(v), \ldots, \phi^n(v), \ldots$ is the sequence of vertices in a reduced infinite path.

By using the same g, w to find as many of $\phi(v), \phi^2(v), \ldots, \phi^n(v)$ as possible, we deduce that $w=\phi^n(v)$ for some n. Hence $g\in\bigcup_{n\geq 0} G_{\phi^n(v)}$. It follows that $\bigcup_{n\geq 0} G_{\phi^n(v)}=G$, and (*b*) holds.

Thus (*a*), (*b*) or (*c*) holds.

Finally, suppose that (*c*) holds. Contracting all edges not in Ge yields a G-tree with exactly one edge orbit and with an edge e translated by an element of G, so no vertex is stabilized by G. By the Structure Theorem 4.1, either $G=B\underset{C}{*}D$ with $B\neq C\neq D$, or $G=B\underset{C}{*}x$. ∎

In case (*c*) somewhat more can be said.

4.13 Proposition. *If some element of G translates some edge of T then there is a unique minimal G-subtree T' of T and ET' consists of all edges translated by elements of G. If G is finitely generated then $G\backslash T'$ is finite.*

Proof. Let E' be the set of edges of T translated by elements of G, and let T' be the subgraph of T with edge set E'. By hypothesis, T' is non-empty; we claim T' is connected and hence is a subtree.

Suppose T_1, T_2 are subtrees of T homeomorphic to $\mathbb{R}$ on which elements g_1, g_2 act by translation, respectively. To show that T' is connected, it suffices to show that each edge in the path p joining T_2 to T_1 lies in E', so we may assume p is nonempty. By replacing g_1 and/or g_2 with its inverse if necessary, we may choose paths p_1, p_2 in T_1, T_2, respectively, so as to have the situation illustrated in Fig. I.2(i). Here each edge of p is translated by g_1g_2 so it lies in E'; see Fig. I.2(ii). Thus T' is a subtree.

Clearly, E' is a G-subset of ET, so T' is a G-subtree of T.

If $g\in G$ translates an edge, then we have a $\langle g\rangle$-subtree T_g of T homeomorphic to $\mathbb{R}$, and $\langle g\rangle$ acts on T by translating T_g. It is then easy to see that T_g is the unique minimal $\langle g\rangle$-subtree of T, so lies in every G-subtree of T. Hence, T' is the unique minimal G-subtree of T.

Now suppose G has a finite generating set S. Let v be a vertex of T'

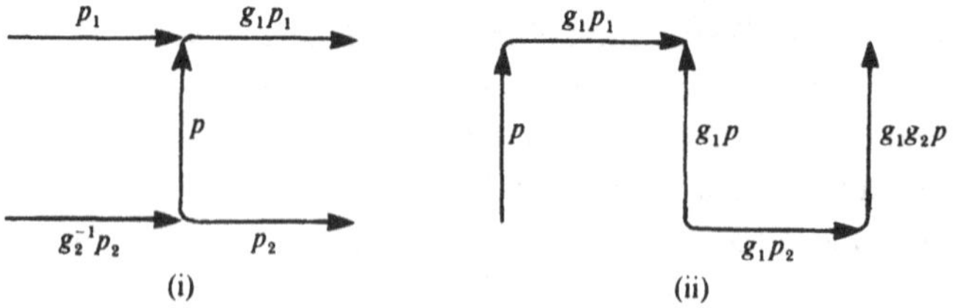

Fig. I.2

and let Y be the subtree of T' generated by $Sv \cup \{v\}$. The union of the $gY, g \in G$, gives a connected G-subgraph of T' which, by minimality, must be all of T'. Since Y is finite, T' is G-finite. ∎

5 Trees for certain automorphism groups

In this section we discuss trees which are acted on by the automorphism group of the free object of rank two in each of the following categories: abelian groups, free groups, vector spaces over principal valuated fields, commutative algebras over a field, and associative algebras over a field.

5.1 Notation. For any commutative ring R and positive integer n, $\mathrm{GL}_n(R)$ denotes the group of all invertible $n \times n$ matrices with entries in R. The centre of $\mathrm{GL}_n(R)$ is the group of invertible scalar matrices, and this can be identified with $UR = \mathrm{GL}_1(R)$, the group of units of R; the quotient group is denoted $\mathrm{PGL}_n(R)$. An $n \times n$ matrix A over R lies in $\mathrm{GL}_n(R)$ if and only if the determinant $\det A$ lies in UR. The subgroup of $n \times n$ matrices with determinant 1 is denoted $\mathrm{SL}_n(R)$, and the centre lies in the centre of $\mathrm{GL}_n(R)$ and the quotient group is denoted $\mathrm{PSL}_n(R)$. ∎

5.2 Examples. (i) Let $G = \mathrm{GL}_2(\mathbb{Z})$. There is a classical G-action on the upper half-plane $\mathscr{H} = \{z \in \mathbb{C} \mid \mathrm{Im}\, z > 0\} = \{x + iy \mid x, y \in \mathbb{R}, y > 0\}$, with

$$g = \begin{pmatrix} a & b \\ c & d \end{pmatrix}$$

acting on $z = x + iy$ to give

$$gz = \frac{az + b}{cz + d}$$

if $ad - bc = 1$, and

$$gz = \frac{a\bar{z} + b}{c\bar{z} + d}$$

if $ad - bc = -1$, where $\bar{z} = x - iy$. Thus

$$\begin{pmatrix} a & b \\ c & d \end{pmatrix}(x + iy) = \frac{(ax + b)(cx + d) + acy^2 + iy}{(cx + d)^2 + c^2y^2}.$$

Let $Y = \{\cos\theta + i\sin\theta \mid \pi/3 \leqslant \theta \leqslant \pi/2\}$, a subset of $\mathscr{H}$ which can be viewed as the geometric realization of a graph with one edge e having $\iota e = i, \tau e = \frac{1}{2}(1 + i\sqrt{3})$. Let $T = GY$, a G-subset of $\mathscr{H}$, which we shall see is

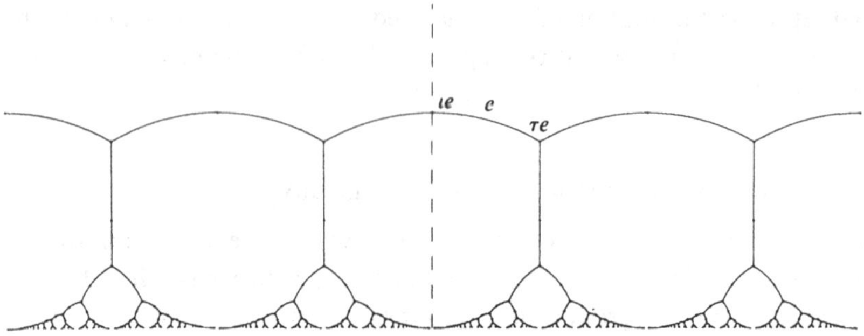

Fig. I.3

a geometric realization of a G-tree with fundamental G-transversal Y. See Fig. I.3.

A straightforward calculation shows that

$$G_{\iota e}=\left\{\pm\begin{pmatrix}1&0\\0&1\end{pmatrix},\pm\begin{pmatrix}0&-1\\1&0\end{pmatrix},\pm\begin{pmatrix}1&0\\0&-1\end{pmatrix},\pm\begin{pmatrix}0&1\\1&0\end{pmatrix}\right\}$$
$$=D_4,$$

$$G_{\tau e}=\left\{\pm\begin{pmatrix}1&0\\0&1\end{pmatrix},\pm\begin{pmatrix}1&-1\\1&0\end{pmatrix},\pm\begin{pmatrix}0&1\\-1&1\end{pmatrix},\right.$$
$$\left.\pm\begin{pmatrix}0&1\\1&0\end{pmatrix},\pm\begin{pmatrix}1&0\\1&-1\end{pmatrix},\pm\begin{pmatrix}-1&1\\0&1\end{pmatrix}\right\}=D_6,$$

$$G_e=\left\{\pm\begin{pmatrix}1&0\\0&1\end{pmatrix},\pm\begin{pmatrix}0&1\\1&0\end{pmatrix}\right\}=D_2.$$

By the Euclidean algorithm, any 2×2 matrix over $\mathbb{Z}$ can be transformed to an upper triangular matrix using the operations of interchanging two rows and adding or subtracting the second row from the first; these operations correspond to left multiplication by

$$t=\begin{pmatrix}0&1\\1&0\end{pmatrix},\quad s=\begin{pmatrix}1&1\\0&1\end{pmatrix}\quad\text{and}\quad s^{-1}=\begin{pmatrix}1&-1\\0&1\end{pmatrix},$$

respectively. It follows that $GL_2(\mathbb{Z})$ is generated by t, s and

$$r=\begin{pmatrix}1&0\\0&-1\end{pmatrix}.$$

Notice that $t\in G_e$, $r\in G_{\iota e}$,

$$srt=\begin{pmatrix}-1&1\\-1&0\end{pmatrix}\in G_{\tau e},$$

so $G_{\iota e}\cup G_{\tau e}$ generates G. Since G permutes the components of T, we see that the component of T containing Y is closed under the action of $G_{\iota e} \cup G_{\tau e}$ so is closed under the action of G, so contains $GY = T$. Thus T is connected.

It remains to verify that T has no simple closed curves. Consider any

$$g = \begin{pmatrix} a & b \\ c & d \end{pmatrix} \in \mathrm{GL}_2(\mathbb{Z})$$

and any $z = x + iy \in Y$. We claim that if $gz \in Y$ then either each point of Y is g-stable, or z is a g-stable endpoint of Y, and thus T is the geometric realization of a G-graph. We claim further that if $\mathrm{Re}(gz) = 0$ then $gz = \iota e$. There are two cases.

Consider first the case where $c^2 \neq d^2$.

Using the fact that $|z| = 1$ one can verify that $|(c^2 - d^2)gz - (ac - bd)| = 1$, that is, gz is on a circle with centre $(ac - bd)/(c^2 - d^2)$ and radius $1/|c^2 - d^2|$.

If $gz \in Y$, then $\mathrm{Im}(gz) \geqslant \sqrt{3}/2 \geqslant \frac{1}{2}$ so $|c^2 - d^2| \leqslant 2$ so $|c^2 - d^2| = 1$. Similarly the centre is limited to 0 or 1 so $ac - bd = 0$ or 1, and it can be verified that either each point of Y is g-stable, or z is a g-stable endpoint of Y. If $\mathrm{Re}(gz) = 0$ then the circle meets the upper half of the imaginary axis and this forces $ac - bd = 0$, from which it follows that each point of Y is g-stable, so $z = gz = \iota e$.

Now suppose that $c^2 = d^2$.

Then $c^2 = d^2 = 1$ and a, b have opposite parity. It can be shown that $2\,\mathrm{Re}(gz) = ac + bd$, an odd integer, so nonzero.

If $gz \in Y$ then $ac + bd = 1$ and $gz = \tau e$, and it follows that τe is g-stable.

Now suppose there is a simple closed path in T. Then there is a path in T from ιe to τe which does not use e. The only edges incident to ιe are e and the edge

$$e' = \begin{pmatrix} 0 & -1 \\ 1 & 0 \end{pmatrix} e.$$

So there is a path from $\tau e'$ to τe which does not pass through $\iota e = i$, but by continuity considerations it must pass through the imaginary axis, a contradiction. Thus T is a G-tree and we deduce $\mathrm{GL}_2(\mathbb{Z}) = D_4 \underset{D_2}{*} D_6$.

It is not difficult to deduce the presentation

$$\mathrm{GL}_2(\mathbb{Z}) = \langle q, r, t \mid q^2, r^2, t^2, (tr)^4, (tr)^2(tq)^3 \rangle$$

where

$$q = \begin{pmatrix} 1 & 0 \\ 1 & -1 \end{pmatrix}, \quad r = \begin{pmatrix} 1 & 0 \\ 0 & -1 \end{pmatrix}, \quad t = \begin{pmatrix} 0 & 1 \\ 1 & 0 \end{pmatrix}.$$

The following three groups also act on T with Y as fundamental transversal, and so have the descriptions indicated.

(ii) $\mathrm{PGL}_2(\mathbb{Z}) = D_2 \underset{D_1}{*} D_3$, where

$$D_2 = \left\{\left\{\pm\begin{pmatrix}1 & 0\\ 0 & 1\end{pmatrix}\right\}, \left\{\pm\begin{pmatrix}0 & -1\\ 1 & 0\end{pmatrix}\right\},\right.$$
$$\left.\left\{\pm\begin{pmatrix}1 & 0\\ 0 & -1\end{pmatrix}\right\}, \left\{\pm\begin{pmatrix}0 & 1\\ 1 & 0\end{pmatrix}\right\}\right\},$$
$$D_3 = \left\{\left\{\pm\begin{pmatrix}1 & 0\\ 0 & 1\end{pmatrix}\right\}, \left\{\pm\begin{pmatrix}1 & -1\\ 1 & 0\end{pmatrix}\right\}, \left\{\pm\begin{pmatrix}0 & 1\\ -1 & 1\end{pmatrix}\right\},\right.$$
$$\left.\left\{\pm\begin{pmatrix}0 & 1\\ 1 & 0\end{pmatrix}\right\}, \left\{\pm\begin{pmatrix}1 & 0\\ 1 & -1\end{pmatrix}\right\}, \left\{\pm\begin{pmatrix}-1 & 1\\ 0 & 1\end{pmatrix}\right\}\right\},$$
$$D_1 = \left\{\left\{\pm\begin{pmatrix}1 & 0\\ 0 & 1\end{pmatrix}\right\}, \left\{\pm\begin{pmatrix}0 & 1\\ 1 & 0\end{pmatrix}\right\}\right\}.$$

(iii) $\mathrm{SL}_2(\mathbb{Z}) = C_4 \underset{C_2}{*} C_6$, where

$$C_4 = \left\{\pm\begin{pmatrix}1 & 0\\ 0 & 1\end{pmatrix}, \pm\begin{pmatrix}0 & -1\\ 1 & 0\end{pmatrix}\right\},$$
$$C_6 = \left\{\pm\begin{pmatrix}1 & 0\\ 0 & 1\end{pmatrix}, \pm\begin{pmatrix}1 & -1\\ 1 & 0\end{pmatrix}, \pm\begin{pmatrix}0 & 1\\ -1 & 1\end{pmatrix}\right\},$$
$$C_2 = \left\{\pm\begin{pmatrix}1 & 0\\ 0 & 1\end{pmatrix}\right\}.$$

(iv) $\mathrm{PSL}_2(\mathbb{Z}) = C_2 * C_3$, where

$$C_2 = \left\{\left\{\pm\begin{pmatrix}1 & 0\\ 0 & 1\end{pmatrix}\right\}, \left\{\pm\begin{pmatrix}0 & -1\\ 1 & 0\end{pmatrix}\right\}\right\},$$
$$C_3 = \left\{\left\{\pm\begin{pmatrix}1 & 0\\ 0 & 1\end{pmatrix}\right\}, \left\{\pm\begin{pmatrix}1 & -1\\ 1 & 0\end{pmatrix}\right\}, \left\{\pm\begin{pmatrix}0 & 1\\ -1 & 1\end{pmatrix}\right\}\right\}.$$

(v) Denote the automorphism group of F_2 by $\mathrm{Aut}(F_2)$. Let F_2 be free on x, y so an automorphism ϕ can be represented by the ordered pair $(\phi x, \phi y)$. We identify F_2 with the group of inner automorphisms, with F_2 acting by left conjugation.

It can be shown that F_2 is generated by $q = (xy, y^{-1})$, $r = (x, y^{-1})$, $t = (y, x)$; see Lyndon and Schupp (1977), Proposition I.4.1. One can compute that $1 = q^2 = r^2 = t^2 = (tr)^4$ and $(tr)^2(tq)^3 = xy$.

Every automorphism of F_2 induces an automorphism on the abelianization $\mathbb{Z}^2$, so there is a homomorphism from $\mathrm{Aut}(F_2)$ to $\mathrm{GL}_2(\mathbb{Z})$, and the preceding paragraph, together with (i), shows that $\mathrm{Aut}(F_2) \to \mathrm{GL}_2(\mathbb{Z})$ is surjective with kernel F_2. Thus $\mathrm{Aut}(F_2) = A \underset{C}{*} B$, where A, B, C are extensions of F_2 by D_4, D_6 and D_2, respectively. See Example IV.1.13 for a further discussion of A, B, C. ▮

5.3 Examples. (i) Let R be a principal valuation ring, K its field of fractions, k the residue field, and t a uniformizer, so $R/tR = k$. For example, R can be the ring of formal power series $k[[t]]$, and then K is the field $k((t))$ of Laurent series.

Let U denote the group of units of R, K^* the group of units of K, and $G = \mathrm{GL}_2(K)$, and view K^* as the centre of G.

We shall construct a G-tree.

Let K^2 be the set of column vectors of length two over K, so K^2 is a G-module under matrix multiplication.

By an *R-lattice* P in K^2 we mean a free R-submodule of K^2 of rank 2, that is, P is generated by a K-basis of K^2. Let $\mathscr{L}$ denote the set of all R-lattices in K^2, so $\mathscr{L}$ is a G-set.

For $P \in \mathscr{L}$ we write $[P] = \{t^i P \mid i \in \mathbb{Z}\} = K^* P$. Then $K^* \backslash \mathscr{L} = \{[P] \mid P \in \mathscr{L}\}$ is again a G-set, since K^* is the centre of G.

If $P, Q \in \mathscr{L}$ then by the theory of matrices over principal ideal domains there exists an R-basis b, b' of P and integers m, n such that $t^m b, t^n b'$ is an R-basis of Q, and then $\{m, n\}$ is independent of the choice of b, b'. We define $d(P, Q) = |m - n|$, and also $d([P], [Q]) = d(P, Q) = |m - n|$, which is easily seen to be well-defined.

Notice $[P] = \{Q \in \mathscr{L} \mid d(P, Q) = 0\}$.

Let $E = \{(p, q) \mid p, q \in K^* \backslash \mathscr{L}$ with $d(p, q) = 1\}$. We claim that E is the edge set of a G-tree T. To construct T we take vertex set $V = \{p, \{p, q\} \mid (p, q) \in E\}$, and incidence functions $\iota(p, q) = p$, $\tau(p, q) = \{p, q\}$. It is clear that T is a G-graph.

If $p, q \in K^* \backslash \mathscr{L}$ then the paths in T from p to q can be identified with the sequences $p = p_0, p_1, \ldots, p_n = q$ such that $d(p_{i-1}, p_i) = 1$ for all $i \in [1, n]$.

Consider any $p, q \in K^* \backslash \mathscr{L}$, and choose representatives $P, Q \in \mathscr{L}$, respectively, and an R-basis b, b' of P such that $t^m b$, $t^n b'$ is an R-basis of Q for some $m, n \in \mathbb{Z}$. Now the sequences $Rt^i b + Rb'$, $i = 0, \ldots, m$, $Rt^m b + Rt^j b'$, $j = 0, \ldots, n$ in $\mathscr{L}$ determine a path in T connecting p to p'. It follows that T is connected.

If $p, q \in K^* \backslash \mathscr{L}$ and $d(p, q) = 1$ then we can choose representatives $P, Q \in \mathscr{L}$, respectively, and an R-basis b, b' of P such that $t^m b, t^{m+1} b'$ is an R-basis of Q for some integer m. Replacing Q by $t^{-m} Q$ we may assume that Q is a maximal submodule of P and tP is a maximal submodule of Q.

If T is not a tree then there exists a sequence $p_0, p_1, \ldots, p_n = p_0$ in $K^*\backslash\mathscr{L}$ such that $p_{i-1} \neq p_{i+1}$ for all $i \in [1, n-1]$, and $d(p_{i-1}, p_i) = 1$ for all $i \in [1, n]$. By the preceding paragraph we can construct a sequence $P_0, P_1, \ldots, P_n = t^m P_0$ in $\mathscr{L}$ such that $tP_{i-1} \neq P_{i+1}$ for all $i \in [1, n-1]$, and for each $i \in [1, n]$, P_i is a maximal submodule of P_{i-1} and tP_{i-1} is a maximal submodule of P_i. Hence, for each $i \in [1, n-1]$, tP_{i-1} and P_{i+1} are distinct maximal submodules of P_i, so $tP_{i-1} + P_{i+1} = P_i$. By reverse induction $tP_{i-1} + P_n = P_i$, and in particular $tP_0 + P_n = P_1$. Now $P_0 \supset P_n = t^m P_0$, so $m \geqslant 1$ and $tP_0 \supseteq P_n$, which means that $tP_0 + P_n = tP_0$. Hence $P_1 = tP_0 + P_n = tP_0 \subset P_1$, a contradiction. Thus T is a G-tree.

Let

$$p = \left[R\begin{pmatrix}1\\0\end{pmatrix} + R\begin{pmatrix}0\\1\end{pmatrix}\right], \quad p' = \left[R\begin{pmatrix}t\\0\end{pmatrix} + R\begin{pmatrix}0\\1\end{pmatrix}\right]$$

in $K^*\backslash\mathscr{L}$, and $e = (p, p')$ in E. Then $\iota e = p$, $\tau e = \{p, p'\}$.

We claim that the subgraph $Y = \{\iota e, e, \tau e\}$ is a G-transversal in T. It is clear that $\iota e, \tau e$ are in different G-orbits, so it suffices to show $E = Ge$. Now $K^*\backslash\mathscr{L} = Gp$ so each edge of T lies in the G-orbit of an edge of the form (p, q), where $d(p, q) = 1$. Here

$$q = \left[R\begin{pmatrix}a\\b\end{pmatrix} + R\begin{pmatrix}c\\d\end{pmatrix}\right]$$

for some $a, b, c, d \in R$ such that $ad - bc = t$. Hence, for a unique r modulo tR,

$$q = \left[R\begin{pmatrix}t\\0\end{pmatrix} + R\begin{pmatrix}r\\1\end{pmatrix}\right] = \begin{pmatrix}1 & r\\0 & 1\end{pmatrix}p'$$

or

$$q = \left[R\begin{pmatrix}0\\t\end{pmatrix} + R\begin{pmatrix}1\\r\end{pmatrix}\right] = \begin{pmatrix}0 & 1\\1 & r\end{pmatrix}p'.$$

Thus $E = Ge$, as desired. We remark that the neighbours of ιe are indexed by two copies of k.

By the Structure Theorem 4.1, $\mathrm{GL}_2(K) = G_{\iota e} \underset{G_e}{*} G_{\tau e}$. Notice that $G_{\iota e} = G_p = \mathrm{GL}_2(R) \cdot K^*$. Also, $p' = gp$, where

$$g = \begin{pmatrix}0 & t\\1 & 0\end{pmatrix},$$

so $G_{p'} = gG_p g^{-1}$, and it follows that

$$G_e = G_{p,p'} = \begin{pmatrix}U & tR\\R & U\end{pmatrix} \cdot K^*.$$

Finally, $G_{\tau e} = G_{\{p,p'\}}$ has index 2 in G_e and contains g which is not an element of $\mathrm{GL}_2(R)$, so

$$G_{\tau e} = \begin{pmatrix} U & tR \\ R & U \end{pmatrix} \cdot K^* \cup \begin{pmatrix} tR & tU \\ U & tR \end{pmatrix} \cdot K^*.$$

In summary, $\mathrm{GL}_2(K) = A \underset{C}{*} B$, where

$$A = \mathrm{GL}_2(R) \cdot K^*,$$

$$B = \begin{pmatrix} U & tR \\ R & U \end{pmatrix} \cdot K^* \cup \begin{pmatrix} R & U \\ t^{-1}U & R \end{pmatrix} \cdot K^*,$$

$$C = \begin{pmatrix} U & tR \\ R & U \end{pmatrix} \cdot K^*.$$

(ii) Let $H = \mathrm{SL}_2(K)$ and view T as an H-tree. By considering determinants it is not difficult to show that $K^* \backslash \mathscr{L}$ has exactly two H-orbits, Hp, Hgp, where

$$g = \begin{pmatrix} 0 & t \\ 1 & 0 \end{pmatrix}.$$

The above argument then shows that E has two H-orbits He, Hge, and one can compute that $\mathrm{SL}_2(K) = A \underset{C}{*} B$, where

$$A = \mathrm{SL}_2(R),$$

$$B = g\,\mathrm{SL}_2(R) g^{-1} = \left\{ \begin{pmatrix} a & tc \\ b/t & d \end{pmatrix} \middle| \begin{pmatrix} a & c \\ b & d \end{pmatrix} \in \mathrm{SL}_2(R) \right\}$$

and

$$C = A \cap B = \left\{ \begin{pmatrix} a & c \\ b & d \end{pmatrix} \in \mathrm{SL}_2(R) \middle| c \in tR \right\}.$$

In fact we can contract all edges in Hge and obtain a new H-tree $\bar{T}$ with vertex set $K^* \backslash \mathscr{L}$ and edge set $\{\{v, w\} \mid v, w \in K^* \backslash \mathscr{L}$ with $d(v, w) = 1\}$. Here $\{v, w\}$ has v, w as its vertices, with the initial vertex being the one lying in Hp. One can think of T as being the barycentric subdivision of $\bar{T}$.

(iii) Consider the case where $K = k(x)$ the field of rational functions in an indeterminate $x, R = \{f/g \mid f, g \in k[x], \deg f \leqslant \deg g\} = k[x^{-1}]_{(x^{-1})}$ and the uniformizer is $t = x^{-1}$. The above $\mathrm{GL}_2(K)$-tree T is then a $\mathrm{GL}_2(k[x])$-tree, and one can verify that

$$\left[R\begin{pmatrix} 1 \\ 0 \end{pmatrix} + R\begin{pmatrix} 0 \\ t^n \end{pmatrix} \right], \quad n \geqslant 0,$$

is a $\mathrm{GL}_2(k[x])$-transversal in $K^* \backslash \mathscr{L}$. The Structure Theorem 4.1 then shows that $\mathrm{GL}_2(k[x]) = \mathrm{GL}_2(k) \underset{T_2(k)}{*} T_2(k[x])$, where T_2 denotes lower triangular 2×2 matrices. ∎

5.4 Examples. Let k be a field and x, y indeterminates. Write $k\langle x, y\rangle$ for the free associative k-algebra in x, y, so x, y commute with the elements of k but not with each other. Write $k[x, y]$ for the polynomial ring over k in x, y, the abelianization of $k\langle x, y\rangle$.

Let $R = k\langle x, y\rangle$ or $k[x, y]$, and let G be the group of all k-algebra automorphisms of R, that is, ring automorphisms of R which act as the identity on k. Such an automorphism ϕ is completely specified by the ordered pair $(\phi x, \phi y)$, so we can use ordered pairs to describe automorphisms.

Let X be the graph whose vertices are the k-subspaces of R and whose edges are the inclusion relations between them. Thus an edge corresponds to an inclusion $v \subseteq w$, where v, w are k-subspaces of R, and we specify that v, w are the initial and terminal vertex, respectively. Clearly X is a G-graph.

Consider the vertices $v = k + kx, w = k + kx + ky$ of X and let e be the edge of X joining them. Let Y be the subgraph $\{v, e, w\}$ of X, and let $T = GY$. Clearly T is a G-subgraph of X. By some rather involved manipulations with leading forms one can show that T is a G-tree; see Cohn (1985). Since T clearly has fundamental G-transversal Y, it follows that $G = A \underset{C}{*} B$, where

$$A = G_v = \{(\lambda + \alpha x, \delta y + f(x)) \mid f(x) \in k[x], \alpha, \delta, \lambda \in k, \alpha\delta \neq 0\},$$

the group of *x-based de Jonquières automorphisms of* R,

$$B = G_w = \{(\lambda + \alpha x + \beta y, \mu + \gamma x + \delta y) \mid \alpha, \beta, \gamma, \delta, \lambda, \mu \in k, \alpha\delta \neq \beta\gamma\},$$

the group of *affine automorphisms* of R, and

$$C = G_e = \{(\lambda + \alpha x, \mu + \gamma x + \delta y) \mid \alpha, \gamma, \delta, \lambda, \mu \in k, \alpha\delta \neq 0\},$$

the group of affine x-based de Jonquières automorphisms of R.

In particular, G is the same for both $R = k\langle x, y\rangle$ and $R = k[x, y]$. ∎

6 The exact sequence for a tree

6.1 Definitions. By a *G-module M* we mean an abelian group M which is a G-set such that G acts by abelian group automorphisms on M. An additive G-map between G-modules is called a *G-linear* map.

For any G-set V we write $\mathbb{Z}V$ or $\mathbb{Z}[V]$ for the free abelian group on V, and write the elements as formal sums $\sum_{v \in V} n_v v$, with $n_v \in \mathbb{Z}$ being

0 for all but finitely many $v \in V$. This is a G-module with G-action $g\left(\sum_{v\in V} n_v v\right) = \sum_{v\in V} n_v gv$. The G-linear map $\varepsilon: \mathbb{Z}V \to \mathbb{Z}$ with $\varepsilon\left(\sum_{v\in V} n_v v\right) = \sum_{v\in V} n_v$ is called the *augmentation map of V*. The kernel of ε is a G-submodule $\omega\mathbb{Z}V$ of $\mathbb{Z}V$ called the *augmentation module* of V. ∎

6.2 Definitions. Let (X, V, E, ι, τ) be a G-graph.

By the *boundary map of X* we mean the G-linear map $\partial: \mathbb{Z}E \to \mathbb{Z}V$ with $\partial(e) = \tau e - \iota e$ for all $e \in E$, that is, $\partial\left(\sum_{e\in E} n_e e\right) = \sum_{e\in E} n_e \tau e - \sum_{e\in E} n_e \iota e$. The sequence

$$(1) \qquad 0 \to \mathbb{Z}E \xrightarrow{\partial} \mathbb{Z}V \xrightarrow{\varepsilon} \mathbb{Z} \to 0$$

is then a *complex*, that is the composite of any two consecutive maps is zero; it is called the *augmented cellular chain complex of X*, or simply the *complex for X*.

The abelian groups in the complex are called the *terms* of the complex. A complex is said to be *exact* at a term if the kernel of the map outward from that term is precisely the image of the map into that term. The complex itself is *exact* if it is exact at each term. ∎

Notice that the complex (1) is exact at $\mathbb{Z}$, that is, ε is surjective, because V is nonempty.

We now examine exactness at each of the terms.

6.3 Lemma. *X is connected if and only if* $\mathbb{Z}E \xrightarrow{\partial} \mathbb{Z}V \xrightarrow{\varepsilon} \mathbb{Z} \to 0$ *is exact.*

Proof. The cokernel of ∂ can be expressed in the form $\mathbb{Z}C$, where C is the G-set obtained from V by identifying $\tau e = \iota e$ for all $e \in E$. Since C can be viewed as the set of components of X, the result is proved. ∎

6.4 Lemma. *X is a forest if and only if* $0 \to \mathbb{Z}V \xrightarrow{\partial} \mathbb{Z}E$ *is exact.*

Proof. If X is not a forest then there is a simple closed path $v_0, e_1^{\varepsilon_1}, \ldots, e_n^{\varepsilon_n}, v_n = v_0$ in X, so $\partial(\varepsilon_1 e_1 + \cdots + \varepsilon_n e_n) = (-v_0 + v_1) + (-v_1 + v_2) + \cdots + (-v_{n-1} + v_n) = -v_0 + v_n = 0$. But $\varepsilon_1 e_1 + \cdots + \varepsilon_n e_n \neq 0$ so ∂ is not injective.

The converse is an easy exercise, and in fact we shall now see how to construct a one-sided inverse for ∂. ∎

6.5 Definitions. Let $T = (T, V, E, \iota, \tau)$ be a G-tree.

For any vertices v, w of T there is a geodesic $v = v_0, e_1^{\varepsilon_1}, \ldots, e_n^{\varepsilon_n}, v_n = w$, and we write $T[v, w]$ for the element $\varepsilon_1 e_1 + \cdots + \varepsilon_n e_n$ of $\mathbb{Z}E$. Notice that $T[w, v] = -T[v, w]$ and $T[v, w] = T[v, u] + T[u, w]$ for all vertices u. Since $\mathbb{Z}V$ is free abelian we can extend the map $T[v, \text{-}]: V \to \mathbb{Z}E$, $w \mapsto T[v, w]$, to a map $\mathbb{Z}V \to \mathbb{Z}E$, and we denote this new map also by $T[v, \text{-}]$. ∎

6.6 Theorem. *If T is a G-tree then there is an exact sequence of G-modules $0 \to \mathbb{Z}ET \xrightarrow{\partial} \mathbb{Z}VT \xrightarrow{\varepsilon} \mathbb{Z} \to 0$; hence $\omega\mathbb{Z}VT \approx \mathbb{Z}ET$ as G-modules.*

Proof. Let v be a vertex of T. For any edge e of T, $T[v, \partial e] = -T[v, \iota e] + T[v, \tau e] = T[\iota e, v] + T[v, \tau e] = T[\iota e, \tau e] = e$. Thus the composite $\mathbb{Z}ET \xrightarrow{\partial} \mathbb{Z}VT \xrightarrow{T[v,\text{-}]} \mathbb{Z}ET$ is the identity, so ∂ is injective. ∎

In Theorem 9.2 we shall see the corresponding exact sequence for a connected G-graph.

In proving that the standard graph is a tree in the next section, we shall use the appropriate part of Lemma 6.4. However, the above proof of Theorem 6.6 provides the actual motivation for the argument. It will be useful to have the following concepts.

6.7 Definitions. Let M be a group on which G acts by group automorphisms. Let us write M additively, although it need not be abelian.

By the *semidirect product* $G \ltimes M$ or $M \rtimes G$ we mean the group with underlying set $G \times M$ and multiplication $(g, m)(g', m') = (gg', m + gm')$, suggestive of 2×2 matrices

$$\begin{pmatrix} G & M \\ 0 & 1 \end{pmatrix}.$$

Now let M be abelian, so M is a G-module.

By a *derivation* $d: G \to M$ we mean a function such that $d(xy) = d(x) + xd(y)$ for all $x, y \in G$.

It is straightforward to verify that a function $d: G \to M$ is a derivation if and only if the function $(1, d): G \to G \ltimes M, g \mapsto (g, dg)$, is a group homomorphism.

For any $m \in M$, the function $\operatorname{ad} m: G \to M, g \mapsto gm - m$, is a derivation; in $G \ltimes M$ it corresponds to right conjugation by $(1, m)$.

A derivation $d: G \to M$ is *inner* if $d = \operatorname{ad} m$ for some $m \in M$; otherwise d is *outer*. ∎

6.8 Example. Let T be a G-tree and v a vertex of T.

The inner derivation $\operatorname{ad} v: G \to \mathbb{Z}VT$ has image lying in $\omega\mathbb{Z}VT \approx \mathbb{Z}ET$. The resulting derivation $G \to \mathbb{Z}ET$ is given by $g \mapsto T[v, gv]$, and is denoted $T[v, \_v]$. ∎

7 The fundamental group and its tree

We now analyze the fundamental group of a graph of groups and verify that the standard graph is a tree. Throughout this section we fix the following terminology introduced in Section 3:

7.1 Notation. Let $(G(-), Y)$ be a graph of groups with connected graph (Y, V, E, ι, τ), vertex groups $G(v), v \in V$, edge groups $G(e) \subseteq G(\iota e), e \in E$, and edge functions $t_e: G(e) \to G(\tau e), g \mapsto g^{t_e}, e \in E$.

Let Y_0 be a maximal subtree of Y and v_0 a vertex of Y.

Let $G = \pi(G(-), Y, Y_0)$, the group presented with

generating set: $\{t_e \mid e \in E\} \vee \bigvee_{v \in V} G(v)$

relations: the relations for $G(v), v \in V$;

$$gt_e = t_e g^{t_e} \quad \text{for all } e \in E, \quad g \in G(e) \subseteq G(\iota e) \quad \text{so } g^{t_e} \in G(\tau e);$$
$$t_e = 1 \quad \text{for all } e \in EY_0.$$

Let $T = T(G(-), Y, Y_0)$, the G-graph presented with generating set Y, relations saying y is $G(y)$-stable, and incidence functions given by $\iota(ge) = g\iota e, \tau(ge) = gt_e\tau e$ for all $g \in G, e \in E$. ∎

7.2 Lemma. *Suppose that H is a group and that there are specified group homomorphisms $\alpha_v: G(v) \to H, v \in VY$, and a function $\alpha: E \to H$, such that $\alpha_{\iota e}(g)\alpha(e) = \alpha(e)\alpha_{\tau e}(g^{t_e})$ for all $e \in E$, $g \in G(e) \subseteq G(\iota e)$. For $v, w \in VY$, define $\alpha(v, w)$ to be the element $\alpha(e_1)^{\varepsilon_1} \cdots \alpha(e_n)^{\varepsilon_n}$ of H where $e_1^{\varepsilon_1}, \ldots, e_n^{\varepsilon_n}$ is the Y_0-geodesic from v to w.*

Then there exists a group homomorphism $\beta: G \to H$ defined on the given generating set as

$$\beta(g) = \alpha(v_0, v)\alpha_v(g)\alpha(v, v_0) \quad \textit{for all } g \in G(v), \quad v \in V,$$

and

$$\beta(t_e) = \alpha(v_0, \bar{\imath}e)\alpha(e)\alpha(\bar{\tau}e, v_0) \quad \text{for all } e \in E.$$

Proof. We need check only that β respects the relations of G.

It is easy to show that, for any $u, v, w \in V$, $\alpha(u, v)\alpha(v, w) = \alpha(u, w)$, and $\alpha(w, v) = \alpha(v, w)^{-1}$.

For $v \in V$, consider the restriction of β to the subset $G(v)$ of the generating set. It is obtained by composing α_v with left conjugation by $\alpha(v_0, v) = \alpha(v, v_0)^{-1}$, so is a group homomorphism, which means that the relations of $G(v)$ are respected.

For each $e \in E$ and $g \in G(e) \subseteq G(\bar{\imath}e)$,

$$\begin{aligned}
\beta(g)\beta(t_e) &= \alpha(v_0, \bar{\imath}e)\alpha_{\bar{\imath}e}(g)\alpha(\bar{\imath}e, v_0)\alpha(v_0, \bar{\imath}e)\alpha(e)\alpha(\bar{\tau}e, v_0) \\
&= \alpha(v_0, \bar{\imath}e)\alpha_{\bar{\imath}e}(g)\alpha(e)\alpha(\bar{\tau}e, v_0) = \alpha(v_0, \bar{\imath}e)\alpha(e)\alpha_{\bar{\tau}e}(g^{t_e})\alpha(\bar{\tau}e, v_0) \\
&= \alpha(v_0, \bar{\imath}e)\alpha(e)\alpha(\bar{\tau}e, v_0)\alpha(v_0, \bar{\tau}e)\alpha_{\bar{\tau}e}(g^{t_e})\alpha(\bar{\tau}e, v_0) \\
&= \beta(t_e)\beta(g^{t_e}).
\end{aligned}$$

For each $e \in EY_0$, $\beta(t_e) = \alpha(v_0, \bar{\imath}e)\alpha(e)\alpha(\bar{\tau}e, v_0) = \alpha(v_0, \bar{\imath}e)\alpha(\bar{\imath}e, \bar{\tau}e)\alpha(\bar{\tau}e, v_0) = \alpha(v_0, v_0) = 1$.

Thus β does respect the relations of G and we have the desired group homomorphism $\beta : G \to H$. ∎

7.3 Definition. Let P be the group presented with

generating set: $\{u_e \,|\, e \in E\} \vee \bigvee_{v \in V} G(v)$

relations: the relations for $G(v)$, $v \in V$;

$gu_e = u_e g^{t_e}$ for all $e \in E$, $g \in G(e) \subseteq G(\bar{\imath}e)$, so $g^{t_e} \in G(\bar{\tau}e)$.

The *fundamental group of* $(G(-), Y)$ *with respect to* v_0, denoted $\pi(G(-), Y, v_0)$, is defined to be the subgroup of P consisting of all elements p for which there exists an expression $p = g_0 u_{e_1}^{\varepsilon_1} u_1 \cdots g_{n-1} u_{e_n}^{\varepsilon_n} g_n$ and a closed path $v_0, e_1^{\varepsilon_1}, v_1, \ldots, e_n^{\varepsilon_n}, v_n = v_0$ in Y at v_0, with $g_i \in G(v_i)$ for all $i \in [0, n]$.

There is a well-defined group homomorphism $P \to G$ sending each u_e to t_e. Using the function $E \to P, e \mapsto u_e$, we can apply Lemma 7.2 to get a group homomorphism $\beta : G \to P$, and the composite $G \xrightarrow{\beta} P \to G$ is then the identity, so β is injective. It is not difficult to show that $\beta G = \pi(G(-), Y, v_0)$, and we have $\pi(G(-), Y, Y_0) = G \approx \beta G = \pi(G(-), Y, v_0)$.

Hence the isomorphism class of $\pi(G(-), Y, Y_0)$ is independent of the choice of Y_0. Where we are dealing with abstract group properties we shall sometimes speak of the *fundamental group* of $(G(-), Y)$ and write $\pi(G(-), Y)$.

Here P is a semidirect product of G by the normal closure of $\{u_e | e \in EY_0\}$. ∎

7.4 Theorem. *If U is a nonempty set such that $|U|$ is uniquely divisible by $|G(v)|$ for each $v \in V$, then there exists a group homomorphism $G \to \operatorname{Sym} U$ such that the composite $G(v) \to G \to \operatorname{Sym} U$ is injective for each $v \in V$.*

Proof. For each $v \in V$, $|G(v)|$ divides $|U|$ so we can partition U into copies of $G(v)$, and hence define a free $G(v)$-action on U; we denote by $\alpha_v : G(v) \to \operatorname{Sym} U$ the resulting injective homomorphism.

Consider any $e \in E$. The free $G(\bar{\iota}e)$ and $G(\bar{\tau}e)$ actions on U induce free $G(e)$-actions on U via the maps $G(e) \subseteq G(\bar{\iota}e)$ and $t_e : G(e) \to G(\bar{\tau}e)$, respectively; let us denote the corresponding $G(e)$-sets as $U_{\bar{\iota}}$, $U_{\bar{\tau}}$. These two free $G(e)$-sets are isomorphic because $|U|$ is uniquely divisible by $|G(e)|$, so there exists a $G(e)$-isomorphism $\alpha(e): U_{\bar{\tau}} \to U_{\bar{\iota}}$. Thus for all $g \in G(e)$, $u \in U$, we have $\alpha(e)(\alpha_{\bar{\tau}e}(g^{t_e})u) = \alpha_{\bar{\iota}e}(g)(\alpha(e)(u))$. Hence $\alpha(e)$ is an element of $\operatorname{Sym} U$ such that for all $g \in G(e)$, $\alpha(e)\alpha_{\bar{\tau}e}(g^{t_e}) = \alpha_{\bar{\iota}e}(g)\alpha(e)$.

By Lemma 7.2, there is a group homomorphism $\beta : G \to \operatorname{Sym} U$ such that for each $v \in VY$, the composite $G(v) \to G \to \operatorname{Sym} U$ is conjugate to α_v, so is injective, since α_v is injective. ∎

7.5 Corollary. *$G(v) \to G$ is injective for each $v \in VY$.*

Proof. Let U be any infinite set such that $|U| > |G(v)|$ for all $v \in VY$. Then $|U|$ is uniquely divisible by $|G(v)|$ for all $v \in VY$. It now follows from Theorem 7.4 that for each $v \in VY$, $G(v) \to G$ is injective. ∎

Henceforth the $G(y)$, $y \in Y$, will be treated as subgroups of G.

We now come to the main result of the chapter which will have many applications in the sequel.

7.6 Theorem. *T is a tree.*

Proof. To see that T is connected we consider the set CT of components of T. There is a natural map $VT \to CT$, $v \mapsto [v]$. This is a G-map, and $[\iota e] = [\tau e]$ for all $e \in ET$. Recall that VT is the G-set generated by V with relations saying that each $v \in V$ is $G(v)$-stable. Hence CT is the G-set generated by V with relations saying that each $[v] \in CT$ is $G(v)$-stable, and additional relations saying that for each $e \in E$, $[\iota e] = [\tau e]$, that is, $[\bar{\iota}e] = [t_e \bar{\tau}e] = t_e[\bar{\tau}e]$. As e ranges over EY_0 the latter relations have the form $[\bar{\iota}e] = [\bar{\tau}e]$ for all $e \in EY_0$, and thus $[v] = [v_0]$ for all $v \in V$, so CT

has only one G-orbit. Now $[v_0]$ is $G(v)$-stable for each $v \in V$, since $[v] = [v_0]$; and $[v_0]$ is t_e-stable for each $e \in EY$, since $[v_0] = [\iota e] = t_e[\tau e] = t_e[v_0]$. Thus $[v_0]$ is G-stable, so CT consists of a single element. Hence T is connected.

By Lemma 6.4 it now sufficies to show that the boundary map $\partial: \mathbb{Z}ET \to \mathbb{Z}VT$ is injective.

Without knowing that T is a tree, we will be able to construct the derivation $T[v_0, \_v_0]: G \to \mathbb{Z}ET$ using the generators and relations of G.

For each $v \in V$, there is an obvious group homomorphism $\alpha_v: G(v) \to G \ltimes \mathbb{Z}ET$, $g \mapsto (g, 0)$. Define a function $\alpha: E \to G \ltimes \mathbb{Z}ET$, $e \mapsto (t_e, e)$.

If $g \in G(e)$, $e \in E$, then $\alpha_{\iota e}(g)\alpha(e) = (g, 0)(t_e, e) = (gt_e, ge) = (t_e g^{t_e}, e) = (t_e, e)(g^{t_e}, 0) = \alpha(e)\alpha_{\tau e}(g^{t_e})$, and the hypotheses of Lemma 7.2 are satisfied.

As in Lemma 7.2, for $v, w \in V$, if $e_1^{\varepsilon_1}, \ldots, e_n^{\varepsilon_n}$ is the Y_0-geodesic from v to w, then $\alpha(v, w) = \alpha(e_1)^{\varepsilon_1} \cdots \alpha(e_n)^{\varepsilon_n} = (1, e_1)^{\varepsilon_1} \cdots (1, e_n)^{\varepsilon_n} = (1, Y_0[v, w])$, where $Y_0[v, w] = \varepsilon_1 e_1 + \cdots + \varepsilon_n e_n \in \mathbb{Z}EY_0 \subseteq \mathbb{Z}ET$. By Lemma 7.2 there exists a group homomorphism $\beta: G \to G \ltimes \mathbb{Z}ET$ such that

$$\begin{aligned}
&\text{if } g \in G(v), \quad v \in V \quad \text{then } \beta(g) = \alpha(v_0, v)\alpha_v(g)\alpha(v, v_0) \\
&\quad = (1, Y_0[v_0, v])(g, 0)(1, Y_0[v, v_0]) = (g, Y_0[v_0, v] + gY_0[v, v_0]); \\
&\text{if } e \in E \quad \text{then } \beta(t_e) = \alpha(v_0, \iota e)\alpha(e)\alpha(\tau e, v_0) \\
&\quad = (1, Y_0[v_0, \iota e])(t_e, e)(1, Y_0[\tau, v_0]) \\
&\quad = (t_e, Y_0[v_0, \iota e] + e + t_e Y_0[\tau e, v_0]).
\end{aligned}$$

It is clear that β has the form $(1, d): G \to G \ltimes \mathbb{Z}ET$, so we have a map $G \to \mathbb{Z}ET$, denoted $T[v_0, \_v_0]$, such that

(1) $T[v_0, \_v_0]$ is a derivation;

(2) if $g \in G(v), v \in V$, then $T[v_0, gv_0] = Y_0[v_0, v] + gY_0[v, v_0]$;

(3) if $e \in E$ then $T[v_0, t_e v_0] = Y_0[v_0, \iota e] + e + t_e Y_0[\tau e, v_0]$.

We claim there is a well-defined additive map $T[v_0, \text{-}]: \mathbb{Z}VT \to \mathbb{Z}ET$ such that

(4) $T[v_0, gv] = T[v_0, gv_0] + gY_0[v_0, v]$ for all $g \in G$, $v \in V$.

Thus suppose $gv = g'v'$ with $g, g' \in G$, $v, v' \in V$. Then $v = v'$ and $g' = gh$ for some $h \in G(v)$ so

$$\begin{aligned}
&T[v_0, g'v_0] + g'Y_0[v_0, v] \\
&\quad = T[v_0, ghv_0] + ghY_0[v_0, v] \\
&\quad = \{T[v_0, gv_0] + gT[v_0, hv_0]\} + ghY_0[v_0, v] \text{ by (1)} \\
&\quad = T[v_0, gv_0] + g\{Y_0[v_0, v] + hY_0[v, v_0]\} + ghY_0[v_0, v] \text{ by (2)} \\
&\quad = T[v_0, gv_0] + gY_0[v_0, v].
\end{aligned}$$

Thus $T[v_0,\text{-}]:\mathbb{Z}VT \to \mathbb{Z}ET$ is well-defined.

For any $g \in G$, $e \in E$ we have

$$\begin{aligned}
T[v_0, \partial ge] &= -T[v_0, g\iota e] + T[v_0, gt_e\tau e] \\
&= \{gY_0[\iota e, v_0] - T[v_0, gv_0]\} + \{T[v_0, gt_e v_0] + gt_e Y_0[v_0, \tau e]\} \\
&\quad \text{by (4)} \\
&= gY_0[\iota e, v_0] - T[v_0, gv_0] + \{T[v_0, gv_0] + gT[v_0, t_e v_0]\} \\
&\quad + gt_e Y_0[v_0, \tau e] \text{ by (1)} \\
&= gY_0[\iota e, v_0] + gT[v_0, t_e v_0] + gt_e Y_0[v_0, \tau e] \\
&= gY_0[\iota e, v_0] + g\{Y_0[v_0, \iota e] + e + t_e Y_0[\tau e, v_0]\} \\
&\quad + gt_e Y_0[v_0, \tau e] \text{ by (3)} \\
&= ge.
\end{aligned}$$

It follows that $\mathbb{Z}ET \xrightarrow{\partial} \mathbb{Z}VT \xrightarrow{T[v_0,\text{-}]} \mathbb{Z}ET$ is the identity map, so ∂ is injective, as desired. ∎

If H is any subgroup of G then T is an H-tree so H is isomorphic to the fundamental group of the graph of groups associated with T with respect to a fundamental H-transversal and connecting elements. If H acts freely on the edges the structure of H is fairly easy to describe.

7.7 Theorem. *If H is a subgroup of G which intersects each G-conjugate of each edge group $G(e)$ trivially then $H = F * \underset{i \in I}{*} H_i$ for some free subgroup F, and subgroups H_i of the form $H \cap gG(v)g^{-1}$ as g ranges over a certain set of double coset representatives in $H \backslash G/G(v)$ and v ranges over VY.* ∎

7.8 The Kurosh Subgroup Theorem. *If H is a subgroup of a free product $\underset{v \in V}{*} G(v)$ then $H = F * \underset{i \in I}{*} H_i$ for some free subgroup F, and subgroups H_i of the form $H \cap gG(v)g^{-1}$ as g ranges over a certain set of double coset representatives in $H \backslash G/G(v)$ and v ranges over V.* ∎

There are even better actions such as those occurring in Propositions 4.5, 4.6 and Corollary 4.9.

7.9 Proposition. *If a subgroup H of G intersects each G-conjugate of each vertex group trivially then H is free.*

For example, if H is torsion-free and the vertex groups are torsion groups then H is free. ∎

7.10 Proposition. *If $G \to A$ is a homomorphism of groups which is injective on each vertex group then the kernel is free.* ∎

7.11 Proposition. *Every finite subgroup of G lies in some conjugate of some vertex group.* ∎

8 Free groups

8.1 Definitions. Let Y be a connected graph, Y_0 a maximal subtree of Y, and v_0 a vertex of Y.

Form the graph of groups $(G(-), Y)$ with $G(y) = 1$ for all $y \in Y$.

In Example 3.5 (i), we defined the *fundamental group of Y with respect to* Y_0 to be $\pi(Y, Y_0) = \pi(G(-), Y, Y_0)$; this is essentially the free group on $EY - EY_0$.

We define the *fundamental group of Y with respect to* v_0 as $\pi(Y, v_0) = \pi(G(-), Y, v_0)$; that is, the subgroup of the free group on EY consisting of all elements $e_1^{\varepsilon_1} e_2^{\varepsilon_2} \cdots e_n^{\varepsilon_n}$, where $e_1^{\varepsilon_1}, e_2^{\varepsilon_2}, \ldots, e_n^{\varepsilon_n}$ is a closed path in Y at v_0. This agrees with the usual notion of the fundamental group of Y at v_0 consisting of homotopy classes of closed paths at v_0.

In Definition 7.3 it was shown that $\pi(Y, Y_0) \approx \pi(Y, v_0)$.

Since the isomorphism type is independent of all choices we agreed to speak of the *fundamental group* of Y and write $\pi(Y)$, thinking of it as the free group of rank $|EY - EY_0|$.

Let $G = \pi(Y, Y_0)$ and write $T = T(G(-), Y, Y_0)$. We treat T as having a distinguished vertex v_0. By Theorem 7.6, T is a tree, and from the construction we see T is a G-free G-tree with $G \backslash T \approx Y$.

Hence the corresponding map $T \to Y$ is an isomorphism on the neighbourhood of each vertex. Any tree with the latter property is called the *universal covering tree* of Y, which agrees with the usual notion of the universal covering space of Y. It is not difficult to show that universal covering trees for Y are unique up to unique isomorphism, as trees with distinguished vertex.

For example, suppose Y has only one vertex, so consists of loops. Here G is free on EY, and T is the Cayley graph of G with respect to EY. The free groups of ranks 1 and 2 were illustrated in Example 2.2(ii), (iv).

We define rank $Y = |EY - EY_0|$, so rank G = rank Y.

For a free group F of finite rank, we define the *Euler characteristic* of F to be $\chi(F) = 1 - \text{rank } F$.

For a finite graph Y the *Euler characteristic of* Y is

$$\chi(Y) = |VY| - |EY| = |VY_0| - |EY| = (1 + |EY_0|) - |EY|$$
$$= 1 - |EY - EY_0| = 1 - \text{rank } Y = 1 - \text{rank } G = \chi(G). \quad ▮$$

Let us repeat one of the above observations and then combine it with Corollary 4.2.

8.2 Theorem. *G is freely generated by a subset S if and only if the Cayley graph $X(G, S)$ is a G-tree.* ▮

8.3 Theorem. *There exists a G-free G-tree if and only if G is a free group.* ▮

The former property is clearly inherited by subgroups.

8.4 The Nielsen–Schreier Theorem. *Every subgroup of a free group is free.* ▮

A closer analysis enables us to describe the ranks of the subgroups.

8.5 The Schreier Index Formula. *If G is a free group of finite rank r and H a subgroup of G of finite index n then H is a free group of rank $1 + n(r - 1)$. In terms of Euler characteristics, $\chi(H) = (G:H)\chi(G)$.*

Proof. Let S be a free generating set of G, and let $T = X(G, S)$ so T is H-free and G-free. By Corollary 4.2, $G \approx \pi(G\backslash T)$, $H \approx \pi(H\backslash T)$. Here $G\backslash T$ is a finite graph, so $\chi(G) = \chi(G\backslash T)$. As ET is G-free, it is $|G\backslash ET|$ copies of G, and hence $(G:H)|G\backslash ET|$ copies of H; thus $(G:H)|G\backslash ET| = |H\backslash ET|$. The analogous result holds for VT, so $H\backslash T$ is a finite graph and $\chi(H\backslash T) = (G:H)\chi(G\backslash T)$. Hence, $\chi(H) = \chi(H\backslash T) = (G:H)\chi(G\backslash T) = (G:H)\chi(G)$. ▮

There are many other results about free groups which can be proved using trees and we conclude this section with a sampling.

8.6 Theorem. *If G is free of finite rank and ϕ an automorphism of G then the subgroup H of elements of G stabilized by ϕ is free of finite rank.*

Proof. Let T be the Cayley graph of G with respect to a free generating set S of G. By Theorem 8.2, T is a G-free G-tree.

Since $VT = G$ we have an H-map $\phi: VT \to VT$, and hence an H-subset

E' of ET consisting of the edges e such that the T-geodesic from $\phi(\iota e)$ to $\phi(\tau e)$ contains e.

We shall show $H\backslash E'$ is finite. Consider any $e \in E'$, so $\iota e = g$, $\tau e = gs$ for some $g \in G$, $s \in S$, and the T-geodesic from $\phi(\iota e) = \phi(g)$ to $\phi(\tau e) = \phi(gs) = \phi(g)\phi(s)$ contains $\iota e = g$. Applying $\phi(g)^{-1}$ we see that the T-geodesic from 1 to $\phi(s)$ contains $\phi(g)^{-1}g$. It is easy to check that the map $H\backslash G \to G$, $Hg \mapsto \phi(g)^{-1}g$, is well-defined and injective. Since S is finite, there are only finitely many pairs (Hg, s) such that $\phi(g)^{-1}g$ lies in the T-geodesic from 1 to $\phi(s)$. Hence, $H\backslash E'$ is finite.

We now ignore the G-action, and view T solely as H-free H-tree.

If $e \in ET - E'$, then $\phi(\iota e)$, $\phi(\tau e)$ lie in the same component of $T - \{e\}$. Hence we can reorient T so that for each $e \in ET - E'$, τe lies in the same component of $T - \{e\}$ as $\phi(\iota e)$, $\phi(\tau e)$. Notice this reorientation respects the H-action.

Also, e is the first edge in the geodesic from ιe to $\phi(\iota e)$. Hence, for any vertex v of T, any edge in $ET - E'$ having v as initial vertex must be the first edge in the T-geodesic from v to $\phi(v)$, but there is at most one such edge. Thus in $T - E'$, each vertex is the initial vertex of at most one edge.

By Corollary 4.2, $H \approx \pi(H\backslash T)$, and it suffices to show that $H\backslash T$ has finite rank. Since $H\backslash E'$ is finite, the graph $(H\backslash T) - (H\backslash E')$ has finitely many components, and it suffices to show that each component has finite rank. But $(H\backslash T) - (H\backslash E') = H\backslash(T - E')$, and here each vertex is the initial vertex of at most one edge. Hence each reduced closed path is oriented cyclically with attached paths pointing in. It follows that no two simple closed paths can ever be attached, and thus each component has rank at most one. ∎

This proves that H is finitely generated.

8.7 Conjecture. If G is free of rank n, and ϕ an automorphism of G, then the subgroup H of elements of G stabilized by ϕ is free of rank at most n.

Discussion. At the time of writing it is not known if rank H can be bounded by a function of n. ∎

8.8 Theorem. *If A and B are finitely generated subgroups of a free group G then $A \cap B$ is finitely generated.*

Proof. Write $C = A \cap B$. By Theorem 8.3 there exists a G-free G-tree T. Let v be a vertex of T, and let T_A, T_B, T_C be the subtrees of T generated by Av, Bv, Cv respectively, so closed under the actions of A, B, C respectively. Clearly T_C is contained in both T_A and T_B. Hence, there is a map $C\backslash T_C \to A\backslash T_A \times B\backslash T_B$, $Ct \mapsto (At, Bt)$, and it is easily seen to be injective.

But $A\backslash T_A$ and $B\backslash T_B$ are finite since A and B are finitely generated. Thus $C\backslash T_C$ is finite, so its fundamental group, C, is finitely generated. ∎

8.9 Conjecture. If A and B are nontrivial finitely generated subgroups of a free group then $\text{rank}(A\cap B) \leqslant 2 - \text{rank}\, A - \text{rank}\, B + \text{rank}(A)\,\text{rank}(B)$, or equivalently $-\chi(A\cap B) \leqslant \chi(A)\chi(B)$.

Discussion. Howson (1954) showed that

$$-\chi(A\cap B) \leqslant 2\chi(A)\chi(B) - \chi(A) - \chi(B) + 2$$

and H. Neumann (1955) improved this to

$$-\chi(A\cap B) \leqslant 2\chi(A)\chi(B).$$

Burns (1969) improved this to

$$-\chi(A\cap B) \leqslant 2\chi(A)\chi(B) + \max\{\chi(A), \chi(B)\},$$

and the matter still rests there. See Nickolas (1985) for more details. ∎

Theorem 8.8 allows us to extend Theorem 8.6 to any finite set of automorphisms.

8.10 Corollary. *For any free group G of finite rank and finitely generated group A of automorphisms of G, the subgroup H of elements of G stabilized by A is free of finite rank.* ∎

9 Groups acting on connected graphs

This section shows that group actions on connected graphs arise from group actions on trees as in Proposition 4.6.

9.1 Notation. Throughout this section let X be a connected G-graph.

Choose a fundamental G-transversal Y in X with subtree Y_0, and denote the incidence functions by $\bar{\iota}, \bar{\tau}$. Choose a vertex v_0 in Y, and for each $e \in EY$ choose an element $t_e \in G$ such that $t_e \bar{\tau} e = \tau e$, with $t_e = 1$ if $e = EY_0$. Let $(G(-), Y)$ be the resulting graph of groups and write $P = \pi(G(-), Y, Y_0)$, $T = T(G(-), Y, Y_0)$. We treat v_0 as an element of Y_0, Y, X and T. ∎

9.2 Structure Theorem for groups acting on connected graphs. *There is a natural extension of groups $1 \to \pi(X) \to P \to G \to 1$.*

Further, $\pi(X)$ acts freely on T, and there is a natural isomorphism of G-graphs $\pi(X)\backslash T \approx X$. In particular, T is the universal covering tree of X.

The action of P on $\pi(X)$ by left conjugation induces a natural G-module structure on $\pi(X)^{ab}$, and there is an exact sequence of G-modules

$$(1) \qquad 0 \to \pi(X)^{ab} \to \mathbb{Z}[EX] \xrightarrow{\partial} \mathbb{Z}[VX] \to \mathbb{Z} \to 0.$$

Proof. Let $v \in VY$. As in the proof of the Structure Theorem 4.1, the paths of length 1 in X starting at v are the sequences of the form $v, gt_e^{\frac{1}{2}(\varepsilon-1)}e^{\varepsilon}, gt_e^{\varepsilon}w$, where v, e^{ε}, w is a path in Y, and $g \in G_v$. Hence, as in the proof of Theorem 4.1, the homomorphism $P \to G$ is surjective. Let N be the kernel, so $G = P/N$.

For each $y \in Y$, the composite $G(y) \to P \to G$ is the natural embedding, so N does not meet any meet any vertex groups. Thus T is N-free and $N \approx \pi(N\backslash T)$ by Corollary 4.2.

The graph $N\backslash T$ is acted on by the group $P/N = G$; moreover, Y is a fundamental G-transversal, the $t_e, e \in EY$, are connecting elements, and the resulting graph of groups agrees with $(G(-), Y)$. As before, the paths of length 1 in $N\backslash T$ starting at v are the sequences of the form $v, gt_e^{\frac{1}{2}(\varepsilon-1)}e^{\varepsilon}, gt_e^{\varepsilon}w$, where v, e^{ε}, w is a path in Y, and $g \in G_v$. It is then not difficult to deduce that $N\backslash T \approx X$ as G-graphs, so $N \approx \pi(X)$. We shall treat this isomorphism as an identification, and we wish to make this precise.

For any element c of N, the path in T from v_0 to cv_0 maps to a closed path in X at v_0 which corresponds to the element of $\pi(X)$ which we identify with c.

We have P acting on $\pi(X)$ by left conjugation, so in the induced action on $\pi(X)^{ab}$, $\pi(X)$ acts trivially, and thus $\pi(X)^{ab}$ has the structure of a module over $P/\pi(X) = G$. The action under $g \in G$ sends an element of $\pi(X)^{ab}$ represented by a closed path q in X at v_0 to the element of $\pi(X)^{ab}$ represented by any *g-translate* of q, that is, a closed path p, gq, p^{-1} where p is a path in X from v_0 to gv_0. This action is independent of all choices.

The function which associates to a path $e_1^{\varepsilon_1}, \ldots, e_n^{\varepsilon_n}$ in X the element $\varepsilon_1 e_1 + \cdots + \varepsilon_n e_n \in \mathbb{Z}[EX]$ induces a natural map $\pi(X)^{ab} \to \mathbb{Z}[EX]$ which is easily seen to be G-linear, and we have a complex (1).

Since X is connected, (1) is exact at $\mathbb{Z}[VX]$ by Lemma 6.3.

Choose a maximal subtree X_0 of X. By Definition 8.1, $\pi(X)$ is the free group on $EX - EX_0$, $\pi(X)^{ab} \approx \mathbb{Z}[EX - EX_0]$, and the natural map to $\mathbb{Z}[EX]$ takes the form $\mathbb{Z}[EX - EX_0] \to \mathbb{Z}[EX], e \mapsto e + X_0[\tau e, \iota e]$, where $X_0[-,-]$ is as in Definition 6.5. This is clearly injective, since composing with the projection onto $\mathbb{Z}[EX - EX_0]$ gives the identity.

It remains to prove exactness at $\mathbb{Z}[EX]$. Suppose $\partial\left(\sum_{e \in EX} n_e e\right) = 0$. We know that $\partial\left(\sum_{e \in EX} n_e(e + X_0[\tau e, \iota e])\right) = 0$ so $\partial\left(\sum_{e \in EX} n_e X_0[\tau e, \iota e]\right) = 0$. But ∂ is injective on $\mathbb{Z}[EX_0]$ by Lemma 6.4, so $\sum_{e \in EX} n_e X_0[\tau e, \iota e] = 0$. Hence $\sum_{e \in EX} n_e e = \sum_{e \in EX} n_e(e + X_0[\tau e, \iota e])$, which is in the image of $\pi(X)^{ab}$, as desired. ∎

We note two consequences of the case where G acts freely on X, so $P=\pi(G\backslash X)$.

9.3 Corollary. *If X is a connected G-free G-graph then there is an extension of groups $1\to\pi(X)\to\pi(G\backslash X)\to G\to 1$.* ∎

9.4 Corollary. *If F is a free group on a set S, N a normal subgroup of F and $G=F/N$ then there is an exact sequence of G-modules*

$$0\to N^{ab}\to\mathbb{Z}[G\times S]\to\mathbb{Z}G\to\mathbb{Z}\to 0.$$

Proof. Let X be the Cayley graph for G with respect to S, so we have an extension of groups $1\to\pi(X)\to\pi(G\backslash X)\to G\to 1$ and an exact sequence of G-modules $0\to\pi(X)^{ab}\to\mathbb{Z}[G\times S]\to\mathbb{Z}G\to\mathbb{Z}\to 0$. Further we can identify $F=\pi(G\backslash X)$ so $N=\pi(X)$. ∎

9.5 Remark. In order to have a complete structure theorem for a group acting on a connected graph, we want an explicit description of $\pi(X)$ as a subgroup of P. An element of $\pi(X)$ corresponds to a unique closed path in X at v_0 and this can be expressed in the form

$$v_0, g_0t_{e_1}^{\frac{1}{2}(\varepsilon_1-1)}e_1^{\varepsilon_1}, g_0t_{e_1}^{\varepsilon_1}v_1, g_0t_{e_1}^{\varepsilon_1}g_1t_{e_2}^{\frac{1}{2}(\varepsilon_2-1)}e_2^{\varepsilon_2}, g_0t_{e_1}^{\varepsilon_1}g_1t_{e_2}^{\varepsilon_2}v_2,\ldots,$$
$$g_0t_{e_1}^{\varepsilon_1}g_1t_{e_2}^{\varepsilon_2}g\cdots g_{n-1}t_{e_n}^{\frac{1}{2}(\varepsilon_n-1)}e_n^{\varepsilon_n}, g_0t_{e_1}^{\varepsilon_1}g_1t_{e_2}^{\varepsilon_2}g_2\cdots g_{n-1}t_{e_n}^{\varepsilon_n}v_n=v_0,$$

where $v_0, e_1^{\varepsilon_1}, v_1, e_2^{\varepsilon_2}, v_2,\ldots, v_{n-1}, e_n^{\varepsilon_n}, v_n=v_0$ is a closed path in Y and $g_i\in G_{v_i}$ for all $i\in[0,n]$. Thdn $g_0t_{e_1}^{\varepsilon_1}g_1t_{e_2}^{\varepsilon_2}g_2\cdots g_{n-1}t_{e_n}^{\varepsilon_n}$ is an element of G_{v_0}, and denoting it by g_n^{-1}, we get an expression $g_0t_{e_1}^{\varepsilon_1}g_1t_{e_2}^{\varepsilon_2}g_2\cdots g_{n-1}t_{e_n}^{\varepsilon_n}g_n$ representing the desired element of P.

For the purpose of presenting G, one wants a set of elements which generate $\pi(X)$ as normal subgroup of P; geometrically this amounts to a set of closed paths at v_0 in X whose G-translates generate all of $\pi(X)$.

For example, if X is the 1-skeleton of a simply connected CW-complex on which G acts cellularly, respecting the orientation of X, then it suffices to take one two-cell from each G-orbit and take the corresponding elements of $\pi(X)$ to present G. The next example illustrates this. ∎

9.6 Example. Let G be the group of symmetries of a cube, so G acts on the graph X in Fig. I.4(i). For any edge e, $\{\iota e, e, \tau e\}$ is a fundamental G-transversal in X, and for definiteness we choose e as indicated.

Let r_1, r_2, r_3 be the reflections in the planes of symmetry π_1, π_2, π_3 indicated in Fig. I.4(ii), (iii), (iv), respectively. It is easy to check that $G_e=\langle r_1\mid r_1^2\rangle\approx D_1$, $G_{\tau e}=\langle r_1, r_2\mid r_1^2, r_2^2, (r_1r_2)^2\rangle\approx D_2$, and $G_{\iota e}=\langle r_1, r_3\mid r_1^2, r_3^2, (r_1r_3)^3\rangle\approx D_3$.

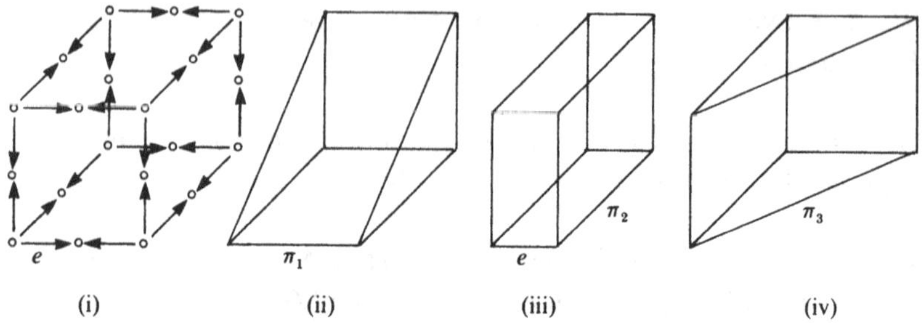

Fig. I.4

By Theorem 9.2, we have an extension of groups

$$1 \to \pi(X) \to D_3 \underset{D_1}{*} D_2 \to G \to 1;$$

here $\pi(X)$ is free of rank $1 - \chi(X) = 1 - 20 + 24 = 5$. (We remark that $D_3 \underset{D_1}{*} D_2 \approx \mathrm{PGL}_2(\mathbb{Z})$ by Example 5.2(ii), and the universal covering tree of X is as in Fig. I.3, page 22; this determines an action of $\mathrm{GL}_2(\mathbb{Z})$ on the cube.)

The path around the bottom of the cube starting with e gives a relation $(r_2 r_3)^4 = 1$, as in Example 2.2(v). The G-translates of this path generate all of $\pi(X)$, and we arrive at the presentation

$$G = \langle r_1, r_2, r_3 \mid r_1^2, r_2^2, r_3^2, (r_1 r_2)^2, (r_1 r_3)^3, (r_2 r_3)^4 \rangle.$$

It is evident that $|G| = |G_e||Ge| = 2 \times 24 = 48$.

In fact, it can be seen by the action on the four diagonals of the cube that $G \approx C_2 \times \mathrm{Sym}\,4$.

The cube has an obvious simply connected CW-structure with X as one-skeleton. There is exactly one G-orbit of faces, and in the above argument the bottom face is chosen as a representative of the G-orbit to find a suitable element generating $\pi(X)$ as normal subgroup of $D_3 \underset{D_1}{*} D_2$.

∎

10 Free products

10.1 Notation. For any set E and equivalence relation $S \subseteq E \times E$ we denote the set of equivalence classes by E/S. The equivalence relation for equality is denoted $\Delta E = \{(e, e) \mid e \in E\}$, called the *diagonal*.

Let X be a graph and $S \subseteq EX \times EX$ an equivalence relation on EX. We define VS to be the equivalence relation on VX generated by $\{(\iota e, \iota f), (\tau e, \tau f) \mid (e, f) \in S\}$. Then the incidence functions $\iota, \tau: EX \to VX$

induce functions $\bar{\iota}, \bar{\tau}: EX/S \to VX/VS$. We denote by X/S the resulting graph. Notice there is a surjective graph map $X \to X/S$.

10.2 Lemma. *If T is a tree and $S \subseteq ET \times ET$ an equivalence relation on ET such that for all $(e, f) \in S$ either $\iota e = \iota f$ or $\tau e = \tau f$ then T/S is a tree.*

Proof. Since the statements concern only finitely many elements at a time, we may assume that T is finite.

Consider now the case where $S = \Delta EX \cup \{(e, f), (f, e)\}$ for distinct edges e, f of T such that $\iota e = \iota f$. Consider the graph $T - \{f\}$ obtained by deleting f from T. Here τe and τf lie in the two different components, so identifying τe and τf attaches together the two components, and we get a tree. It is not difficult to see that this graph is isomorphic to T/S, so T/S is a tree.

The general case is obtained by a finite repetition of such constructions, and the result follows by induction. ∎

10.3 Theorem. *Let N be a normal subgroup of G, and $\bar{T}$ a G/N-tree such that $E\bar{T}$ is G/N-free. If there exists a G-tree T with G-free edge set, and a surjective map of G-trees $T \to \bar{T}$, then there exists such a T with $N\backslash T \approx \bar{T}$ as G-trees.*

Proof. Let us denote the map $T \to \bar{T}$ by $t \mapsto \bar{t}$. Here we have a surjective map of G/N-graphs, $N\backslash T \to \bar{T}$, $Nt \mapsto \bar{t}$. Suppose it is not injective. Since $\bar{T}$ is a tree, it follows that $N\backslash ET \to E\bar{T}$, $Ne \mapsto \bar{e}$, is not injective. This is a map of free G/N-sets, so two G/N-orbits get identified. Thus there exist $e, f \in ET$ with $Ge \neq Gf$, $\bar{e} = \bar{f}$. Hence we can construct a path $p = e_1^{\varepsilon_1}, \ldots, e_n^{\varepsilon_n}$ in T with $Ge_1 \neq Ge_n$, $\bar{e}_1 = \bar{e}_n$.

We claim we can choose such a path p with $n = 2$. Clearly $n \geqslant 2$. Since $\bar{p} = \bar{e}_1^{\varepsilon_1}, \ldots, \bar{e}_n^{\varepsilon_n}$ is a path in $\bar{T}$ with $\bar{e}_1 = \bar{e}_n$, there is some $i \in [1, n-1]$ with $\bar{e}_{i+1}^{\varepsilon_{i+1}} = \bar{e}_i^{-\varepsilon_i}$. If $Ge_{i+1} \neq Ge_i$ then we can take $p = e_i^{\varepsilon_i}, e_{i+1}^{\varepsilon_{i+1}}$ and achieve $n = 2$. It remains to consider the case $Ge_{i+1} = Ge_i$. Here $n \geqslant 3$ and $e_{i+1} = ge_i$ for some $g \in G$. If $i = 1$ or $n - 1$ we can delete the first or last edge of p, respectively, and reduce n by 1. Thus we may assume $n \geqslant 4$ and $i \in [2, n-2]$. Since $\bar{e}_{i+1} = g\bar{e}_i = g\bar{e}_{i+1}$ and $E\bar{T}$ is G/N-free we see $g \in N$. Thus $g\bar{e}_1 = \bar{e}_1 = \bar{e}_n$ and $Gge_1 = Ge_1 \neq Ge_n$, while the path $p' = ge_1^{\varepsilon_1}, \ldots, ge_i^{\varepsilon_i}, e_{i+1}^{\varepsilon_{i+1}}, \ldots, e_n^{\varepsilon_n}$ in T is not a reduced path. Hence we can delete $ge_i^{\varepsilon_i}, e_{i+1}^{\varepsilon_{i+1}}$ from p' and reduce n. This proves the claim that we may assume $n = 2$.

Here $e_1^{\varepsilon_1}, e_2^{\varepsilon_2}$ is a path in T with $\bar{e}_1 = \bar{e}_2$, $Ge_1 \neq Ge_2$. Since $\bar{e}_1^{\varepsilon_1}, \bar{e}_2^{\varepsilon_2}$ is a path in $\bar{T}$ with $\bar{e}_1 = \bar{e}_2$ we see that $\varepsilon_1 = -\varepsilon_2$. Since $e_1^{\varepsilon_1}, e_2^{\varepsilon_2}$ is a path in T either $\iota e_1 = \iota e_2$ or $\tau e_1 = \tau e_2$. Let $S = \Delta E \cup \{(ge_1, ge_2), (ge_2, ge_1) \mid g \in G\}$. It is easy to see that this is an equivalence relation on EX since $Ge_1 \neq Ge_2$ and e_1, e_2 have trivial stabilizers. By Lemma 10.2, it follows that the G-graph $T' = T/S$ is a tree. Further T' has G-free edge set and there is a surjective map of G-trees $T' \to \bar{T}$.

By Zorn's Lemma there is a largest equivalence relation S on ET such that the graph $T' = T/S$ is still a G-tree with G-free edge set having a surjective map of G-trees $T' \to \bar{T}$. The preceding argument shows that, by maximality, $N \backslash T' \to \bar{T}$ must be an isomorphism, as desired. ∎

10.4 Theorem. *Let I be a set and $(\alpha_i : G_i \to \bar{G}_i | i \in I)$ a family of surjective group homomorphisms. Denote the free products by $G = \underset{i \in I}{*} G_i$, $\bar{G} = \underset{i \in I}{*} \bar{G}_i$, and the resulting map by $\alpha : G \to \bar{G}$. For any subgroup H of G such that $\alpha(H) = \bar{G}$, there exist subgroups H_i, $i \in I$, of H such that $\alpha(H_i) = \bar{G}_i$ and $H = \underset{i \in I}{*} H_i$.*

Proof. Clearly we may assume that I is nonempty. Let Y be a tree with vertex set I, and incidence functions $\bar{\iota}, \bar{\tau}$; for example, Y can be a star with one specified element of I being the initial vertex of each edge.

Let $(G(-), Y)$ be the graph of groups such that, for all $i \in VY = I$, $G(i) = G_i$, and for all $e \in EY$, $G(e) = 1$ so $G(e) \subseteq G(\bar{\iota}e)$ and there is a unique group homomorphism $t_e : G(e) \to G(\bar{\tau}e)$. Then $G = \pi(G(-), Y, Y)$ acts on the tree $T = T(G(-), Y, Y)$. Here ET is G-free.

Define $(\bar{G}(-), Y)$ and $\bar{T} = T(\bar{G}(-), Y)$ similarly, so $E\bar{T}$ is $\bar{G}$-free.

There is a surjective graph map $T \to \bar{T}$; in fact, $\bar{T} = (\text{Ker } \alpha) \backslash T$.

We view $T, \bar{T}$ as H-trees in the obvious way. Then $T \to \bar{T}$ is a surjective map of H-trees, ET is H-free and $E\bar{T}$ is H/N-free, where $N = H \cap \text{Ker } \alpha$.

By Theorem 10.3 there exists an H-tree T' with H-free edge set such that $N \backslash T' = \bar{T}$. The natural copy of Y in $\bar{T}$ is a $\bar{G}$-transversal, and $\bar{G} = \alpha(H) = H/N$. Thus we can lift the tree Y in $N \backslash T'$ back to a subtree Z in T' and Z is a fundamental H-transversal in T'. Since ET' is H-free, the Structure Theorem 4.1 shows that $H = \underset{v \in VZ}{*} H_v$. Further, for each $v \in VZ$,

$$\bar{G}_{\bar{v}} = \alpha(H)_{\bar{v}} = \alpha(H_{\bar{v}}) = \alpha(NH_v) = \alpha(H_v), \text{ as desired.} \quad ∎$$

10.5 Theorem. *Let I be a set, and $(G_i | i \in I)$ a family of groups. For any free group F and surjective homomorphism $\alpha : F \to \underset{i \in I}{*} G_i$, there exist subgroups F_i, $i \in I$, of F such that $F = \underset{i \in I}{*} F_i$ and $\alpha(F_i) = G_i$.*

Proof. Consider the natural map $\beta : \underset{i \in I}{*} \alpha^{-1}(G_i) \to F$. The image of β is the subgroup of F generated by the $\alpha^{-1}(G_i)$, $i \in I$, which is $\alpha^{-1}(G) = F$. As β is a surjective map to a free group, there exists a homomorphism $\gamma: F \to \mathop{*}_{i\in I} \alpha^{-1}(G_i)$ such that $\beta\gamma$ is the identity on F; in particular, γ is injective. Hence we have a copy γF of F in $\mathop{*}_{i\in I} \alpha^{-1}(G_i)$ and it is easy to check that the restriction of the natural surjection $\mathop{*}_{i\in I} \alpha^{-1}(G_i) \to \mathop{*}_{i\in I} G_i$ to the copy of F is the given surjection α. The result now follows from Theorem 10.4. ∎

10.6 The Grushko–Neumann Theorem. *For any set I and family $(G_i \mid i \in I)$ of groups,* $\operatorname{rank}(\mathop{*}_{i\in I} G_i) = \sum_{i\in I} \operatorname{rank} G_i$.

Proof. It is clear that $\operatorname{rank}(\mathop{*}_{i\in I} G_i) \leqslant \sum_{i\in I} \operatorname{rank} G_i$.

Let F be a free group with $\operatorname{rank}(F) = \operatorname{rank}(\mathop{*}_{i\in I} G_i)$, so there is a surjection $\alpha: F \to \mathop{*}_{i\in I} G_i$. By Theorem 10.5, there are subgroups F_i, $i \in I$, of F such that $F = \mathop{*}_{i\in I} F_i$ and $\alpha(F_i) = G_i$, so $\sum_{i\in I} \operatorname{rank} G_i \leqslant \sum_{i\in I} \operatorname{rank} F_i$. But the F_i are free groups by the Nielsen–Schreier Theorem 8.4, so $\sum_{i\in I} \operatorname{rank} F_i = \operatorname{rank} F = \operatorname{rank}(\mathop{*}_{i\in I} G_i)$, and we have the reverse inequality. ∎

Notes and comments

The book of Serre (1977), written with the collaboration of Bass, forms the foundation of this chapter.

The first three sections contain little more than notation, not all of which is standard, as noted by Rota (1986). For example, what we call a *graph* is usually called a directed multigraph, and what we call a *tree* is usually called an oriented tree. We felt at liberty to give the short names to the concepts which occurred most frequently in this work.

Free products with amalgamation were introduced by Schreier (1927); the HNN extension take its name from the initials of the authors of Higman, Neumann and Neumann (1949), where the concept was first studied.

Section 4 is taken from Serre (1977). The Structure Theorem 4.1 is due to Bass and Serre; the proof can be used to obtain a normal form, which in turn can be used as a rather cumbersome substitute for a tree. Theorem 4.12 is due to Tits, and Proposition 4.7 is classical.

In Example 5.2, the tree is from Serre (1977), where the action of $SL_2(\mathbb{Z})$ is described; the extension to $GL_2(\mathbb{Z})$ was pointed out to us by Paul Gerardin. The group-theoretic conclusions are essentially well-known.

Example 5.3 is due to Serre (1977), and the interested reader can find many more details there. The group-theoretic conclusions in (iii), (ii) are a theorem of Nagao (1959) and a generalization of a theorem of Ihara (1966), respectively.

Example 5.4 is a condensed version of the survey we wrote for Cohn (1985); the interested reader will find there details of the arguments, and the numerous attributions.

Theorem 7.6 is due to Bass and Serre, see Serre (1977), and the proof here follows Dicks (1980). Other, more topological, proofs can be found in Chiswell (1979), and Scott and Wall (1979). Theorem 7.7 is a generalization of a result of H. Neumann (1948), which in turn generalizes Theorem 7.8, of Kurosh (1937).

Theorem 8.2 has long been known. Theorem 8.3 seems to have been first stated explicitly in Serre (1977), but is essentially contained in Reidemeister (1932), Section 4, 20. Nielsen (1921) proved Theorem 8.4 for finitely generated subgroups, and Schreier (1927) proved the general case and Theorem 8.5.

Theorem 8.6 is due to Gersten (1984); the elegant proof given here is extracted from Goldstein and Turner (1986) where more is proved; Conjecture 8.7 is due to G.P. Scott.

Theorem 8.8 is due to Howson (1954), whose proof mentions trees. Conjecture 8.9 is from H. Neumann (1953).

Theorem 9.2 is from Serre (1977), and the explicit version in Remark 9.5 was written out by Brown (1984). Corollary 9.4 is due to Lyndon (1950).

Theorem 10.3 is based on results of Chiswell (1976) and Stallings (1965). Theorem 10.4 is due to Higgins (1966). Theorem 10.5 goes back to Wagner (1957), and the argument given here was shown to us by E. Formanek. Theorem 10.6 was proved independently by Grushko (1940) and B.H. Neumann (1943).

II

Cutting graphs and building trees

Section 1 gives a useful characterization of trees in terms of vertices acting as functions on the edge set. Section 2 introduces the concept of the Boolean ring of a connected G-graph, and associates with it an inverse limit of G-trees. This is used in Section 3 to determine the infinite finite-valency distance-transitive graphs, and will also be important in the next chapter.

1 Tree sets

We begin with terminology and notation which will be used frequently throughout the chapter.

1.1 Definitions. Let E, V be G-sets and A a nonempty set.

(i) The set of all functions from E to A will be denoted (E, A); this is a G-set with $(gv)(e) = v(g^{-1}e)$ for all $v \in (E, A)$, $g \in G$, $e \in E$.

If there is specified a G-map $V \to (E, A)$ we denote it by $v \mapsto v|E$ and write $V|E$ for the image. The value of $v|E$ at e will be denoted simply $v(e)$. There is then a dual G-map $E \to (V, A)$, denoted $e \mapsto e|V$, and the same notation applies; thus for $e \in E$, $v \in V$, $e(v) = v(e)$.

(ii) Since $\mathbb{Z}_2$ has a ring structure, $(V, \mathbb{Z}_2)$ is a ring under pointwise addition and multiplication. The 0 and 1 are the obvious constant functions.

If $b \in (V, \mathbb{Z}_2)$ then b^* denotes $1 - b$, or equivalently $1 + b$. For any subset F of $(V, \mathbb{Z}_2)$, F^* denotes $\{f^* | f \in F\}$.

Let $a, b \in (V, \mathbb{Z}_2)$. If $ab^* = 0$ we write $a \leqslant b$; this defines a partial order on $(V, \mathbb{Z}_2)$. We denote by $a \square b$ the family consisting of ab, a^*b, ab^*, a^*b^*; these are then four distinct elements of $a \square b$ even though some of them may be equal as elements of $(V, \mathbb{Z}_2)$. We say that a and b are *nested* if one of the elements of $a \square b$ is 0, or equivalently one of $b \leqslant a^*$, $b \leqslant a$, $a \leqslant b$, $b^* \leqslant a$ holds.

By a *nested* subset of $(V, \mathbb{Z}_2)$ we mean a subset whose elements are pairwise nested.

There is a natural bijection between the elements of $(V, \mathbb{Z}_2)$ and the subsets of V, with a function s corresponding to the set $s^{-1}(1)$. Addition, multiplication, $*$, and $\leqslant$ correspond to symmetric difference, intersection, complement and inclusion, respectively. Where there is little risk of confusion we treat elements of $(V, \mathbb{Z}_2)$ as subsets of V.

(iii) Let $v, v' \in (E, A)$. We write $v \triangledown v' = \{e \in E \mid v(e) \neq v'(e)\}$; if $A = \mathbb{Z}_2$ then $v \triangledown v'$ is the symmetric difference of the corresponding sets, or equivalently (the set corresponding to) $v + v'$. We say that v and v' are *almost equal*, denoted $v =_a v'$, if $v \triangledown v'$ is a finite set.

Almost equality is an equivalence relation on (E, A), and the equivalence classes are called *almost equality classes.* It is clear that the G-action preserves almost equality, so G acts on the set of almost equality classes of (E, A); any element stabilized by this action is called a *G-stable almost equality class.* The elements of a G-stable almost equality class are said to be *almost G-stable.* It is easy to see that an almost equality class is G-stable if and only if it contains a G-orbit, or equivalently an almost G-stable element. ∎

1.2 Definitions. Let T be a G-tree.

There is a function $VT \times ET \to \mathbb{Z}_2$ given by

$$(v, e) \mapsto \langle v, e \rangle = \begin{cases} 1 & \text{if } e \text{ points to } v \\ 0 & \text{if } e \text{ points away from } v. \end{cases}$$

By the *structure map* of T we mean the G-map $ET \to (VT, \mathbb{Z}_2)$, $e \mapsto e|VT: v \mapsto \langle v, e \rangle$. This associates to each $e \in ET$ the set of vertices of the component of $T - \{e\}$ containing τe.

For any $e, e' \in ET$ the elements of $(e|VT) \square (e'|VT)$ correspond to the components of $T - \{e, e'\}$, and 0. Thus $ET|VT$ is nested.

The *costructure map* of T is the dual G-map $VT \to (ET, \mathbb{Z}_2)$, $v \mapsto v|ET: e \mapsto \langle v, e \rangle$ which associates to each vertex v the set of edges pointing to v.

If $v, v' \in VT$ then $(v|ET) \triangledown (v'|ET)$ is the set of edges in the T-geodesic from v to v', so is finite. Thus each $v|ET$ is almost G-stable, and $VT|ET$ lies in a G-stable almost equality class $\widetilde{V}T$ of $(ET, \mathbb{Z}_2)$.

It is easy to see that both the structure map and the costructure map are injective. ∎

1.3 Definition. Let U be a nonempty set.

By a *tree set* in $(U, \mathbb{Z}_2)$ we mean a nested subset E of $(U, \mathbb{Z}_2)$ which does not contain any constant functions and such that the dual, $U|E$, lies in an almost equality class of $(E, \mathbb{Z}_2)$. An element v of $(E, \mathbb{Z}_2)$ is said to *respect the nesting of* E if for all $e, f \in E$, $\varepsilon, \xi \in \{1, *\}$, if $e^\varepsilon f^\xi = 0$ in $(U, \mathbb{Z}_2)$ then $v(e)^\varepsilon v(f)^\xi = 0$ in $\mathbb{Z}_2$. Equivalently, v extends to $E \cup E^* \to \mathbb{Z}_2$ in such a way that for all $e, f \in E \cup E^*$, $v(e^*) = v(e)^*$, and $ef = 0$ implies $v(e)v(f) = 0$.

Let E be a tree set in $(U, \mathbb{Z}_2)$. Since E is nested and contains no constant functions, for all $e, f \in E$ exactly one of $f < e$, $f = e$, $f > e$, $f^* < e$, $f^* = e$, $f^* > e$ holds. Thus there are well-defined maps $\iota, \tau: E \to (E, \mathbb{Z}_2)$ such that

$$\iota e(f) = \begin{cases} 1 & \text{if } f > e \text{ or } f > e^* \\ 0 & \text{if } f \leqslant e \text{ or } f \leqslant e^* \end{cases} \qquad \tau e(f) = \begin{cases} 1 & \text{if } f \geqslant e \text{ or } f > e^* \\ 0 & \text{if } f < e \text{ or } f \leqslant e^* \end{cases}$$

for all $e, f \in E \subseteq (U, \mathbb{Z}_2)$. We write $T(E)$ for the graph having edge set E, vertex set $\iota(E) \cup \tau(E)$, and incidence functions ι, τ.

Theorem 1.5 will show that $T(E)$ is a tree with costructure map $VT(E) \subseteq (E, \mathbb{Z}_2)$, and $VT(E)$ contains $U|E$. Thus for all $v, v' \in VT(E)$, $v \nabla v'$ is the set of edges in the $T(E)$-geodesic joining v, v'.

One way of picturing the double dual $U \to U|E \subseteq VT(E)$ is to represent the elements of U as points, and each vertex v of $T(E)$ as a circle surrounding the elements of U such that $u|E = v$. Each edge e then points to the elements of U which it contains. ∎

1.4 Example. Let $U = \{u_1, u_2, u_3, u_4\}$. Let E be the nested subset of $(U, \mathbb{Z}_2)$ consisting of $e_1 = \{u_1\}$, $e_2 = \{u_1, u_3, u_4\}$, $e_3 = \{u_1, u_2\}$, $e_4 = \{u_3, u_4\}$, $e_5 = \{u_1, u_2, u_3\}$ and $e_6 = \{u_3\}$.

The nesting relations equate $e_1e_2^*$, $e_1e_3^*$, e_1e_4, $e_1e_5^*$, e_1e_6, $e_2^*e_3^*$, $e_2^*e_4$, $e_2^*e_5^*$, $e_2^*e_6$, e_3e_4, $e_3^*e_4^*$, $e_3e_5^*$, e_3e_6, $e_4^*e_5^*$, $e_4^*e_6$, $e_5^*e_6$, to zero.

The elements of $(E, \mathbb{Z}_2)$ which respect the nesting are $v_1 = \{e_1, e_2, e_3, e_5\}$, $v_2 = \{e_3, e_5\}$, $v_3 = \{e_2, e_4, e_5, e_6\}$, $v_4 = \{e_2, e_4\}$, $v_5 = \{e_2, e_3, e_5\}$, $v_6 =$

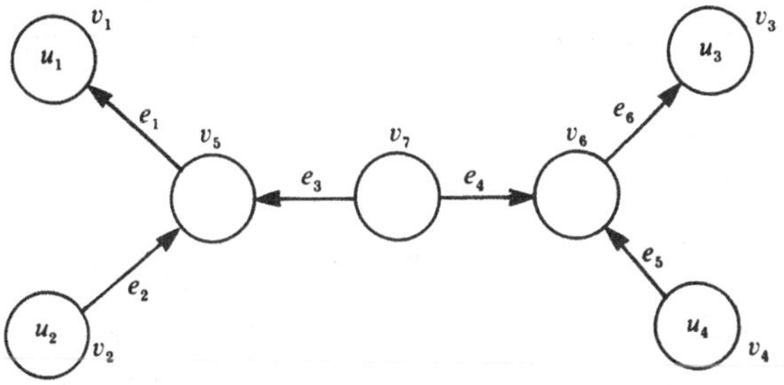

Fig. II.1

$\{e_2, e_4, e_5\}$. The double dual map $U \to (E, \mathbb{Z}_2)$ carries u_i to v_i, $i \in [1, 4]$. The tree $T(E)$ is illustrated in Fig. II.1. It has a vertex $v_7 = \{e_2, e_5\}$ which respects all the nesting relations except $e_3^* e_4^* = 0$. ∎

1.5 Theorem. *Let U be a nonempty G-set and E a tree G-set in $(U, \mathbb{Z}_2)$.*

Then $T(E)$ is a G-tree and the costructure map is the inclusion $VT(E) \to (E, \mathbb{Z}_2)$.

Further, if $\tilde{U}$ is the almost equality class containing $U|E$ then $VT(E)$ contains all elements of $\tilde{U}$ which respect the nesting of E. In particular, $VT(E)$ contains $U|E$.

Proof. We shall treat the elements of E as subsets of U, and the elements of $(E, \mathbb{Z}_2)$ as subsets of E. Also, we shall ignore until the end, both the definition of $T(E)$ and the G-action.

For $e \in E$, and $v \in (E, \mathbb{Z}_2)$ define

$$e[v] = \begin{cases} e \text{ if } v(e) = 1 \\ e^* \text{ if } v(e) = 0, \end{cases} \quad e^*[v] = (e[v])^* = \begin{cases} e^* \text{ if } v(e) = 1 \\ e \text{ if } v(e) = 0. \end{cases}$$

If $u \in U$ then $e[u|E]$ is whichever of e, e^* contains u, and we abbreviate it to $e[u]$.

If $v \in (E, \mathbb{Z}_2)$ then v respects the nesting of E if and only if for all $e, e' \in E$, $e[v]e'[v] \neq 0$ in $(U, \mathbb{Z}_2)$. If $u \in U$, then $u|E$ respects the nesting, since u lies in $e[u]e'[u]$.

To simplify the exposition let us consider first the case where

(1) $\qquad E \cap E^* = \varnothing$;

recall that $E^* = \{e^* | e \in E\}$.

Let $V = \{v \in \tilde{U} | v \text{ respects the nesting}\}$. Clearly $U|E \subseteq V \subseteq \tilde{U}$.

Let T be the labelled graph with $VT = V$ and ET the set of $(v', v) \in V^2$ such that $v = v' \cup \{e\}$ for some $e \in E$ with $e \notin v'$. For any such (v, v') we have $v + v' = \{e\}$ for a unique e, and $v'(e) = 0$, and we assign to (v', v) initial vertex v', terminal vertex v, and label e. It will turn out that the labelling $ET \to E$ is bijective.

Let $v, v' \in V$.

We claim that the following holds:

(2) $\qquad$ for any distinct $e, e' \in v + v'$, either $e'[v] < e[v]$ or $e[v] < e'[v]$.

Here $e[v]e'[v] \neq 0$ since v respects the nesting, and $e^*[v]e'^*[v] = e[v']e'[v'] \neq 0$ since v' respects the nesting. Since $e[v]$, $e'[v]$ are nested by hypothesis, one of $e^*[v]e'[v]$, $e[v]e'^*[v]$ is zero; moreover both cannot be zero since $e \neq e'$, and by (1), $e \neq e'^*$. Thus $e'[v] \leqslant e[v]$ or $e[v] \leqslant e'[v]$, but not both. This proves (2).

Since $v, v' \in \tilde{U}$, we know $v =_a v'$ and it follows from (2) that there is a unique expression $v + v' = \{e_1, \ldots, e_n\}$ such that

(3) $\qquad e_1[v] < \cdots < e_n[v].$

Define $v_1 = v + \{e_1\} \in (E, \mathbb{Z}_2)$.

We claim that $v_1 \in V$. Since $v_1 =_a v$, we see that $v_1 \in \tilde{U}$, and it remains to show that v_1 respects the nesting. Let e, e' be distinct elements of E. We wish to show that $e[v_1]e'[v_1] \neq 0$. If e and e' are different from e_1, then $e[v_1]e'[v_1] = e[v]e'[v] \neq 0$. It remains to show that if $e \neq e_1$ then $e[v_1]e_1[v_1] \neq 0$. If $e \notin v + v'$ then $e[v_1]e_1[v_1] = e[v']e_1[v'] \neq 0$; if $e \in v + v'$ then $e = e_i$ for some $i \in [2, n]$, so by (3) $e_1[v] < e[v]$, and thus $0 < e[v]e_1^*[v] = e[v_1]e_1[v_1]$. Hence $v_1 \in V$.

Now v, v_1 are joined by an edge labelled e_1, and $v_1 + v' = \{e_2, \ldots, e_n\}$. It follows by induction on n that there is a path in T from v to v' and therefore T is connected.

Let p be a reduced path in T. There is a corresponding sequence v_0, $e_1, v_1, e_2, v_2, \ldots, e_n, v_n$ where the v_i are vertices of T and the e_i are elements of E that label the edges. Here $v_0 + v_1 = \{e_1\}$, and $v_1 + v_2 = \{e_2\}$ so $v_0 + v_2 = \{e_1\} + \{e_2\}$. If $e_1 = e_2$ then $v_0 = v_2$ and the underlying edges are the same, which contradicts p being reduced; hence e_1, e_2 are distinct elements of $v_0 + v_2$. If $e_2[v_0] < e_1[v_0]$ then $0 = e_1^*[v_0]e_2[v_0] = e_1[v_1]e_2[v_1]$, which contradicts v_1 respecting the nesting; so by (2) $e_1[v_0] < e_2[v_0] = e_2[v_1]$. By induction there is an ascending sequence $0 < e_1[v_0] < e_2[v_1] < \cdots < e_n[v_{n-1}] < 1$. In particular, $e_1 \neq e_n$.

Hence T is a tree, for otherwise we could construct a reduced path consisting of a simple closed path followed by its first edge, and get a contradiction.

The labelling $ET \to E$ is injective, for otherwise there is a reduced path joining two distinct edges having the same label, which gives a contradiction.

The labelling is surjective, since for any $e \in E$, e is not constant so there exist elements u, u' of U such that $e(u) \neq e(u')$, so $e \in (u|E) + (u'|E)$, and the reduced path in T joining $u|E$ to $u'|E$ contains an edge labelled e.

Thus we may identify $ET = E$.

Consider any $e \in E$, $u \in U$. If e points to $u|E$ in T, that is, e does not lie in the T-geodesic from τe to $u|E$, then by the above analysis, $e \notin (u|E) + \tau e$, so $u(e) = \tau e(e)$, that is, $e(u) = 1$, so $u \in e$. Similarly, $e(u) = 0$ if e points away from $u|E$.

Let us turn now to the general situation where (1) may fail. Express E in the form $E' \cup E''^*$, where $E' \cap E'^* = \emptyset$ and E'' is a subset of E'. Clearly E' is nested and contains no constant functions, and $U|E'$ lies in an almost equality class $\tilde{U}|E'$. Thus the above construction applies to give a tree T' with $U|E' \subseteq VT' \subseteq (E', \mathbb{Z}_2)$ such that if $e \in E'$ and $u \in U$ then e points to $u|E'$ if and only if $u \in e$.

Expand T' to a new tree T with edge set $E = E' \cup E''^*$ and vertex set $V = VT' \cup E''$, intuitively constructed by adding to each $e \in E''$ a vertex in the middle and an arrowhead at the other end. The new edge with the old arrowhead is to be called e, and the other new edge is to be called e^*.

Corresponding to T there is an injective costructure map $V \to (E, \mathbb{Z}_2)$. We have a map $U \to U|E' \subseteq VT' \subseteq V$, and it is clear from the construction of T that if $e \in E$ and $u \in U$ then $u \in e$ if and only if the edge e of T points to the image of u in $VT' \subseteq V$. Thus the composite $U \to V \to (E, \mathbb{Z}_2)$ is $u \mapsto u|E$.

We want to show that if $e, f \in E$ then $\iota e(f)$, $\tau e(f)$ are as described in Definition 1.3. If $f = e$ then $\iota e(f) = 0$, $\tau e(f) = 1$. If $f = e^*$ then by construction $\iota e = \iota f$, so f points away from ιe, τe and $\iota e(f) = \tau e(f) = 0$. If $f > e$ then for any $u \in e$, $u' \in f e^*$ we see f points to $u|E$, $u'|E$, and e is in the geodesic joining $u|E$, $u'|E$, so f points to ιe, τe, and thus $\iota e(f) = \tau e(f) = 1$. Similarly for the remaining three cases. Thus $T = T(E)$.

It is clear from Definition 1.3 that the graph structure of $T(E)$ is respected by the G-action, so $T(E)$ is a G-tree. ∎

1.6 Remarks. (i) If E is a tree subset of $(U, \mathbb{Z}_2)$ then any subset E' of E is again a tree set, and the tree $T(E')$ can be obtained from the tree $T(E)$ by contracting all edges in $E - E'$, and the resulting vertex map $VT(E) \to VT(E')$ is given by the restriction map $(E, \mathbb{Z}_2) \to (E', \mathbb{Z}_2)$.

(ii) If E is any nested subset of $(U, \mathbb{Z}_2)$ not containing any constant functions we can form the graph $T(E)$ and it is a forest. If we add to the vertex set of this graph all functions respecting the nesting, then the resulting structure can be identified with the inverse limit of the inverse system of contracting maps for the finite trees $T(E')$, E' a finite subset of E. See also Section IV.6. ∎

It is sometimes useful to have a more abstract version of the above result. We sketch the steps, leaving the details to the reader.

1.7 Definition. Let E be a G-subset of a G-set with a partial order $\leqslant$ and an order-reversing involution $*$ all respected by G.

We say that E is *nested* if for all $e, f \in E$, exactly one of the six possibilities $e < f, e = f, e > f, e < f^*, e = f^*, e > f^*$ holds.

In this event there is G-graph $T(E)$ as in Definition 1.3.

An element v of $(E, \mathbb{Z}_2)$ *respects the nesting* if v extends to $E \cup E^* \to \mathbb{Z}_2$ so that for all $e, f \in E \cup E^*$, $v(e^*) = v(e^*)$ and $e \leqslant f$ implies $v(e)v(f)^* = 0$. ∎

1.8 Theorem. *Suppose that E is a nested G-subset of a G-set with a partial order and an order-reversing involution all respected by G, and that $VT(E)$ lies in an almost equality class $\tilde{U}$ of $(E, \mathbb{Z}_2)$.*

Then $T(E)$ is a tree with costructure map given by the inclusion map $VT(E) \to (E, \mathbb{Z}_2)$.

Further, $VT(E)$ contains all elements of $\tilde{U}$ which respect the nesting of E.

Proof. Let $U = VT(E) - \{\iota e | e \in E \cap E^*\} \subseteq (E, \mathbb{Z}_2)$, and denote the dual map $E \to (U, \mathbb{Z}_2)$ by $e \mapsto e|U$. It is a straightforward exercise to show that for all $e, f \in E$ if $e < f, e < f^*, e = f^*, e > f^*$ then $(e|U) < (f|U)$, $(e|U) < (f|U)^*$, $(e|U) = (f|U)^*$, $(e|U) > (f|U)^*$, respectively. Now Theorem 1.5 applies. ∎

The results of this section can be phrased and proved more easily but applied with greater difficulty in the following setting.

1.9 Definition. By an *unoriented G-tree T* we mean a G-graph T with no loops, such that for each edge e of T there is a unique edge e^* with $\iota e^* = \tau e$, $\tau e^* = \iota e$, and for all vertices v, w of T there is a unique reduced path in T of the form $v = v_0, e_1, v_1, \ldots, e_n, v_n = w$; for $v \neq w$ we then write $e_1 \geqslant e_n$, so ET is a partially ordered set with involution *. This definition is simply a formal device and we think of the pair e, e^* as a single undirected edge joining the two vertices involved.

We can choose an orientation to make T into a G-tree if and only if there is no edge e with $ge = e^*$ for some g in G; in the latter event we can subdivide the offending edges as in the proof of Theorem 1.5. ∎

Theorem 1.8 easily gives the following characterization of the edge set of an unoriented tree.

1.10 Corollary. *A G-set E with a partial order $\leqslant$ and an order-reversing involution * respected by G arises as the edge set of an unoriented G-tree if and only if for all $e, f \in E$ exactly one of $e < f$, $e = f$, $e > f$, $e < f^*$, $e = f^*$,*

$e > f^*$ holds, and if $e < f$ then there are only finitely many $g \in E$ such that $e < g < f$.

Moreover, the unoriented G-tree is uniquely determined. ∎

2 The Boolean ring of a graph

Throughout this section let X be a connected G-graph and $n \geqslant 0$ an integer.

2.1 Definitions. For any function s on VX the *coboundary* of s is defined as $\delta s = \{e \in EX \mid s(\iota e) \neq s(\tau e)\}$; the elements of δs are said to be *cut* by s. If one thinks of s as a potential, then δs locates the places where the potential changes. By the *components* of s we mean the vertex sets of the components of the graph $X - \delta s$. Thus s is constant on each component of s.

A *Boolean ring* is a commutative ring with 1 such that each element x is idempotent, that is, $x^2 = x$.

For example, for any set V, the ring $(V, \mathbb{Z}_2)$ is a Boolean ring. As usual, the elements of $(V, \mathbb{Z}_2)$ will be treated both as functions and as subsets of V. In particular, $\cup$ and $\cap$ will be used as binary operations, although for simple expressions we shall use juxtaposition rather than $\cap$. Let E be a finite subset of $(V, \mathbb{Z}_2)$. The inclusion map $E \to (V, \mathbb{Z}_2)$ has a dual $V \to (E, \mathbb{Z}_2)$, and the dual of the surjection $V \to V|E$ can be viewed as an embedding $(V|E, \mathbb{Z}_2) \subseteq (V, \mathbb{Z}_2)$. Specifically, an element s of $(V, \mathbb{Z}_2)$ lies in $(V|E, \mathbb{Z}_2)$ if and only if, for all $v, w \in V$, if $s(v) = s(w)$ then $e(v) = e(w)$ for all $e \in E$; in other words, E partitions V at least as finely as s. Let B be the subring of $(V, \mathbb{Z}_2)$ generated by E. Then B is a Boolean ring, and its atoms, that is, the smallest nonzero elements, form a partition of V. But the latter partition is easily seen to be the partition determined by E. It follows that B can be viewed as the set of all subsets of $V|E$, that is $B = (V|E, \mathbb{Z}_2) \subseteq (V, \mathbb{Z}_2)$. This means that a finitely generated Boolean ring is essentially the set of all subsets of a finite set. If $E = \{e_1, \ldots, e_n\}$ then the atoms of B are the nonzero elements of the form $e_1^{\varepsilon_1} \cdots e_n^{\varepsilon_n}$ where $\varepsilon_i \in \{1, *\}$, and each element of B can be written uniquely as a sum of distinct atoms. A subring of $(V, \mathbb{Z}_2)$ generated by an infinite set is the directed union of the finite subrings generated by the finite subsets.

The *Boolean ring* of X, $\mathscr{B}X$, is defined to be the set of all $s \in (VX, \mathbb{Z}_2)$ such that δs is finite. Here s can be thought of as a set of vertices of X such that there are only finitely many edges joining an element of s to an element not in s. It is clear that if $s, s' \in \mathscr{B}X$ then $\delta s = \delta s^*$ and $\delta(ss') \subseteq \delta s$. Thus $\mathscr{B}X$ is a Boolean subring of $(VX, \mathbb{Z}_2)$ since it is closed under forming * and products.

An element $s \in \mathscr{B}X$ is said to be *connected* if the set corresponding to s is a component of s, that is, for any two vertices v, v' of X, if $s(v) = s'(v) = 1$ then s takes the value 1 on every vertex in some path in X from v to v'.

Let $\mathscr{B}_n X$ denote the subring of $\mathscr{B}X$ generated by the elements whose coboundaries have at most n elements. An element s of $\mathscr{B}X$ is said to be n-thin if $|\delta s| = n$ and $s \notin \mathscr{B}_{n-1}X$; an element is *thin* if it is m-thin for some m.

Suppose s is a thin element of $\mathscr{B}_n X$. Then $|\delta s| \leqslant n$. Furthermore, s is connected, since s lies in the ring generated by its components, and for any proper component s' of s, $\delta s' \subset \delta s$. Also, s^* is thin and connected. These facts will be used frequently. ∎

Clearly $\mathscr{B}X$, $\mathscr{B}_n X$ are G-subsets of $(VX, \mathbb{Z}_2)$.

In this section we will prove by induction that there exists a tree G-subset E of $\mathscr{B}_n X$ consisting of thin elements such that E generates $\mathscr{B}_n X$ as Boolean ring.

If $n = 0$ we take E to be empty.

2.2 Remark. It is instructive to look at the case $n = 1$. Let E be the set of edges e of X such that $X - \{e\}$ is not connected, and define $e|VX$ to be the vertex set of the component of $X - \{e\}$ containing τe. Then E is easily seen to be the edge set of a tree whose vertices are the components of $X - E$. Thus $E|VX$ is a tree G-subset of $\mathscr{B}_1 X$ consisting of thin elements such that $E|VX$ generates $\mathscr{B}_1 X$ as Boolean ring. ∎

2.3 Notation. Suppose then that $n \geqslant 2$, and we are given a tree G-set E_{n-1} in $\mathscr{B}_{n-1}X$ consisting of thin elements and generating $\mathscr{B}_{n-1}X$.

By Zorn's Lemma there is a maximal nested G-set E in $\mathscr{B}_n X$ consisting of thin elements and containing E_{n-1}. It will be shown in Corollary 2.6 that E is a tree set. Let B be the subring of $\mathscr{B}_n X$ generated by E. We wish to show that $B = \mathscr{B}_n X$, so we suppose that $\mathscr{B}_n X - B$ is nonempty and try to obtain a contradiction.

Since $\mathscr{B}_{n-1}X \subseteq B$, an element s of $\mathscr{B}_n X - B$ is thin if and only if $|\delta s| = n$; let E' be the set of all such s. Clearly E' is a G-set. Also, E' is nonempty since $B \cup E'$ contains all s with $|\delta s| \leqslant n$ and therefore generates $\mathscr{B}_n X$. ∎

We shall carefully choose $s \in E'$ such that $E \cup Gs$ is nested, which will contradict the maximality of E and the proof will be complete. We begin by deriving a sequence of facts about E and E' that will be used in the construction of s.

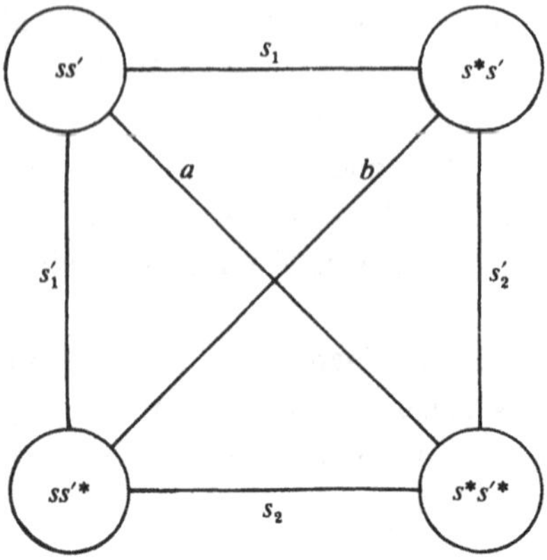

Fig. II.2

2.4 Lemma. (i) *If* $s \in E$ *and* $s' \in E'$ *then some element of* $s \square s'$ *lies in* E'.

(ii) *If* $s, s' \in E'$ *and no element of* $s \square s'$ *lies in* B *then all elements of* $s \square s'$ *lie in* E'.

(iii) *If* $s, s' \in E'$ *and either* $ss'^* \in B$ *or* $\delta ss' \cap \delta s^*s'^* \neq \emptyset$ *then* $ss', s^*s'^* \in E'$.

Proof. Let $s, s' \in E \cup E'$. We partition $\delta s \cup \delta s'$ into six sets, as shown schematically in Fig. II.2:

$$s_1 = \delta ss' \cap \delta s^*s', \quad s_2 = \delta ss'^* \cap \delta s^*s'^*, \quad a = \delta ss' \cap \delta s^*s'^*,$$
$$b = \delta ss'^* \cap \delta s^*s', \quad s'_1 = \delta ss' \cap \delta ss'^*, \quad s'_2 = \delta s^*s' \cap \delta s^*s'^*.$$

The individual partitions are then

$$\delta ss' = s_1 \vee a \vee s'_1; \quad \delta s^*s' = s_1 \vee b \vee s'_2; \quad \delta ss'^* = s'_1 \vee b \vee s_2;$$
$$\delta s^*s'^* = s_2 \vee a \vee s'_2; \quad \delta s = s_1 \vee a \vee b \vee s_2; \quad \delta s' = s'_1 \vee a \vee b \vee s'_2.$$

It is straightforward to deduce that

(1) $\qquad |\delta s| + |\delta s'| = |\delta ss'| + |\delta s^*s'^*| + 2|\delta ss'^* \cap \delta s^*s'|,$

(1*) $\qquad |\delta s| + |\delta s'| = |\delta s^*s'| + |\delta ss'^*| + 2|\delta ss' \cap \delta s^*s'^*|.$

Notice also

(2) $\qquad \sum_{r \in s \square s'} |\delta r| = 2|\delta s \cup \delta s'| \leqslant 2|\delta s| + 2|\delta s'|,$

since the left hand expression counts twice each element of $\delta s \cup \delta s'$.

With this notation we now proceed with the proof.

(i) Suppose $s \in E$, $s' \in E'$. Since $s' \notin B$, we cannot have three elements of $s \square s'$ lying in B, so there exist two elements $r, r' \in s \square s'$ which do not lie in B. In particular, r and r' each cut at least n edges.

Let us say that elements of $s\square s'$ joined by a line labelled s_1, s_2, s'_1 or s'_2 are *adjacent*; and elements of $s\square s'$ joined by a line labelled a or b are *opposite*. Let r^{op} be the unique element of $s\square s'$ which is opposite r, and similarly for r'.

Since $s\in E$, s is an m-thin cut for some $m\leqslant n$. We claim that r^{op} or r'^{op} cuts at least m edges. Suppose not. Then $|\delta r^{\text{op}}|\leqslant m-1<n\leqslant|\delta r'|$, so $r'\neq r^{\text{op}}$; hence r^{op} is adjacent to r'^{op}. But r^{op} and r'^{op} lie in $\mathscr{B}_{m-1}X$, so s or s' lies in $\mathscr{B}_{m-1}X\subseteq B$, a contradiction. Thus the claim is proved and we may assume that r^{op} cuts at least m edges.

Now by (1), (1*), $m+n\leqslant|\delta r^{\text{op}}|+|\delta r|\leqslant|\delta s|+|\delta s'|=m+n$, and it follows that $|\delta r|=n$. Hence $r\in E'$.

(ii) Let $s, s'\in E'$. If no element of $s\square s'$ lies in B then every element cuts at least n edges. By (2), $4n\leqslant\sum_{r\in s\square s'}|\delta r|\leqslant 2|\delta s|+2|\delta s'|=4n$. Hence, equality holds throughout, and for each $r\in s\square s', |\delta r|=n$, so $r\in E'$.

(iii) Let $s, s'\in E'$. Suppose $ss'^*\in B$. Since $s, s'\in E'$, we see $ss', s^*s'^*$ cannot lie in B, so cut at least n edges. Using (1) we have

$$n+n=|\delta s|+|\delta s'|=|\delta ss'|+|\delta s^*s'^*|+2|\delta ss'^*\cap\delta s^*s'|$$
$$\geqslant n+n+2|\delta ss'^*\cap\delta s^*s'|.$$

Thus $\delta ss'^*\cap\delta s^*s'=\varnothing$ and $|\delta ss'|=n$, $|\delta s^*s'^*|=n$, so $ss', s^*s'^*\in E'$.

Now suppose $\delta ss'\cap\delta s^*s'^*\neq\varnothing$. By the contrapositive of the preceding paragraph, with s'^* in place of s', we see $ss'\notin B$. Similarly, taking s^* in place of s we see $s^*s'^*\notin B$. Now the same argument shows $ss', s^*s'^*\in E'$. ∎

2.5 Lemma. *If $u, v\in VX$ then every descending chain in $\{s\in E\cup E'\,|\,s(u)=1\}$ is finite, and every chain in $\{s\in E'\cup E\,|\,s(u)=1,\ s(v)=0\}$ is finite.*

Proof. Suppose $s_1>s_2>\cdots$ is an infinite descending chain of elements of $E\cup E'$ containing u, so u belongs to the set $s_\omega=\bigcap_{i\geqslant 1}s_i$.

Since s_1 is connected, there exists an edge e_1 joining $s_1s_\omega^*$ to s_ω, and hence joining $s_1s_{i_1}^*$ to s_ω for some $i_1>1$. Thus $e_1\in\bigcap_{i\geqslant i_1}\delta s_i$. Similarly s_{i_1} is connected so there exists an edge e_2 joining $s_{i_1}s_{i_2}^*$ to s_ω for some $i_2>i_1$. Notice e_2 has both vertices in s_{i_1} so $e_2\neq e_1$. Continuing in this way we arrive at an i_{n+1} and $n+1$ distinct edges $e_1,\ldots,e_{n+1}$ lying in $\delta s_{i_{n+1}}$, as illustrated in Fig. II.3. But $s_{i_{n+1}}\in E\cup E'$, so $|\delta s_{i_{n+1}}|\leqslant n$, a contradiction. This proves the first part. By applying * to an ascending chain, we can use the same argument to show that $\{s\in E'\cup E\,|\,s(u)=0\}$ has the ascending chain condition. The second result now follows from the fact that any chain with the ascending and descending chain condition is finite. ∎

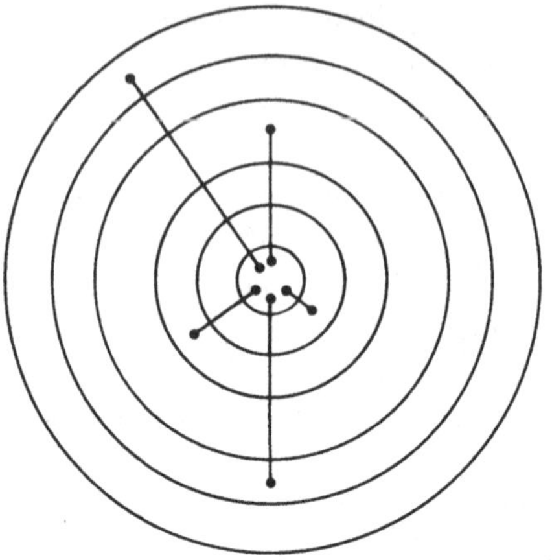

Fig. II.3

2.6 Corollary. *E is a tree set.*

Proof. Let $u, u' \in VX$ and write $C = \{e \in E \mid e(u) = 1,\ e(u') = 0\}$. For any $e, e' \in C$, $u \in ee'$, $u' \in e^*e'^*$; but E is nested so either $e \geqslant e'$ or $e' \geqslant e$. Thus C is a chain. By Lemma 2.5, C is finite. By interchanging u, u' we see that $(u \mid E) \nabla (u' \mid E) = \{e \in E \mid e(u) \neq e(u')\}$ is finite. Thus E is a tree set. ∎

2.7 Lemma. *Each edge of X is cut by only finitely many elements of $E \cup E'$.*

Proof. Let e be any edge of X.

The set of elements of E which cut e is $(\iota e \mid E) \nabla (\tau e \mid E)$, which is finite by the previous result.

It remains to consider E', and by symmetry it suffices to show that the set $L = \{s \in E' \mid s(\iota e) = 0,\ s(\tau e) = 1\}$ is finite.

The following argument shows that L is closed under intersection and union. Let $s, s' \in L$. Then $ss'(\iota e) = 0$, $ss'(\tau e) = 1$, $s^*s'^*(\iota e) = 1$, $s^*s'^*(\tau e) = 0$ so $e \in \delta ss' \cap \delta s^*s'^*$. It follows from Lemma 2.4(iii) that ss' and $s^*s'^*$ lie in E'. Hence, $ss' \in L$ and $(s^*s'^*)^* \in L$, that is $s \cap s', s \cup s' \in L$.

By Lemma 2.5 every chain in L is finite.

Let s be any element of L, and t_i, $i \in [1, \gamma]$, the smallest elements of L strictly containing s, indexed by an ordinal γ. The intersection of any two distinct t_i lies in L so must be s. Thus the $t_i s^*$ are pairwise disjoint, and we can construct a chain $t_1 < t_1 \cup t_2 < t_1 \cup t_2 \cup t_3 < \cdots$ in L indexed by γ. Hence γ is finite.

Suppose L has infinitely many elements. Let s_0 be the least element of

L. Among the finitely many next smallest elements of L we can choose an s_1 contained in infinitely many elements of L. In this way we construct an infinite chain $s_0 < s_1 < s_2 < \cdots$ in L, a contradiction. Thus L is finite. ∎

The last two paragraphs of the above proof amount to the usual verification of the finiteness of a distributive lattice having the ascending and descending chain conditions.

Throughout, we shall use *almost all* to mean 'for all but finitely many'.

2.8 Lemma. *Each $s \in E \cup E'$ is nested with almost all elements of $E \cup E'$.*

Proof. Let Y be a finite connected subgraph of X containing δs. By Lemma 2.7, for almost every $s' \in E \cup E'$, s' does not cut any edge of Y. Consider any such s'. Then Y is a connected subgraph of $X - \delta s'$, so lies in one of the two components. Let X' denote the component of $X - \delta s'$ disjoint from Y, so $X' \subseteq X - Y \subseteq X - \delta s$. Here VX' is s' or s'^*, and by symmetry we may assume $VX' = s'$. If $u, u' \in s'$ then u, u' are joined by a path in $X' \subseteq X - \delta s$, so $s(u) = s(u')$. Thus s is constant on s', so s, s' are nested. ∎

With Lemmas 2.4(i), 2.4(ii), 2.5, 2.8 at our disposal, we have no further need for the edges of X, and are interested only in the vertices.

2.9 Lemma. *Some element of E' is nested with every element of E.*

Proof. For any $s \in E'$, let $n(s)$ be the number of elements of E which are not nested with s; by Lemma 2.8, $n(s)$ is finite.

Choose $s \in E'$ with $n(s)$ minimal. It remains to show that $n(s) = 0$. Suppose $n(s) \geqslant 1$. Then there exists $e \in E$ which is not nested with s. By Lemma 2.4(i), some $r \in s \square e$ belongs to E'. We claim $n(r) < n(s)$. Since e is nested with r and not with s, it suffices to show that for any $e' \in E$, if e' is nested with s then e' is nested with r. For this argument we are free to replace s with s^*, e' with e'^* so without loss of perspective we may assume that s being nested with e' takes the form $s \geqslant e'$; we are also free to replace e with e^* and so assume that e being nested with e' takes the form either $e \geqslant e'$ or $e' \geqslant e$. But $e' \geqslant e$ would imply $s \geqslant e$, which contradicts s not being nested with e; hence, $e \geqslant e'$. Thus $se \geqslant e'$, so either $r = se \geqslant e'$ or $r^* \geqslant se \geqslant e'$ and r, e' are nested. Thus $n(r) < n(s)$ as claimed, and the result is proved. ∎

2.10 Notation. Let s_0 be an element of E' such that $E \cup \{s_0\}$ is nested.

Since adding an element does not affect almost equality, $E \cup \{s_0\}$ is a tree set. By Theorem 1.5, with $G = 1$, $T(E \cup \{s_0\})$ is a tree. Here $s_0 \notin E$, so $T(E)$ is obtained from $T(E \cup \{s_0\})$ by contracting s_0 and both its vertices to some vertex v of $T(E)$. This vertex v of $T(E)$ will remain fixed throughout the rest of the section.

The vertices $\iota s_0, \tau s_0$ are functions on $E \cup \{s_0\}$ which agree on E, and their common restriction to E is v.

Let $\mathscr{B}_v X = \{s \in \mathscr{B}_n X \mid \forall e \in E$, either $e[v] \geqslant s$ or $e[v] \geqslant s^*\}$, where $e[v]$ is e if $v(e) = 1$, and is e^* if $v(e) = 0$. Using the fact that $\mathscr{B}_v X$ is closed under $*$, and that if $e[v] \geqslant s$ and $e[v] \geqslant s'$ then $e[v] \geqslant s \cup s'$, one can show without difficulty that $\mathscr{B}_v X$ is a subring of $\mathscr{B}_n X$. The remainder of the proof will be centred on this ring. ∎

In Fig. II.4 the elements of VX are depicted as dots within the vertices of $T(E)$, and s_0 is depicted first as a dashed line that separates the elements of s_0, s_0^* and then as an edge in $T(E \cup \{s_0\})$.

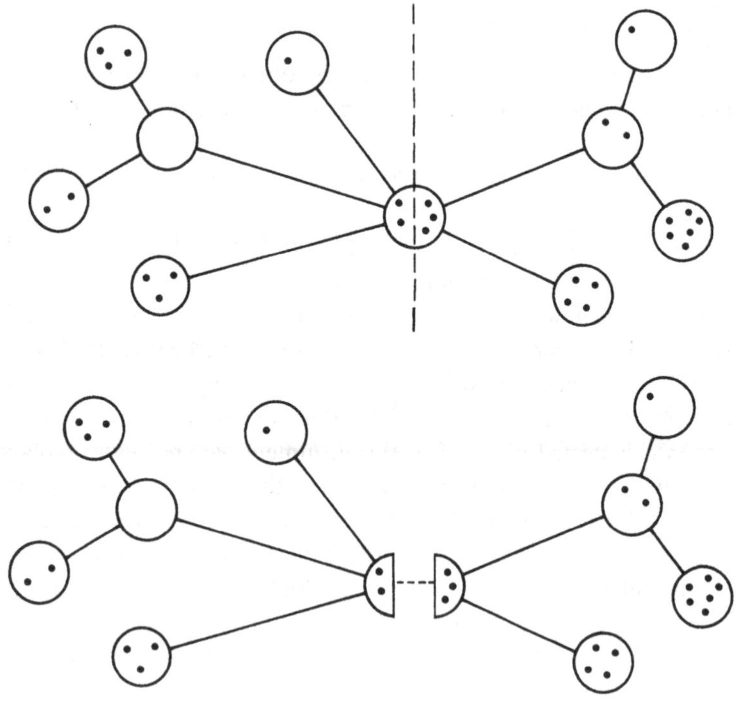

Fig. II.4

The next result shows that the elements of $\mathscr{B}_v X$ are essentially the subsets that can be depicted as in Fig. II.4.

2.11 Lemma. *If $s \in \mathscr{B}_n X$ then $s \in \mathscr{B}_v X$ if and only if either s is constant or $E \cup \{s, s^*\}$ is a tree set, and some vertex of s or s^* in $T(E \cup \{s, s^*\})$ agrees with v on E.*

Proof. We may assume that s is not constant.

Suppose $E \cup \{s, s^*\}$ is a tree set and that some vertex of s in $T(E \cup \{s, s^*\})$ agrees with v on E. By Definition 1.3, for all $e \in E \cup \{s, s^*\}$, $e[\iota s] \geqslant s$ or $e[\iota s] \geqslant s^*$, and $e[\tau s] \geqslant s$ or $e[\tau s] > s^*$. Whichever vertex of s agrees with v on E, we still have, for each $e \in E$, either $e[v] \geqslant s$ or $e[v] \geqslant s^*$. Hence, $s \in \mathscr{B}_v X$.

Conversely, suppose $s \in \mathscr{B}_v X$. It follows from the definition of $\mathscr{B}_v X$ that s is nested with each $e \in E$, so $E \cup \{s, s^*\}$ is nested. Adding two elements cannot affect almost equality, so $E \cup \{s, s^*\}$ is a tree set. Consider any $e \in E - \{s, s^*\}$. In the tree $T(E \cup \{s, s^*\})$, e does not lie between s and ιs, τs so either $e[\iota s] = e[\tau s] > s$ or $e[\iota s] = e[\tau s] > s^*$, which forces $e[\iota s] = e[\tau s] = e[v]$, and similarly for s^*. Thus the three vertices ιs, $\tau s = \tau s^*$, ιs^* agree with v on $E - \{s, s^*\}$. It remains to show that one of them agrees with v on all of E. If neither s nor s^* lies in E, then obviously all three agree with v on E. If $s \in E$ and $s^* \notin E$ then either $v(s) = 0$ and ιs agrees with v on E, or $v(s) = 1$ and τs agrees with v on E. Similarly, if $s \notin E$, $s^* \in E$. Finally if $s, s^* \in E$ then v is a vertex of $T(E \cup \{s, s^*\})$ with no edge other than s, s^* between v and the three vertices ιs, $\tau s = \tau s^*$, ιs^*. Hence, v is one of these three vertices. ∎

Let $B_v = B \cap \mathscr{B}_v X$ and let $E_v = \text{star}(v) \subseteq E$ in $T(E)$.

2.12 Lemma. *E_v generates B_v.*

Proof. Let $e \in E_v$. Then e is incident to v, so, by Lemma 2.11, e lies in $\mathscr{B}_v X$, so e lies in B_v.

Let $s \in B_v$. Since E generates B there is a finite subset E_0 of E such that for all $u, u' \in VX$, if $e(u) = e(u')$ for all $e \in E_0$ then $s(u) = s(u')$. For each $e \in E_0$, let e_v be the first edge in the geodesic in $T(E)$ from v to e; these e_v, $e \in E_0$, give a finite subset of E_v. Suppose $u, u' \in VX$ with $s(u) \neq s(u')$, so $s^*(u) \neq s^*(u')$. Then $e(u) \neq e(u')$ for some $e \in E_0$. As $s \in \mathscr{B}_v X$, we can apply Lemma 2.11. In the tree $T(E \cup \{s, s^*\})$ the geodesic between $u | E \cup \{s, s^*\}$ and $u' | E \cup \{s, s^*\}$ passes through s and s^*, and by Lemma 2.11, on con-tracing down to $T(E)$, we see that the geodesic between $u|E$ and $u'|E$ passes through v. But e occurs in this geodesic, and hence so does e_v. Thus $e_v(u) \neq e_v(u')$. By the observations made in Definition 2.1, we see that E_v generates B_v. ∎

2.13 Notation. Let $e \in E_v$. For each $s \in \mathscr{B}_v X$ either $s \geqslant e^*[v]$ or $s^* \geqslant e^*[v]$, so $e^*[v]$ is an atom in $\mathscr{B}_v X$. The $e^*[v]$, $e \in E_v$, will be called the *v-atoms*. Thus for each $e \in E_v$ either e or e^* is a v-atom; with a little care one could even reorient so that the elements of E_v are the v-atoms. Let I_v be the ideal of $\mathscr{B}_v X$ generated by the v-atoms; thus the v-atoms are a $\mathbb{Z}_2$-basis of I_v, and I_v actually lies in B_v. In fact, if $B_v \neq 0$ then I_v is a maximal ideal of B_v. Notice that if $q \in \mathscr{B}_v X$ and $r \in I_v$ and $q \leqslant r$ then $q \in I_v$.

Graphically, one can think of deleting v from $T(E)$ to get a disconnected graph having one component for each element of star(v), and the v-atoms correspond to the subsets of VX which lie in the different components.

Let $E'_v = E' \cap \mathscr{B}_v X$. The original definition of v arose from an $s_0 \in E'$ such that $E \cup \{s_0\}$ is a tree set and $\iota s_0|E = \tau s_0|E = v$. Thus $s_0 \in E'_v$, so E'_v is nonempty.

We now want to find some s in E'_v such that $E \cup Gs$ is nested. Here each gs is nested with $gE = E$, so the only difficulties arise where gs, s are not nested. If $gv \neq v$ then there is some edge e of $T(E)$ with $e(v) \neq e(gv)$, so $e^*[v] = e[gv] = g(g^{-1}e[v])$ and then $(s \text{ or } s^*) \geqslant e^*[v] = g(g^{-1}e[v]) \geqslant (gs \text{ or } gs^*)$. Thus s, gs are nested. By the contrapositive, if s, gs are not nested then $g \in G_v$.

Let $H = G_v$. It suffices to find $s \in E'_v$ such that $E \cup Hs$ is nested, for then $E \cup Gs$ is nested and we have our contradiction. It is not difficult to show that $\mathscr{B}_v X$ is closed under the action of H.

Our interest is now entirely in $E_v \cup E'_v$, so each element u of VX could be replaced with its image $u|(E_v \cup E'_v)$. There would then be two types of elements. For those $u \in VX$ such that $u|E = v$ we could call $u|E_v \cup E'_v$ an *interior* element. Those $u \in VX$ with $u|E \neq v$ become associated with the appropriate edge of star(v), and we could call these *exterior* elements. Each v-atom then consists of a single exterior element. We shall not formalize this viewpoint but shall use it to illuminate some of the arguments.

We now introduce a set that is even better than E'_v. Let E''_v denote the set of all $s \in E'_v$ such that for each $h \in H$, some element of $s \square hs$ lies in I_v. ∎

2.14 Lemma. $E''_v \neq \varnothing$.

Proof. For some $u \in VX$ the set $\{s \in E'_v | s(u) = 1\}$ is nonempty. By Lemma 2.5 it has a least element s.

Let $h \in H$.

We claim that some element of $s \square hs$ lies in B. Suppose not. Then, by Lemma 2.4(ii), each element of $s \square hs$ lies in E', so lies in E'_v. Let r be the element of $s \square hs$ with $r(u) = 1$, so $r \leqslant s$. Since $r \in E'_v$, the minimality condition shows that $r = s$, which means that 0 is an element of $s \square hs$, and this certainly lies in B. This contradiction proves the claim.

We now have an element $q \in s \square hs$ with $q \in B$. Thus $q \in B_v$. If $q \in I_v$ we are finished. If $q \notin I_v$ then $q^* \in I_v$. Consider any $p \in s \square hs$ with $p \neq q$. Then $p \leqslant q^*$ and $p \in \mathscr{B}_v X$ so $p \in I_v$. ∎

2.15 Notation. We define a function $\mathscr{B}_v X \to \mathbb{Z} \cup \{\infty\}$, $s \mapsto \|s\|$, as follows. For $s \in \mathscr{B}_v X$, either $s \notin I_v$, and we set $\|s\| = \infty$, or $s \in I_v$, and then s can be expressed uniquely as a sum of distinct v-atoms, and we define $\|s\|$ to be the number of these atoms. Here s consists of $\|s\|$ exterior elements.

For $s, s' \in \mathscr{B}_v X$ we write $s \leqslant_v s'$ to mean $\infty > \|ss'^*\| \leqslant \|s^*s'\|$. This is equivalent to the existence of a finite permutation of exterior elements which carries s to s'.

As usual, $s =_v s'$ means $s \leqslant_v s'$ and $s' \leqslant_v s$, while $s <_v s'$ will mean $s \leqslant_v s'$ but not $s' \leqslant_v s$.

Obviously if $s < s'$ then $s <_v s'$.

Notice that if ss'^* lies in I_v then $\|ss'^*\|$ is finite and s and s' are comparable under $\leqslant_v$. In particular, if $s \in E''_v$, and $q, r \in Hs \cup Hs^*$, then one of qr, qr^* q^*r, q^*r^* lies in I_v, so $q \leqslant_v r$ or $q \leqslant_v r^*$ or $q^* \leqslant_v r$ or $q^* \leqslant_v r^*$.

For each element s of E''_v we write $m(s)$ for the number of elements of Hs which are not nested with s; this is finite by Lemma 2.8.

For the remainder of the section we assume that s has been chosen in E''_v to minimize $m(s)$, and we want to show $m(s) = 0$. Suppose then that $m(s) \geqslant 1$. By replacing s with s^* if necessary we may assume there is some $s' \in Hs \cup Hs^*$ such that $s' \leqslant_v s$ and s is not nested with s'. By Lemma 2.8 there are only finitely many possibilities for s', so we may assume that s' has further been chosen minimal under $\leqslant_v$. Let $K = \{h \in H | hs' =_v s'\}$ and set $s'' = s' \cap \cap Ks$, where $\cap Ks = \bigcap_{k \in K} ks$.

This notation will remain fixed throughout the remainder of the proof. ∎

It is clear that $s'' \in \mathscr{B}_v X$, but is not clear that s'' lies in E''_v or even E'. We shall show $s'' \in E''_v$ and that $m(s'') < m(s') = m(s)$, and we will have our desired contradiction.

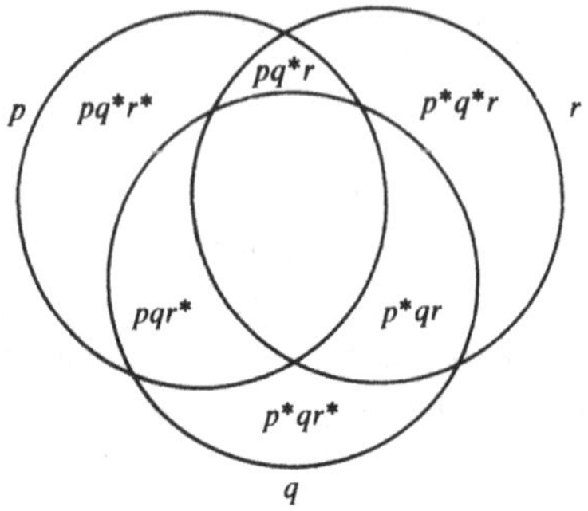

Fig. II.5

2.16 Lemma. *The relation* $\leqslant_v$ *is transitive on* $\mathscr{B}_v X$.

Proof. Suppose p, q, r are elements of $\mathscr{B}_v X$ such that $p \leqslant_v q$ and $q \leqslant_v r$.

Thus $\infty > \|pq^*\| = \|pq^*r\| + \|pq^*r^*\|$, and $\infty > \|qr^*\| = \|pqr^*\| + \|p^*qr^*\|$, so $\infty > \|pqr^*\| + \|pq^*r^*\| = \|pr^*\|$. If $\|p^*r\| = \infty$ then $p \leqslant_v r$, and it remains to consider the case $\infty > \|p^*r\| = \|p^*qr\| + \|p^*q^*r\|$.

Here $\|pr^*\| = \|pqr^*\| + \|pq^*r^*\| = \|qr^*\| - \|p^*qr^*\| + \|pq^*\| - \|pq^*r\|$. Similarly, $\|p^*r\| = \|q^*r\| - \|pq^*r\| + \|p^*q\| - \|p^*qr^*\|$. Hence, $\|p^*r\| - \|pr^*\| = \|q^*r\| - \|qr^*\| + \|p^*q\| - \|pq^*\| \geqslant 0$. Thus $p \leqslant_v r$, as desired. ∎

2.17 Lemma. $s'' \in E''_v$.

Proof. If $k \in K$ then $k^{-1}s' =_v s' \leqslant_v s$, so $s' \leqslant_v ks$; here ks is nested with s' only if $s' \leqslant ks$.

By Lemma 2.8, s' is nested with almost all $ks \in Ks$, so $s' \leqslant ks$ for almost all $ks \in Ks$. Hence, the elements $s'' = s' \cap \bigcap Ks$ can be expressed as a finite intersection $s'' = s' \cap k_1 s \cap \cdots \cap k_m s$ with $k_1, \ldots, k_m \in K$, and $s' \leqslant_v k_i s$.

We show by induction that if $i \in [0, m]$ then the intersection $s_i = s' \cap k_1 s \cap \cdots \cap k_i s$ lies in E'. Here $s_0 = s' \in E'$. Suppose $i \in [1, m]$ and $s_{i-1} \in E'$. Then $s_{i-1} \leqslant s' \leqslant_v k_i s$, so $s_{i-1} \cap k_i s^*$ is a finite sum of v-atoms, so lies in B. Now, by Lemma 2.4 (iii), $s_{i-1} \cap k_i s \in E'$, that is, $s_i \in E'$. By induction, $s'' \in E'$. Hence, $s'' \in E'_v$.

Notice $s' \cap s''^* = \bigcup_{i \in [1,m]} (s' \cap k_i s^*) \in I_v$, and $s'^* \cap s'' = 0$ lies in I_v. Let $h \in H$. To show that $s'' \in E''_v$ we must show that some element of $s'' \square hs''$ lies in I_v. Since $s' \in E'_v$ we know that an element of $s' \square hs'$ lies in I_v and now the relations

$$s'' \cap hs'' \leqslant (s'^* \cap s'') \cup (s' \cap hs') \cup h(s'^* \cap s''),$$
$$s''^* \cap hs'' \leqslant (s' \cap s''^*) \cup (s'^* \cap hs') \cup h(s'^* \cap s''),$$

$$s'' \cap hs''^* \leqslant (s'^* \cap s'') \cup (s' \cap hs'^*) \cup h(s' \cap s''^*),$$
$$s''^* \cap hs''^* \leqslant (s' \cap s''^*) \cup (s'^* \cap hs'^*) \cup h(s' \cap s''^*)$$

in $\mathscr{B}_v X$ show that the corresponding element of $s'' \square hs''$ lies in I_v. Thus $s'' \in E''_v$. ∎

2.18 Lemma. *Let $r \in Hs \cup Hs^*$. If $r <_v s'$ then $s'r \leqslant s''$; hence, if $r < s'$ then $r \leqslant s''$.*

Proof. Suppose $r <_v s'$. If $k \in K$ then $kr <_v ks' =_v s'$, so, by the minimality hypothesis on s', kr is nested with s, and the only way this can happen is for $kr < s$. Thus $r \leqslant \cap Ks$ and $s'r \leqslant s' \cap \cap Ks = s''$.

If $r < s'$ then $r = s'r \leqslant s''$ by the first part, since $r <_v s'$. ∎

2.19 Lemma. $m(s'') < m(s') = m(s)$.

Proof. Since $s' \in Hs \cup Hs^*$ there exists $h_0 \in H$ such that $s' = h_0 s$ or $s' = h_0 s^*$, and it is clear that $m(s) = m(s')$.

If k is any element of $H_{s'}$ then $ks' = s'$ so $k \in K$ and $ks'' = ks' \cap \cap kKs = s' \cap \cap Ks = s''$. Thus $H_{s'} \subseteq H_{s''}$, and we have a surjective map $Hs' \to Hs''$, $hs' \mapsto hs''$. We shall show that if hs' is nested with s' then hs'' is nested with hs'', while $h_0 s'$ is not nested with s' and $h_0 s''$ is nested with s''. This will prove $m(s'') < m(s)$.

We know that s' is not nested with s or s^* so s' is not nested with $h_0^{-1} s'$. Thus s' is not nested with $h_0 s'$.

We next show that s'' and $h_0 s''$ are nested. By the construction of s'' we have $s'' \leqslant s'$, $s'' \leqslant s$. If $s' = h_0 s^*$ then $h_0 s''^* \geqslant h_0 s^* = s' \geqslant s''$, so we may assume $s' = h_0 s$. If $h_0 \in K$, then $s'' = \cap Ks$, so $h_0 s'' = s''$, and we may assume $h_0 \notin K$. Here $s' \neq_v h_0 s' = s$ so $s' <_v s$ and $h_0 s' <_v h_0 s = s'$. By Lemma 2.18, $s' \cap h_0 s' \leqslant s''$. Also $h_0 s'' \leqslant h_0(s' \cap s) = h_0 s' \cap s' \leqslant s''$. This completes the proof of the claim.

Let $h \in H$. It remains to show that if hs' and s' are nested then hs'' and s'' are nested. We distinguish five cases.

If $hs' = s'$ then $hs'' = s''$.

Suppose $hs' < s'$. By Lemma 2.18, $hs' \leqslant s''$. But $s'' \leqslant s'$ so $hs'' \leqslant hs' \leqslant s''$.

Suppose $hs' > s'$. Then $h^{-1}s' < s'$, so by the previous case, $h^{-1}s'' \leqslant s''$, and $hs'' \geqslant s''$.

Suppose $hs'^* \geqslant s'$. Then $hs''^* \geqslant hs'^* \geqslant s' \geqslant s''$, so $hs''^* \geqslant s''$.

Suppose $hs'^* < s'$. Then $s'^* < h^{-1}s'$ so $s' > h^{-1}s'^*$. Now $hs'^* < s'$ and $h^{-1}s'^* < s'$, so, by Lemma 2.18, $hs'^* \leqslant s''$ and $h^{-1}s'^* \leqslant s''$. Thus we have $hs' \geqslant s''^*$. We next show $\cap hKs \geqslant s''^*$. Let $k \in K$. By the construction of s'',

$ks \geqslant s'' \geqslant h^{-1}s'^*$, so $hks^* \leqslant s'$. To see that equality does not hold, observe that $s' \leqslant_v s$ by choice of s', so $hks^* \leqslant_v hks'^* =_v hs'^* < s'$. Thus $hks^* < s'$, and, by Lemma 2.18, $hks^* \leqslant s''$ or equivalently $hks \geqslant s''^*$, as desired. Now $hs'' = hs' \cap \cap hKs \geqslant s''^*$. ∎

This contradicts the minimality of $m(s)$, so the underlying hypothesis that E' is nonempty is therefore false. Thus E is a tree G-set which generates $\mathscr{B}_n X$ and consists of thin elements. This completes the proof of the main result of this section.

2.20 Theorem. *If X is a connected G-graph and n a natural number then $\mathscr{B}_n X$ contains a tree G-subset E_n which generates $\mathscr{B}_n X$ as Boolean ring.* ∎

2.21 Remarks. (i) Intuitively we have the following situation. By Theorem 1.5 there is a G-tree $T_n = T(E_n)$ and a G-map $VX \to VT_n$. Deleting an edge from T_n divides VT_n into two parts, and the resulting division of VX into two parts cuts at most n edges of X. Further, using all the subsets of VX arising from this division, and the operations union, intersection and complement, one can arrive at any partition of VX which arises by deleting at most n edges from X.

(ii) Theorem 2.20 is the result most useful for our applications, but we have proved more. We have constructed an ascending chain $E_0 \subseteq E_1 \subseteq E_2 \subseteq \cdots$ of tree G-sets of thin elements such that each E_n generates $\mathscr{B}_n X$.

If $n \geqslant 1$ we may assume that E_n contains the set E_1 constructed in Remark 2.2. This will be needed in one application.

The union E_ω of all the E_n is a nested G-set of thin elements which generates $\mathscr{B}X$. The $T(E_n)$ form an inverse directed system of trees and contracting maps, and E_ω is the edge set of the inverse limit $T_\omega = \varprojlim T(E_n)$. Notice that T_ω need not be a tree since there can be infinitely many edges 'between' two vertices. ∎

2.22 Examples. (i) If X is a tree then $\mathscr{B}X = \mathscr{B}_1 X$ and we can take $E = E_1$ as in Remark 2.2.

(ii) Let G be the free abelian group on a, b, and X the Cayley graph $X(G, \{a, b\})$, so X is a lattice in the plane with vertex set $\mathbb{Z}^2$. It is not difficult to show that $\mathscr{B}X = \mathscr{B}_4 X$, and that we can choose as our tree G-subset $E \subseteq \mathscr{B}_4 X \subseteq (VX, \mathbb{Z}_2)$, the set $E = \{\{v\} \mid v \in VX\}$. There is a natural embedding $VX \subseteq (E, \mathbb{Z}_2)$, and then $VT(E) = VX \cup \{0\} \subseteq (E, \mathbb{Z}_2)$, and for each edge of the form $e = \{v\}$ we have $\iota e = 0$, $\tau e = v$. Thus $T(E)$ is simply a star.

(iii) Let $G = \langle a_1, b_1, a_2, b_2, a_3, b_3, \ldots | b_1 = [a_2, b_2], b_2 = [a_3, b_3], \ldots \rangle$ and X the Cayley graph $X(G, \{a_1, b_1, a_2, b_2, a_3, b_3, \ldots\})$. We leave it as an exercise for the interested reader to use the arguments of Example IV.7.2 to show that no tree G-subset generates $\mathscr{B}X$. Of course, $\mathscr{B}X$ has a nested G-subset which generates it, by Remark 2.21. ∎

3 Distance-transitive graphs

In this section we use the result of the previous section to characterize the infinite finite-valency distance-transitive graphs.

Throughout this section let $X = (X, V, E, \iota, \tau)$ be a graph and G its automorphism group.

3.1 Definitions. We say that X is *unoriented* if there are no loops and for each edge e of X there is a unique edge e^* with $\iota e^* = \tau e$, $\tau e^* = \iota e$. This definition is simply a formal device, and we think of the pair e, e^* as a single undirected line joining the two vertices involved. Here E is completely specified by a subset of the set of two-element subsets of V. For example, if E corresponds to the set of all two-elements subsets of V we get the *complete graph on* V.

Suppose X is unoriented. For vertices u, v of X the *distance*, $d(u, v)$, is the least possible value of i such that there is a path of length i from u to v in X, or ∞ if no such i exists, that is, u, v lie in different components of X. By a *geodesic* in X we mean a path of the form $v_0, e_1, v_1, \ldots, e_i, v_i$ in X with $d(v_0, v_i) = i$. Clearly in this event $d(v_k, v_j) = j - k$ for all $k \leqslant j$ in $[0, i]$.

We say that X is *distance-transitive* if X is connected, unoriented and for all u, v, u', v' in VX, if $d(u, v) = d(u', v')$ then there exists $g \in G$ such that $gu = u'$, $gv = v'$. Here one can take $u = v, u' = v'$, so G must act transitively on VX; in particular all vertices have the same valency, called the *valency* of X. Similarly, G acts transitively on EX. ∎

3.2 Examples. (i) The one-skeleton of a regular solid is a finite distance-transitive graph.

(ii) For any integers $m, n \geqslant 2$ we denote by $X(m, n)$ the graph consisting of countably many complete m-vertex graphs attached in the most general possible way so that each vertex belongs to exactly n of the m-vertex graphs. Thus replacing each of the complete m-vertex graphs with an $(m + 1)$-vertex star in a natural way gives a tree, namely the standard tree for the free product $C_m * C_n$. For instance, $X(2, n)$ is a tree with each vertex having valency n, and the graph $X(4, 2)$ is depicted in Fig. II.6, with gaps

Fig. II.6

indicating crossings. It is not difficult to show that $X(m, n)$ is an infinite distance-transitive graph with valency $(m-1)n$; the main theorem of this section is the converse, that every infinite finite-valency distance-transitive graph is of the form $X(m, n)$ for some $m, n \geqslant 2$. ∎

3.3 Lemma. *If X is an infinite finite-valency distance-transitive graph then $\mathscr{B}X$ contains an element which is not almost constant.*

Proof. Consider any geodesic $v_0, e_1, v_1, \ldots, e_i, v_i$ in X. We claim we can extend this to a longer geodesic. Since X is locally finite there are only finitely many vertices of X within distance i of v_0, but X is infinite, so there is some vertex u of X with $d(v_0, u) > i$. Choose a geodesic from v_0 to u, and let u_i, f_i, u_{i+1} be the $(i+1)$th edge in the geodesic. Then $d(v_0, u_i) = i$, so by distance transitivity there is a $g \in G$ with $gv_0 = v_0$, $gu_i = v_i$. Set $e_{i+1} = gf_{i+1}$, $v_{i+1} = gu_{i+1}$. Then $d(v_0, v_{i+1}) = d(gv_0, gu_{i+1}) = d(v_0, u_{i+1}) = i+1$, so the original geodesic can be extended, as claimed. By induction we can extend it to an infinite geodesic $v_0, e_1, v_1, e_2, v_2, \ldots$, that is, $d(v_0, v_i) = i$ for all $i \geqslant 0$. We fix such an infinite geodesic for the remainder of the proof.

For each $i \geqslant 0$, let $b_i = \{w \in VX \mid d(w, v_0) = 1,\ d(w, v_i) = i+1\}$ and $c_i = \{w \in VX \mid d(v_0, w) = 1,\ d(w, v_i) = i-1\}$. Notice that by distance-transitivity the cardinality of these sets does not depend on the choice of infinite geodesic.

If $w \in b_{i+1}$ then $d(w, v_0) = 1$, $d(w, v_{i+1}) = i + 2$ and we have

$$1 + i = d(w, v_0) + d(v_0, v_i) \geqslant d(w, v_i) \geqslant d(w, v_{i+1}) - d(v_i, v_{i+1})$$
$$= (i + 2) - 1,$$

which means that $d(w, v_i) = i + 1$, so $w \in b_i$. Hence, $b_0 \supseteq b_1 \supseteq \cdots$.

Similarly, if $w \in c_i$ then $d(w, v_0) = 1$, $d(w, v_i) = i - 1$ and we have

$$(i - 1) + 1 = d(w, v_i) + d(v_i, v_{i+1}) \geqslant d(w, v_{i+1}) \geqslant d(v_0, v_{i+1}) - d(v_0, w)$$
$$= (i + 1) - 1$$

which means that $d(w, v_{i+1}) = i$, so $w \in c_{i+1}$. Hence, $c_0 \subseteq c_1 \subseteq \cdots$.

Since v_0 has finite valency there exists some k such that $c_k = c_{k+1} = c_{k+2} = \ldots$, $b_k = b_{k+1} = b_{k+2} = \cdots$. As we remarked before, the cardinality of these sets is independent of the choice of infinite geodesic, so the same value of k will work for any infinite geodesic.

Let $s = \{u \in VX \mid k + 1 \leqslant d(v_0, u) = 1 + d(v_1, u)\}$. Since it contains v_{k+1}, $v_{k+2}, \ldots$, the set s is infinite.

Here v_0, v_1 are arbitrary neighbours in X, so on interchanging them we get another infinite set $\{u \in VX \mid k + 1 \leqslant d(v_1, u) = 1 + d(v_0, u)\}$. The latter set is disjoint from s, so s^* is infinite. Thus, viewed as an element of $(VX, \mathbb{Z}_2)$, is s is not almost constant. It remains to show that $s \in \mathscr{B}X$, that is, $|\delta s| < \infty$.

Let us analyze an edge $e \in \delta s$, say e has vertices u, v with $u \in s$, $v \in s^*$. Write $i = d(v_0, u)$. Here $d(v_1, u) = i - 1 \geqslant k$ because $u \in s$.

We claim that $d(v_0, v) = 1 + d(v_1, v)$, or equivalently $d(v_0, v) \geqslant 1 + d(v_1, v)$. Observe that

$$i - 1 = d(v_0, u) - d(v, u) \leqslant d(v_0, v) \leqslant d(v_0, v_1) + d(v_1, v)$$
$$= 1 + d(v_1, v) \leqslant 1 + d(v_1, u) + d(u, v) = 1 + (i - 1) + 1 = 1 + i.$$

We consider three cases.

Suppose $d(v_0, v) = i - 1$. Thus $d(v_0, u) = i = d(v_0, v) + d(v, u)$ so there is a geodesic of length i from v_0 to u finishing with v, e, u. Since $i - 1 \geqslant k$,

$$\{w \in VX \mid d(v_0, w) = 1,\ d(w, v) = i - 2\}$$
$$= \{w \in VX \mid d(v_0, w) = 1,\ d(w, u) = i - 1\}.$$

But v_1 belongs to the latter set so belongs to the former set, that is, $d(v_1, v) = i - 2 = d(v_0, v) - 1$, as desired.

Suppose next that $d(v_1, v) = i$. Here $d(v_1, v) = i = d(v_1, u) + d(u, v)$ so there is a geodesic of length i from v_1 to v finishing with u, e, v. Since $i - 1 \geqslant k$,

$$\{w \in VX \mid d(w, v_1) = 1,\ d(w, u) = i\}$$
$$= \{w \in VX \mid d(w, v_1) = 1,\ d(w, v) = i + 1\}.$$

But v_0 belongs to the former set, so belongs to the latter set, that is, $d(v_0, v) = i + 1 = d(v_1, v) + 1$, as desired.

This leaves the case where $d(v_0, v) \geqslant i$ and $d(v_1, v) \leqslant i - 1$, and here $d(v_1, v) + 1 \leqslant i \leqslant d(v_0, v)$, as desired. The proof of the claim that $d(v_0, v) = 1 + d(v_1, v)$ is complete.

The fact that $v \in s^*$ means that $k + 1 \leqslant d(v_0, v)$ fails, so $k \geqslant d(v_0, v)$. Hence there are only finitely many possibilities for e, which proves that $|\delta s| < \infty$.

∎

3.4 Theorem. *If X is an infinite finite-valency distance-transitive graph then $X = X(m, n)$ for some $m, n \geqslant 2$.*

Proof. By Lemma 3.3, $\mathscr{B}X$ has an element which is not almost constant, so for some k, $\mathscr{B}_k X$ has an element which is not almost constant. By Theorem 2.20, there exists a tree G-subset E of $\mathscr{B}_k X$ which generates $\mathscr{B}_k X$. In particular, E must contain an element s which is not almost constant, and clearly Gs is a tree G-subset of $\mathscr{B}X$. Let $T = T(Gs)$, as in Definition 1.3.

The first step is to find $g \in G$ and $v_0 \in s$ such that v_0, gv_0 are neighbours and $s \supset gs$.

For $v \in VX$ and $r \subseteq VX$ let us write $d(v, r) = \min\{d(v, u) \mid u \in r\}$. Since s, s^* are infinite and δs is finite, there exist $u, v \in VX$ such that $d(u, s^*) \geqslant 2$, $d(v, s) \geqslant 2$. Hence there exists a geodesic $v_0, e_1, v_1, \ldots, v_{i-1}, e_i, v_i$ in X with $d(v_0, s^*) \geqslant 2$, $d(v_i, s) \geqslant 2$. Notice that $v_0, v_1 \in s$, $v_{i-1}, v_i \in s^*$.

The sequence of non-negative integers $d(v_0, s), d(v_1, s), \ldots, d(v_i, s)$ changes by at most one at each step, and starts at 0 and ends above 1, so there is some step where it changes from 1 to 2. Omitting the segment after this step does not diminish the required properties of the path, so we may further assume that $d(v_{i-1}, s) \neq d(v_i, s)$.

Now $d(v_0, v_{i-1}) = i - 1 = d(v_1, v_i)$, so by distance transitivity there exists $g \in G$ with $gv_0 = v_1$, $gv_{i-1} = v_i$.

Here $v_1 = gv_0 \in s \cap gs$ and $v_i = gv_{i-1} \in s^* \cap gs^*$, but Gs is nested, so either $gs \subseteq s$ or $gs \supseteq s$. Also $d(v_i, s) \neq d(v_{i-1}, s) = d(gv_{i-1}, gs) = d(v_i, gs)$, so $gs \neq s$. Thus $s \supset gs$ or $gs \supset s$.

In summary, either $s \supset gs$, $gv_0 = v_1$ or $s \supset g^{-1}s$, $g^{-1}v_1 = v_0$. By relabelling we may assume we have the fomer situation, which is what we wanted.

Let u, v be vertices of X, and let $d = d(u, v)$. We now show that the T-geodesic from $u|Gs$ to $v|Gs$ contains at least d edges pointed in the same direction.

Since Gs is a tree set, the chain $s \supset gs \supset g^2 s \supset \cdots$ must have empty intersection. As $v_0 \in s$ there is a unique integer $a \geqslant 0$ such that $v_0 \in g^a s$ and

$v_0 \notin g^{a+1}s$; further, $v_0, e_1, gv_0, ge_1, g^2v_0, g^2e_1, \ldots$ is an infinite path in X with no repeated vertices. Since X is locally finite, $\lim_{j \to \infty} d(v_0, g^j v_0) = \infty$ and there exists $c \geqslant d$ with $d(v_0, g^c v_0) = d = d(u, v)$.

For each $j \in [a+1, a+c]$, we have $(g^j s)(v_0) = 0 \neq 1 = (g^{j-c}s)(v_0) = (g^j s)(g^c v_0)$, so $g^j s$ lies in the T-geodesic joining $v_0 | Gs$ and $g^c v_0 | Gs$. Thus the T-geodesic from $v_0 | Gs$ to $g^c v_0 | Gs$ contains at least $c \geqslant d$ edges all pointing in the same direction. It follows by distance transitivity that the T-geodesic from $u | Gs$ to $v | Gs$ contains at least d edges pointing in the same direction, as claimed.

One consequence is that the map $VX \to VT$ is injective, and we view VX as a G-orbit in VT.

Another consequence is that, if $u, v \in VX$ and the T-geodesic from u to v does not contain two edges pointing in the same direction, then u and v are neighbours in X.

Since ET has one G-orbit, VT has either one or two G-orbits.

If VT has one G-orbit then $VX = VT$. Let u, v be the vertices of s in T. By the above, u and v are neighbours in X. Since G acts transitively on EX it follows that $X = T$. Thus X is a tree, so is of the form $X(2, n)$, where n is the valency of X.

This leaves the case where $VX \subset VT$. Here $VT - VX$ is one G-orbit. All edges of T join VX to $VT - VX$, and all incident edges of T point in opposite directions. It is a simple matter to pick two distinct elements $u, v \in VX$ such that the T-geodesic from u to v consists of two edges pointing in opposite directions. By the above, u, v must be neighbours in X. By distance transitivity, EX joins up all pairs of elements of VX which are at distance 2 in T. Since X is locally finite it follows easily that T is locally finite, and that X is of the form $X(m, n)$ for some $m, n \geqslant 2$. ∎

Notes and comments

In Section IV.6 we shall discuss ends of groups which were studied by Fredenthal (1931, 1942, 1944), Hopf (1943) and Specker (1949, 1950). This last paper notes a connection with Boolean rings; the next breakthrough in this area came from Stallings (1968), and the accompanying article by Bergman (1968) makes explicit the relationship with the Boolean ring of a graph.

The seminal results of Stallings (1968, 1971) which will be given in Section III.2, together with the theory of Bass and Serre (Serre, 1977) given in Chapter I, indicated there was a strong connection with trees; the connection was given substance in Dunwoody (1979). One of the key results there is stated here, mostly for historical interest, as Corollary 1.10. The novelty of Theorem 1.5 lies in the concrete description of vertices as functions; this is fundamental in many subsequent sections.

Bergman's result for the Boolean ring of a locally finite graph was refined, and extended to arbitrary graphs, by Dunwoody (1982). Theorem 2.20 further advances the theory to provide an extremely close link with trees, which will be useful in Chapter III.

The results of Section 3 are due to Macpherson (1982); the proofs given here simplify the original arguments. See Brouwer, Cohen and Neumaier (1988) for further information on distance-transitive graphs.

III

The Almost Stability Theorem

This chapter is devoted to the proof of a theorem which is motivated and stated in Section 1 and will have several applications in the next chapter. As most of the consequences do not require the full force of the result, we digress in Section 2 to give a case which is of historical interest and is adequate for certain applications.

1 Motivation

To develop the motivation we make some preliminary observations.

1.1 Definition. We say that U is a *G-retract* of V if U, V are G-sets and there exists an injective G-map $U \to V$, and a G-map $V \to U$. Obviously the latter map can then be chosen so that the composite $U \to V \to U$ is the identity.

Thus a G-subset U of V is a G-retract if and only if for each subgroup H of G, if H stabilizes an element of V then H stabilizes an element of U. ∎

1.2 Remarks. There is a very close connection between retracts and subtrees.

(i) If T is a G-tree and T' a G-subtree of T then contracting every edge of T not in T' contracts each component of $T - ET'$ to the unique vertex of T' which it contains. Hence VT' is a G-retract of VT.

(ii) Conversely, if T' is a G-tree, and VT' is a G-retract of a G-set V then T' can be extended to a G-tree with vertex set V as follows. View VT' as a G-subset of V, and let $\phi: V \to VT'$ be a G-map. Extend T' to a tree T with vertex set V, by adjoining, for each $v \in V - VT'$, an edge e_v

with $\iota e_v = v$, $\tau e_v = \phi(v)$. For all $g \in G$ and $v \in V - VT'$, we have $\iota e_{gv} = gv$, $\tau e_{gv} = \phi(gv) = g\phi(v)$, so the G-action on T' extends to a G-action on T with $ge_v = e_{gv}$. Thus T' has been extended to a G-tree T with vertex set V.

For example, if T' consists of one vertex v_0 stabilized by G then the edge set of T is formed by joining the elements of $V - \{v_0\}$ to v_0. ∎

1.3 The fundamental example. Let T be a G-tree, with vertex set V and edge set E, and let v be a vertex of T. Let $V \to (E, \mathbb{Z}_2)$ be the costructure map of T, so $v|E$ is the set of edges which point to v. Then $V|E$ lies in a G-stable almost equality class $\tilde{V} = \tilde{V}T$ in $(E, \mathbb{Z}_2)$.

Let $v \in V$, and consider any $u \in \tilde{V}$. Then $v + u$ is a finite set of edges $\{e_1, \ldots, e_n\}$. For any $g \in G_u$, $\{e_1, \ldots, e_n\} + \{ge_1, \ldots, ge_n\} = v + u + gv + gu = v + gv$ since $gu = u$. Thus the distance from v to gv in T is at most $2n$. By Proposition I.4.7, G_u stabilizes a vertex of T. This shows that V is a G-retract of $\tilde{V}$, so by Remark 1.2(ii), T extends to a G-tree $\tilde{T}$ with vertex set $\tilde{V}$.

The interested reader can show that $\tilde{V}$ is isomorphic as G-set to the set of all finite subsets of V with an odd number of elements. ∎

This example motivates the statement of the main theorem of the chapter.

The Almost Stability Theorem (Theorem 8.5). *If E is a G-set with finite stabilizers, A a nonempty set, and V a G-stable almost equality class in (E, A), then there exists a G-tree with finite edge stabilizers and vertex set V.* ∎

Conceptually the project is quite simple. It is observed in Lemma 5.1 that the complete graph on V has finite edge stabilizers, and all we want is a G-stable maximal subtree. Nonetheless, the proof is rather long and technical, and the result is quite powerful, as will be seen from the applications in the next chapter.

Example IV.1.4 gives a two-generator group G, an infinite cyclic subgroup C, and a G-stable almost equality class in $(G/C, \mathbb{Z}_2)$ that cannot be the vertex set of a G-tree; thus the finite stabilizer condition on E is of some importance.

2 Digression: the classical approach

For historical interest we recall the proof of one of the first results in this area. Although the result itself will not be used, the proof illustrates some of the more important ideas that are needed, containing, for example, the historical basis of Section II.2.

2.1 Stallings' Ends Theorem. *If G is a finitely generated group and there exists an almost G-stable function on G which is not almost constant then there exists a finite subgroup C such that G can be expressed either in the form* $B \underset{C}{*} D$ *with* $B \neq C \neq D$ *or in the form* $B \underset{C}{*} x$.

Proof. Since G is finitely generated, we can form the Cayley graph X with respect to a finite generating set. Here X is G-free, G-finite and connected. Let v_0 be a vertex of X, and write $V = VX = Gv_0$.

For any function s on V, recall that the *coboundary* of s is $\delta s = \{e \in EX \mid s(\iota e) \neq s(\tau e)\}$. The first step in the proof will be to show that there is some $s \in (V, \mathbb{Z}_2)$ such that s is not almost constant and δs is finite (so $s \in \mathscr{B}X$).

By hypothesis, there exists a nonempty set A and a function $\tilde{v}: G \to A$ which is almost G-stable, and not almost constant.

Denote the G-map $V \to (G, A)$, $gv_0 \mapsto g\tilde{v}$, by $v \mapsto v|G$. Since $\tilde{v}$ is almost G-stable, $v|G =_a \tilde{v}$ for all $v \in Gv_0 = V$. There is a dual G-map $G \to (V, A)$ denoted $g \mapsto g|V$. Let $\tilde{s}$ denote the image of 1.

For any edge e of X, since $\iota e|G =_a \tau e|G$ in (G, A), there are only finitely many $g \in G$ such that $\iota e(g) \neq \tau e(g)$, that is, $g^{-1}\iota e(1) \neq g^{-1}\tau e(1)$, or equivalently, $g^{-1}e$ lies in $\delta\tilde{s}$. Thus $Ge \cap \delta\tilde{s}$ is finite. As EX is G-finite we see that $\delta\tilde{s}$ is finite. Hence $X - \delta\tilde{s}$ has only finitely many components so $\tilde{s}$ takes only finitely many values. As $v_0|G = \tilde{v}$ is not almost constant, so $\tilde{s}|Gv_0$ is not almost constant, and hence $\tilde{s}$ is not almost constant. Thus there exists a component of $X - \delta\tilde{s}$ with vertex set $s \in (V, \mathbb{Z}_2)$ which is not almost constant. Further, $\delta s \subseteq \delta\tilde{s}$, so δs is finite.

In summary, we have found an element $s \in (V, \mathbb{Z}_2)$ which is not almost constant and δs is finite. We now forget all the preceding notation since we shall want another way of assigning to each vertex of X a function on G.

Let m be the smallest value of $|\delta s|$ as s ranges over all $s \subseteq V$ which are not almost constant.

Choose $s \subseteq V$ not almost constant, with $|\delta s| = m$. Clearly the same condition is satisfied by all elements of $Gs \cup Gs^*$.

Since s is not almost constant, one of the finitely many components s' of s with $s' \leqslant s$ is such that s' is not almost constant. By minimality of $|\delta s|$, we see that $|\delta s'| \geqslant m$. But $\delta s' \subseteq \delta s$, so $\delta s' = \delta s$, and thus $s' = s$. So s is connected. Similarly, s^* is connected.

Replacing s with s^*, if necessary, we may assume that s contains v_0.

We claim we can choose s minimal under inclusion among all $s \subseteq V$ such that s is not almost constant, $|\delta s| = m$, and $v_0 \in s$.

Suppose not, so there exists an infinite descending chain $s_1 > s_2 > \cdots$ of connected subsets of V with $|\delta s_i| = m$, and containing v_0. Let $s_\omega = \bigcap_{i \geqslant 1} s_i$. Since s_1 is connected, there exists an edge e_1 joining $s_1 s_\omega^*$ to s_ω, and hence joining $s_1 s_{i_1}^*$ to s_ω for some $i_1 > 1$. Thus $e_1 \in \bigcap_{i \geqslant i_1} \delta s_i$. Similarly s_{i_1} is connected so there exists on edge e_2 joining $s_{i_1} s_{i_2}^*$ to s_ω for some $i_2 > i_1$. Notice e_2 has both vertices in s_{i_1} so $e_2 \neq e_1$. Continuing in this way we arrive at an i_{m+1} and $m+1$ distinct edges $e_1, \ldots, e_{m+1}$ lying in $\delta s_{i_{m+1}}$, as illustrated in Fig. II.3, page 58. But $|\delta s_{i_{m+1}}| \leqslant m$, a contradiction. This proves the claim.

Let $g \in G$. We claim that some $r \in s \square gs$ is almost 0. Suppose not, so each $r \in s \square gs$ is not almost constant. By minimality, $m \leqslant |\delta r|$ for all $r \in s \square gs$, so, as in Fig. II.2, page 56,

$$4m \leqslant \sum_{r \in s \square gs} |\delta r| \leqslant 2|\delta s| + 2|\delta gs| = 4m,$$

and hence $|\delta r| = m$ for all $r \in s \square gs$. Consider the unique $r \in s \square gs$ containing v_0. Then $r \leqslant s$, and by minimality under inclusion, $r = s$, which forces some element of $s \square gs$ to be 0, and the claim is proved.

Let Σ be the set of all permutations of V which move only finitely many elements. Then Σ acts in the usual way on $(V, \mathbb{Z}_2)$, and we have a map $(V, \mathbb{Z}_2) \to \Sigma \backslash (V, \mathbb{Z}_2)$, $s \mapsto \Sigma s$.

Since Σ is a normal subgroup of the group of all permutations of V, the G-action on $(V, \mathbb{Z}_2)$ induces a G-action on $\Sigma \backslash (V, \mathbb{Z}_2)$. The involution $*$ on $(V, \mathbb{Z}_2)$ induces an involution, again denoted $*$, on $\Sigma \backslash (V, \mathbb{Z}_2)$.

The partial order $\leqslant$ on $(V, \mathbb{Z}_2)$ arising from inclusion induces a partial order, again denoted $\leqslant$, on $\Sigma \backslash (V, \mathbb{Z}_2)$. Explicitly, if $e, f \in (V, \mathbb{Z}_2)$ then $\Sigma e \leqslant \Sigma f$ if and only if $\sigma e \leqslant f$ for some $\sigma \in \Sigma$.

Let $e = \Sigma s \in \Sigma \backslash (V, \mathbb{Z}_2)$, and let $E = Ge$. The facts that for all $g \in G$ one element of $s \square gs$ is almost 0, and s, gs are not almost constant imply that E is nested, and we can form a G-graph $T = T(E)$ as in Definition II.1.7.

The surjection $G \to E$, $g \mapsto ge$, allows us to view $(E, \mathbb{Z}_2)$ as a subset of $(G, \mathbb{Z}_2)$, and we have $VT \subseteq (E, \mathbb{Z}_2) \subseteq (G, \mathbb{Z}_2)$.

We have a map $G \to (V, \mathbb{Z}_2)$, $g \mapsto gs$, and its dual $V \to (G, \mathbb{Z}_2)$, $v \mapsto v|G$. Let $\tilde{V}$ be the almost equality class of $v_0|G$ in $(G, \mathbb{Z}_2)$. For each edge e of X, there are only finitely many $g \in G$ such that $g^{-1} e \in \delta s$. Thus $(gs)(\iota e) = (gs)(\tau e)$ for almost all $g \in G$, so $\iota e|G =_a \tau e|G$. Since X is connected, $V|G$ lies in $\tilde{V}$. Moreover, $\tilde{V}$ is G-stable.

Hence $VT \subseteq (E, \mathbb{Z}_2) \subseteq (G, \mathbb{Z}_2) \supseteq \tilde{V}$; we claim that $\tilde{V}$ contains VT.

Choose $v \in s$, $v' \in s^*$.

We shall show that for almost all $g \in G$, $\tau e(g) \in \{v(g), v'(g)\}$ and since $v|G =_a v'|G \in \tilde{V}$ it follows that $\tau e|G \in \tilde{V}$.

Let Y be a finite connected subgraph of X containing δs. Since δs and Y are finite and EX is G-free, for almost all $g \in G$, $g\delta s$ does not meet Y. Consider any such g.

Then Y is a connected subgraph of $X - g\delta s$, so lies in one of the two components. Let X' denote the component of $X - g\delta s$ disjoint from Y, so $X' \subseteq X - Y \subseteq X - \delta s$. Here VX' is gs or gs^*. If $VX' = gs$ then, for any $u, u' \in gs$, there is a path joining them in $X' \subseteq X - \delta s$, so $s(u) = s(u')$. Thus s is constant on gs, so s, gs are nested. Similarly, if $VX' = gs^*$ then s, gs are nested. So in any event s, gs are nested. Since

$$\tau e(g) = \tau e(ge) = \begin{cases} 1 & \text{if } ge \geqslant e \text{ or } ge > e^* \\ 0 & \text{if } ge < e \text{ or } ge \leqslant e^*, \end{cases}$$

one of the following holds:

$$\begin{array}{lll} gs \supseteq s \text{ so } ge \geqslant e & \text{and} & gs(v) = 1 = \tau e(ge), \\ gs \supset s^* \text{ so } ge > e^* & \text{and} & gs(v') = 1 = \tau e(ge), \\ gs \subset s \text{ so } ge < e & \textit{and} & gs(v') = 0 = \tau e(ge), \\ gs \subseteq s^* \text{ so } ge \leqslant e^* & \text{and} & gs(v) = 0 = \tau e(ge). \end{array}$$

In summary, for almost all $g \in G$, $\tau e(g) = \tau e(ge) \in \{gs(v), gs(v')\} = \{v(g), v'(g)\}$, as claimed. Thus $\tau e|G \in \tilde{V}$. Similarly, $\iota e|G \in \tilde{V}$, and, since $\tilde{V}$ is a G-subset, $VT \subseteq \tilde{V}$ as desired.

By Theorem II.1.8, T is a G-tree with costructure map given by the embedding $VT \subseteq (E, \mathbb{Z}_2) \subseteq (G, \mathbb{Z}_2)$.

Here $ET = E = Ge$, and G_e is finite, since for any $g \in G_e$, $\iota e(g) = \iota e(ge) = \iota e(e) \neq \tau e(e) = \tau e(ge) = \tau e(g)$, and $\iota e|G =_a \tau e|G$, so there are only finitely many such g.

Since $s = s|V = s|Gv_0$ is not almost constant, $v_0|G$ is not almost constant, so no element of $\tilde{V} \supseteq VT$ is stabilized by G. The result now follows by the Structure Theorem I.4.1. ∎

In the case where G is torsion-free, the finite subgroup must be trivial, and here the Grushko–Neumann Theorem I.10.6 applies.

2.2 Corollary. *If G is a finitely generated torsion-free group and there exists an almost G-stable function on G which is not almost constant then either G is infinite cyclic or $G = B * D$ with* rank $B <$ rank G, rank $D <$ rank G. ∎

2.3 Remark. In several of the arguments in the next chapter the above results can be substituted for the Almost Stability Theorem as follows.

(i) It follows from Proposition IV.2.3 that if G is finitely generated and $H^1(G, AG) \neq 0$ for some abelian group A, then G is either a nontrivial free product amalgamating a finite subgroup or an HNN extension over a finite subgroup. This is the usual statement of Stallings' Ends Theorem.

The corresponding result for infinitely generated groups, given in Theorem IV.6.10, requires the full force of the Almost Stability Theorem.

(ii) If G is a finitely generated torsion-free group of cohomological dimension at most one over a nonzero ring R then G is free. For ωRG is an RG-summand of AG for some free R-module A. The map $G \to \omega RG \subseteq AG, g \mapsto g - 1$, then gives rise to a nonzero element of $H^1(G, AG)$. By the preceding remark, either G is infinite cyclic or $G = B * D$ with rank $B <$ rank G, rank $D <$ rank G. By Proposition IV.3.7, B, D have cohomological dimension at most one over R, so are free by induction. Hence G is free.

For $R = \mathbb{Z}$, if G is a finitely generated group of cohomological dimension one, then by Proposition IV.3.8, G is torsion-free, so G is free.

(iii) If G is finitely generated torsion-free with a free subgroup of finite index then G is free. For by Lemma IV.3.18(ii), G has cohomological dimension at most one over $\mathbb{Q}$, so G is free by the preceding remark. ∎

3 Blowing up

We begin with some general background.

3.1 Definition. If H is a subgroup of G and U an H-set we write $G \otimes_H U$ for the set obtained from $G \times U$ by identifying $(gh, u) = (g, hu)$ for all $g \in G$, $h \in H$, $u \in U$. The image of (g, u) will be denoted $g \otimes u$. There is a natural G-set structure on $G \otimes_H U$ by $g'(g \otimes u) = (g'g) \otimes u$.

For example, if $U = H/K$ for some subgroup K of H then $G \otimes_H U = G/K$. ∎

3.2 Definition. Let T be a G-tree and U a G-transversal in VT and suppose that for each $u \in U$, there is given a G_u-tree T_u; for example, T_u could consist of the single vertex $\{u\}$. Let Z be the G-forest $\bigcup_{u \in U} G \otimes_{G_u} T_u$, and $\psi: VZ \to VT$ the G-map which sends VT_u to u for each $u \in U$. Suppose further that there are given G-maps $\phi_\iota, \phi_\tau: ET \to VZ$ such that for all $e \in ET$, $\psi\phi_\iota(e) = \iota e$, $\psi\phi_\tau(e) = \tau e$.

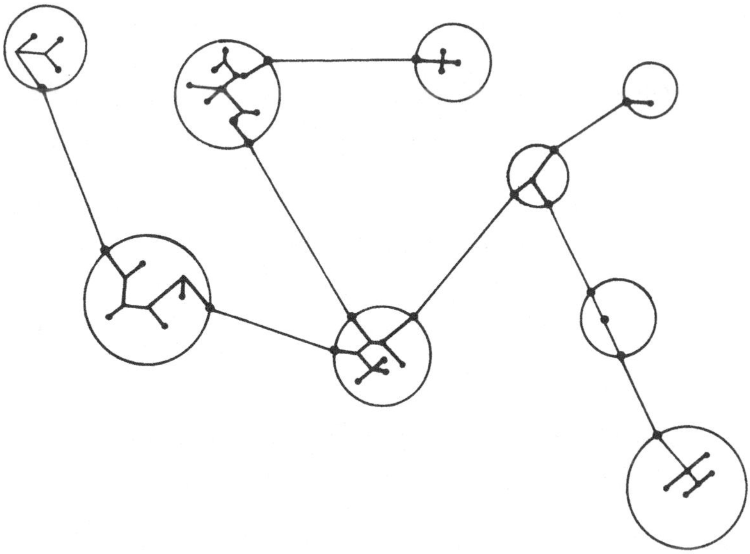

Fig. III.1

Let $\tilde{T}$ be the G-graph obtained by adjoining the set of edges ET to the forest Z, where the additional incidence functions are determined by ϕ_ι and ϕ_τ. The next proposition shows that $\tilde{T}$ is a tree; it is called the *fibre* G-tree with *base* T, *fibre* T_u *over* $u \in U$, and *attaching maps* ϕ_ι, ϕ_τ. The fibre T_u is said to be *trivial* if $T_u = \{u\}$.

Here we are reversing the procedure of contracting edges, since T can be obtained from $\tilde{T}$ by contracting certain edges. One can have in mind the sort of picture given in Fig. III.1. ∎

3.3 Proposition. *$\tilde{T}$ is a tree.*

Proof. Suppose that in $\tilde{T}$ we contract every edge lying in Z. The resulting vertex set is $\bigcup_{u \in U} G \otimes_{G_u} \{u\} = VT$ and the remaining edge set is ET, and the incidence functions are those of T. Thus the resulting graph is T. In particular, $\tilde{T}$ is connected. Consider any simple closed path in $\tilde{T}$; contracting the edges of Z leaves a closed path in T with no repeated edges, so is contracted to a single vertex because T is a tree. Thus the simple closed path lies in a single component of Z, which is impossible. Thus $\tilde{T}$ is a tree. ∎

3.4 Lemma. *If U, V are G-sets, $\phi: V \to U$ a G-map and U_0 a G-transversal in U then there is a natural identification of G-sets $V = \bigcup_{u \in U_0} G \otimes_{G_u} \phi^{-1}(u)$.*

Proof. There is an obvious G-map $\bigcup_{u\in U_0} G\otimes_{G_u}\phi^{-1}(u)\to V$.

The map is surjective, since for any $v\in V$, $\phi(v)=gu$ for some $g\in G$, $u\in U_0$, which gives $g^{-1}v\in\phi^{-1}(u)$ and $g\otimes g^{-1}v\mapsto v$.

To see that the map is injective, suppose that $g\otimes v$ and $g'\otimes v'$ have the same image, with $v\in\phi^{-1}(u)$, $v'\in\phi^{-1}(u')$. Then $gv=g'v'$ and $\phi(v)=u$, $\phi(v')=u'$ so $gu=g\phi(v)=\phi(gv)=\phi(g'v')=g'\phi(v')=g'u'$. Since U_0 is a G-transversal, $u=u'$ and $g^{-1}g'\in G_u$, and thus $g\otimes v=g'\otimes v'$ in $G\otimes_{G_u}\phi^{-1}(u)$. ∎

The crucial property of finite edge stabilizers is that they allow attaching maps to be constructed as follows.

3.5 Definition. Let T be a G-tree with finite edge stabilizers and U a G-transversal in VT, and suppose that for each $u\in U$, there is given a G_u-tree T_u.

Let Z be the G-forest $\bigcup_{u\in U} G\otimes_{G_u}T_u$, and let $\psi: VZ\to VT$ be the G-map which sends VT_u to u for each $u\in U$.

Applying Lemma 3.4 to the G-map $\iota: ET\to VT$ we see $ET=\bigcup_{u\in U} G\otimes_{G_u}\iota^{-1}(u)$. Let $u\in U$. For each $e\in\iota^{-1}(u)$, G_e is finite, so stabilizes a vertex of T_u; thus there exists a G_u-map $\iota^{-1}(u)\to VT_u$, and we can construct a G-map

$$ET=\bigcup_{u\in U} G\otimes_{G_u}\iota^{-1}(u)\xrightarrow{\phi_\iota}\bigcup_{u\in U} G\otimes_{G_u}VT_u$$

which carries $\iota^{-1}(u)$ to VT_u for each $u\in U$. Hence, $\psi\phi_\iota$ carries $\iota^{-1}(u)$ to u.

Similarly, there exists a G-map $\phi_\tau: ET\to\bigcup_{u\in U} G\otimes_{G_u}VT_u$ such that $\psi\phi_\tau$ carries $\tau^{-1}(u)$ to u for each $u\in U$.

Thus we can form the fibred G-tree $\tilde{T}$ with base T, fibre T_u over $u\in U$, and attaching maps ϕ_ι, ϕ_τ. We then call $\tilde{T}$ a *fibred G-tree $\tilde{T}$ with base T, and fibre T_u over $u\in U$*.

If the edge stabilizers of the T_u are finite then the edge stabilizers of $\tilde{T}$ are again finite. ∎

4 Notation

4.1 Notation. For the remainder of this chapter we fix a G-set E with finite stabilizers, a nonempty set A, a G-stable almost equality class V in (E, A), an element v_0 of V, and a subgroup H of G. ∎

We wish to construct a G-tree T with vertex set V. All our arguments will take place in the complete graph on V and we are looking for a G-stable maximal subtree inside this graph.

We ignore the trivial case where A has only one element; in Section 6, we shall assume there is a copy of $\mathbb{Z}_2$ contained in A.

4.2 Notation. For any subset E' of E we define $V(E') = \{v \in V \mid v \nabla v_0 \subseteq E'\}$, so $v_0 \in V(E')$.

Dually, if V' is a subset of V containing v_0, we define $E(V') = \bigcup_{v \in V'} v \nabla v_0$.

We write $E_H = E(Hv_0) = \bigcup_{h \in H} v_0 \nabla hv_0$ and $V_H = V(E_H) = \{v \in V \mid v \nabla v_0 \subseteq E_H\}$.

∎

It is sometimes helpful to speak loosely of $E(V')$ as the set of edges which connect v_0 to V', and $V(E')$ as the set of vertices connected to v_0 using only edges in E', or the component containing v_0 after deleting $E - E'$.

4.3 Lemma *Let E' be a subset of E, and V' a subset of V containing v_0.*

(i) *$E(-)$, $V(-)$ preserve inclusions.*

(ii) $V' \subseteq VE(V')$.

(iii) $EV(E') = E'$.

(iv) $G_{V'} \subseteq G_{VE(V')} \subseteq G_{E(V')}$.

(v) $G_{V(E')} \subseteq G_{E'}$.

(vi) *There is an identification of $G_{E'}$-sets $V = V|(E - E') \times V|E'$, which induces an identification of $G_{V(E')}$-sets $V(E') = v_0|(E - E') \times V|E' = V|E'$.*

Proof. (i) and (ii) are straightforward.

(iii) Since V is an almost equality class and A has at least two elements, there exists, for each element e of E, some $v \in V$ such that $v \nabla v_0 = \{e\}$. Hence, $E' \subseteq EV(E')$, and the reverse inclusion is clear from the definitions.

(iv), (v), (vi). If $g \in G_{V'}$ then

$$gE(V') = \bigcup_{v \in V'} gv \nabla gv_0 \subseteq \bigcup_{v \in V'} (gv \nabla v_0 \cup gv_0 \nabla v_0) \subseteq E(V')$$

which proves

(1) $\qquad G_{V'} \subseteq G_{E(V')}$.

Replacing V' with $VE(V')$ in (1) we have $G_{VE(V')} \subseteq G_{EVE(V')} = G_{E(V')}$ by (iii), so the second part of (iv) holds.

Replacing V' with $V(E')$ in (1) we have $G_{V(E')} \subseteq G_{EV(E')} = G_{E'}$ by (iii), so (v) holds.

Since V is an almost equality class, there is a natural identification $V = V|(E-E') \times V|E'$, and these are $G_{E'}$-sets. Hence there is induced an identification $V(E') = v_0|(E-E') \times V|E'$ of $G_{V(E')}$-sets so (vi) holds.

It follows that $G_{V(E')} = \{g \in G_{E'} | gv_0 \nabla v_0 \subseteq E'\}$. Replacing E' with $E(V')$ we have $G_{VE(V')} = \{g \in G_{E(V')} | gv_0 \nabla v_0 \subseteq E(V')\}$ and the latter clearly contains $G_{V'}$, which completes the proof of (iv). ∎

4.4 Corollary. *$H \subseteq G_{V_H} \subseteq G_{E_H}$, and V_H is an H-retract of V which is identified with an H-stable almost equality class in (E_H, A) by restriction. Also $E_H \subseteq E_G$, $V_H \subseteq V_G$ so $GE_H \subseteq E_G$, $GV_H \subseteq V_G$.* ∎

In (vi) we are identifying the act of restricting a function to a subset with the act of putting a function equal to v_0 on the complement of the subset. This corresponds to identifying a $G_{V(E')}$-subtree and the $G_{V(E')}$-tree obtained by contracting the complementary edges.

In the situation of Example 1.3, E_H is the edge set of the H-tree T_H generated by Hv_0, while V_H meets VT in VT_H.

5 Preliminaries

5.1 Lemma. *The complete G-graph on V has finite edge stabilizers, that is, if v, v' are distinct elements of V then $G_{vv'}$ is finite.*

Proof. As $v \neq v'$, there exists some $e \in v \nabla v'$. For each $g \in G_{vv'}$, $v(ge) = (g^{-1}v)(e) = v(e) \neq v'(e) = (g^{-1}v')(e) = v'(ge)$, which shows that $G_{vv'}e$ lies in the finite set $v \nabla v'$. Since G_e is finite, $G_{vv'}$ is finite. ∎

5.2 Definitions. We say that G is *finitely generated over H* if there exists a finite subset $\{g_1, \ldots, g_m\}$ of G such that G is generated by $H \cup \{g_1, \ldots, g_m\}$. *Countably generated over H* is defined similarly. ∎

The next section is devoted to proving that if G is finitely generated over H and $gE_H \cap E_H = \emptyset$ for all $g \in G - H$, then for any G-finite G-subset W of V_G and any H-tree T_H with vertex set V_H, the G-graph $W \cup GT_H$ embeds in a G-tree T with finite edge stabilizers for which there exists a G-map $VT \to V_G$. This is the most important step in the proof and is rather involved. On a first reading it might be preferable to concentrate on the case $H = 1$ and $W = \emptyset$, which can be used in the arguments of the next chapter to determine the finitely generated groups with a free subgroup of finite index and the finitely generated groups of cohomological dimension one over an arbitrary ring.

The following will be used frequently.

5.3 Lemma. *Suppose that X is a G-graph, E' a G-set with finite stabilizers, and $E' \to (VX, A)$ a G-map such that the image of the dual map $VX \to (E', A)$ lies in an almost equality class. Then for any G-transversal S' in E' and any G-finite G-subset F of EX, $\bigvee_{s \in S'} (F \cap \delta(s|VX))$ is finite.*

Proof. Let e be an edge of X.

Then $\iota e|E' =_a \tau e|E'$ so there are only finitely many $s \in E'$ such that $\iota e(s) \neq \tau e(s)$, that is, $e \in \delta(s|VX)$. Hence $\bigvee_{s \in E'} (\{e\} \cap \delta(s|VX))$ is finite.

Since E' has finite stabilizers, the set $\bigvee_{(g,s) \in G \times S'} (\{e\} \cap g\delta(s|VX))$ is finite. Thus $\bigvee_{(g,s) \in G \times S'} (g^{-1}\{e\} \cap \delta(s|VX))$ is finite, so $\bigvee_{s \in S'} (Ge \cap \delta(s|VX))$ is finite.

Since F is G-finite, $\bigvee_{s \in S'} (F \cap \delta(s|VX))$ is finite. ∎

6 The main argument

We fix the following throughout this section.

6.1 Notation. Suppose that G is finitely generated over H, that $gE_H \cap E_H = \varnothing$ for all $g \in G - H$, and that T_H is an H-tree with vertex set V_H.

Let W be a G-finite G-subset of V_G.

Let $H \cup \{g_1, \ldots, g_b\}$ generate G, and let $\{w_1, \ldots, w_a\}$ be a G-transversal for W.

Let X be the G-subgraph of the complete graph on V_G consisting of $GT_H \cup W$ together with the G-set generated by edges joining v_0 to $w_1, \ldots, w_a$, $g_1 v_0, \ldots, g_b v_0$ omitting loops.

By Lemma 5.1, X has finite edge stabilizers.

Write $Y = GT_H$ and $F = X - (Y \cup W)$. Thus Y is a G-subgraph of X, F is a G-finite set of edges in X, and $X = Y \cup W \cup F$.

Let S_H, ST_H be H-transversals in E_H, ET_H, respectively. By hypothesis no two elements of S_H lie in the same G-orbit, so we can extend S_H to a G-transversal S_G in E_G.

By Corollary 4.4, V_H is identified with its image in $V|E_H$ and there is an H-retraction $VX \to V_H$, $v \mapsto v|E_H$. Viewing $\mathbb{Z}_2$ as a subset of A, we have an H-map $(V_H, \mathbb{Z}_2) \to (VX, \mathbb{Z}_2) \subseteq (VX, A)$. Now the structure map for T_H gives an H-map $ET_H \to (V_H, \mathbb{Z}_2) \to (VX, A)$, $s \mapsto s|VX$. Explicitly, $s(v) =$

$1 \in A$ if and only if s points to $v|E_H$ in the tree T_H. We write δs for $\delta(s|VX)$. Define $F_0 = \bigcup_{s \in S_H} (F \cap \delta s)$.

Since $VX \subseteq (E_G, A)$, there is a dual G-map $E_G \to (VX, A)$, $s \mapsto s|VX$, and for $s \in E_G$ we write δs for $\delta(s|VX)$. ∎

6.2 Lemma. *X is connected.*

Proof. Imagine all the edges of X contracted and let $\bar{v}_0$ be the image of v_0. Since T_H is connected, all the G-orbits in Y contract into $G\bar{v}_0$ and H stabilizes $\bar{v}_0$; also W is contracted into this orbit and $\bar{v}_0$ is stabilized by $H \cup \{g_1, \ldots, g_b\}$. Hence, $\bar{v}_0$ is stabilized by G. Thus X contracts down to a single vertex, so is connected. ∎

6.3 Lemma. *For each $g \in G - H$ the H-retraction $VX \to V_H$, $v \mapsto v|E_H$, collapses gV_H to a single vertex.*

Proof. By Corollary 4.4, $V \to V|E_H$, $v \mapsto v|E_H$, is an H-retraction onto V_H.

Let $g \in G - H$. By hypothesis, $g^{-1}E_H \subseteq E_G - E_H$. By definition, if $v \in V_H$ then $v|E_G - E_H = v_0|E_G - E_H$ so $v|g^{-1}E_H = v_0|g^{-1}E_H$ and hence $gv|E_H = gv_0|E_H$. Thus all of gV_H is mapped to $gv_0|E_H$. ∎

We will make frequent use of the effect on the edges for this H-retraction.

6.4 Corollary. *$Y = GT_H$ is a forest and $gET_H \cap ET_H = \emptyset$ for all $g \in G - H$. Hence ST_H is a G-transversal in EY, and $EY = G \otimes_H ET_H$, and the H-map $ET_H \to (VX, A)$ extends to a G-map $EY \to (VX, A)$, $s \mapsto s|VX$.*

Proof. If Y is not a forest then there is a simple closed path involving an edge of T_H. By Lemma 6.3, the retraction $VX \to VT_H$ contracts this simple closed path to a nonempty closed path of T_H with no repeated edges, which is impossible.

If $g \in G - H$ then under the retraction $VX \to VT_H$ any edge in $gET_H \cap ET_H$ is both collapsed to a vertex and left fixed, which is absurd. Thus $gET_H \cap ET_H = \emptyset$ for all $g \in G - H$. It follows that there is a natural identification $EY = G \otimes_H ET_H$. ∎

We now have a G-map $EY \cup E_G \to (VX, A)$, $s \mapsto s|V$, and we write δs for $\delta(s|VX)$. We use these to cut up X and build some nice trees.

Recall $F_0 = \bigcup_{s \in S_H} (F \cap \delta s)$.

6.5 Lemma. *$HF_0 = \bigcup_{s \in E_H} (F \cap \delta s)$ and F_0 is finite.*

Proof. Since $HS_H = E_H$, $HF_0 = \bigcup_{s \in E_H} (F \cap \delta s)$.

Since F is G-finite, $\bigcup_{s \in S_G} (F \cap \delta s)$ is finite by Lemma 5.3, so the subset $F_0 = \bigcup_{s \in S_H} (F \cap \delta s)$ is finite. ∎

6.6 Lemma. *If $s \in ST_H$ then $s \in \delta s \subseteq \{s\} \cup HF_0$.*

Proof. Recall that EX is the union of the following sets: ET_H, gET_H for $g \in G - H$, and F.

If $e \in ET_H$ then in T_H, $\iota e | V_H = \iota e$ and $\tau e | V_H = \tau e$, and s lies between these if and only if $e = s$; thus $\delta s \cap ET_H = \{s\}$.

If $g \in G - H$ then $gV_H | E_H$ is a single vertex of T_H by Lemma 6.3, so $s | gV_H$ is constant and thus $\delta s \cap gET_H = \varnothing$.

If $f \in F \cap \delta s$ then in T_H, s lies between $\iota f | E_H$ and $\tau f | E_H$. In particular, $\iota f | E_H \neq \tau f | E_H$, so $f \in \delta e$ for some $e \in E_H$ and thus $f \in HF_0$.

This proves $s \in \delta s \subseteq \{s\} \cup HF_0$. ∎

6.7 Lemma. *$\bigvee_{s \in ST_H} (HF_0 \cap \delta s)$ is finite.*

Proof. The dual of the H-map $ET_H \to (VT_H, A) \to (VX, A)$ is $VX \to VT_H \to (ET_H, A)$ which has image in the almost equality class $\tilde{V}T_H$, and ET_H has finite stabilizers. By Lemma 5.3 applied to the group H, $\bigvee_{s \in ST_H} (HF_0 \cap \delta s)$ is finite. ∎

6.8 Corollary. *The set $\{e \in EY \mid \delta e \neq \{e\}\}$ is G-finite, and for all $e \in EY$, δe is finite and $\delta e \cap EY = \{e\}$. Hence there exists some integer n such that $|\delta e| \leqslant n$ for all $e \in EY$.*

Proof. For all $s \in ST_H$, $s \in \delta s \subseteq \{s\} \cup HF_0$ by Lemma 6.6, so by Lemma 6.7, for all $s \in ST_H$, δs is finite, and for almost all $s \in ST_H$, $\delta s = \{s\}$.

Since $EY = GST_H$, the set $\{e \in EY \mid \delta e \neq \{e\}\}$ is G-finite, and for all $e \in EY$, δe is finite and $\delta e \cap EY = \{e\}$. ∎

We can now glue together the G-forest Y to get a G-tree.

6.9 Theorem. *There exists a G-tree T_Y having a map of G-graphs $Y \cup W \to T_Y$ which is bijective on edge sets.*

Proof. Let

$$R = \{r \in (VX, \mathbb{Z}_2) | \exists e, e' \in EY, r \in (e|VX) \square (e'|VX), \delta r \cap EY = \emptyset\}.$$

The idea is that the elements of R are the obstruction to $EY|VX$ being nested, so for the map $VX \to VX|R \subseteq (R, \mathbb{Z}_2)$ the fibre over $v_0|R$ has the behaviour we want, and the base $VX|R$ can be handled separately.

Clearly R is a G-subset of $(VX, \mathbb{Z}_2)$, and for each $r \in R$, $|\delta r| \leqslant |\delta e \cup \delta e'| \leqslant 2n$. Form a new graph $X|R$ with vertex set $VX|R$ and edge set EX and the incidence functions given by composition $EX \xrightarrow{\iota, \tau} VX \to VX|R$. The double-dual map $R \to (VX|R, \mathbb{Z}_2)$ allows each $r \in R$ to be viewed as a function on $VX|R$, and δr has the same meaning as it did for X, since the edge sets are identical, as are the values at the vertices. By Theorem II.2.20 there is a tree G-subset E_R of $(VX|R, \mathbb{Z}_2)$ which generates $\mathscr{B}_{2n}(X|R)$. Let $T_R = T(E_R)$, which is a G-tree by Theorem II.1.5. Since $R|(VX|R)$ lies in the Boolean ring generated by E_R, the map $VX|R \to (R_R, \mathbb{Z}_2)$ is injective, and we have $VX \to VX|R \subseteq VT_R \subseteq (E_R, \mathbb{Z}_2)$.

This puts $VX|R$ where we want it, and we now look at the fibre over $v_0|R$. Let G_0 be the stabilizer of $v_0|R$.

Let e be an edge of Y. For all $r \in R$, $e \notin \delta r$ so $\iota e|R = \tau e|R$. Hence each component of Y maps to a single vertex of $X|R$. Let Y_0 be the G_0-subgraph of Y consisting of the components mapping to $v_0|R$, so $T_H \subseteq Y_0$ and $H \subseteq G_0$. Since $Y = GT_H = GY_0$, we see $Y = G \otimes_{G_0} Y_0$ by Lemma 3.4. Let VX_0 be the set of vertices of X mapping to $v_0|R$, so $VX_0 \supseteq VY_0$.

Let $e \neq e' \in EY_0$. Here $\delta e \cap EY = \{e\}$ and $\delta e' \cap EY = \{e'\}$ so $e \in \delta e - \delta e'$, $e' \in \delta e' - \delta e$. In terms of Fig. II.2, page 56, e and e' do not lie in the diagonals, nor in a parallel pair. It follows that there is an element r of $(e|VX) \square (e'|VX)$ with $e, e' \notin \delta r$; therefore $\delta r \cap EY = \emptyset$, that is, $r \in R$. Hence $r|VX_0$ is constant; in fact $r|VX_0$ must be zero since $e|VX_0$, $e'|VX_0$ are not constant. Thus $e|VX_0$, $e'|VX_0$ are nested, and we have proved that $EY_0|VX_0$ is a nested G_0-subset of $(VX_0, \mathbb{Z}_2)$.

We want to show that $VX_0|EY_0$ lies an almost equality class. By Lemma 6.6, if $s \in ST_H$ then $\delta s \subseteq HF_0 \cup \{s\}$. If $e \in EX$ then

$$\bigvee_{s \in ST_H} (Ge \cap \delta s) \subseteq \bigvee_{s \in ST_H} (Ge \cap \{s\}) \vee \bigvee_{s \in ST_H} (HF_0 \cap \delta s);$$

the second union is finite since no two elements of ST_H are in the same G-orbit, and the third union is finite by Lemma 6.7. Hence there are only finitely many pairs $(g, s) \in G \times ST_H$ such that $ge \in \delta s$ or equivalently $e \in \delta g^{-1} s$. Now if $v, v' \in VX_0$ we can choose a path p in X joining v, v', and there are only finitely many $gs \in EY$ such that δgs meets p, so $v(gs) = v'(gs)$ for almost all $gs \in EY_0$.

Thus $EY_0 | VX_0$ is a tree G_0-subset of $(VX_0, \mathbb{Z}_2)$.

By Theorem II.1.5, $T_0 = T(EY_0 | VX_0)$ is a G_0-tree. This has edge set EY_0, and a G_0-map $VX_0 \to VT_0$. For each edge e of Y_0, $e|VX_0$ is the unique element of $EY_0|VX_0$ cutting e; the images of $\iota e, \tau e$ in VT_0 are therefore joined by $e|VX_0$ in T_0, by Theorem II.1.5. By the above definition of $e|VX$, $e(\tau e) = 1$, so the orientation is correct. In summary, we have a map of G_0-graphs $Y_0 \cup VX_0 \to T_0$ which is bijective on edge sets.

Form a fibred G-tree T_Y with base T_R and fibre T_0 over $v_0|R$ (and trivial fibre for all other G-orbits) as in Definition 3.5. We then have a map of G-graphs

$$\begin{aligned} Y \cup VX &= (VX - G \otimes_{G_0} VX_0) \cup (G \otimes_{G_0} (Y_0 \cup VX_0)) \\ &\to (VT_R - G(v_0|R)) \cup (G \otimes_{G_0} T_0) \subseteq T_Y, \end{aligned}$$

which on edge sets is injective, with image $ET_Y - ET_R$. Contracting all the edges of T_Y which lie in ET_R leaves us with a G-tree $\bar{T}_Y$ having a map of G-graphs $Y \cup VX \to \bar{T}_Y$ which is bijective on edge sets, as desired. ∎

We shall use T_Y as the base for a fibred tree, and the fibres will be provided by the next result.

6.10 Theorem. *There exists a G-tree T_W with finite edge stabilizers such that there are G-maps $VY \cup W \to VT_W \to V_G$ whose composite is the inclusion map.*

Proof. Let $E_1 = \{e \in EY | \delta e = \{e\}\}$. Since $EX - EY = F$ is G-finite, and $EY - E_1$ is G-finite by Corollary 6.8, we see $EX - E_1$ is G-finite. By Lemma 5.3, the set $\delta = \bigvee_{s \in S_G} ((EX - E_1) \cap \delta s)$ is finite. Let $m = \max\{1, |\delta|\}$.

By Theorem II.2.20 there exists a tree G-subset E_m of $(VX, \mathbb{Z}_2)$ such that E_m generates $\mathscr{B}_m X$. Since $m \geqslant 1$, by Remark II.2.21(ii) we may assume that $E_1 | VX$ belongs to E_m.

By Theorem II.1.5, $T_m = T(E_m)$ is a G-tree; we shall show that T_m has all the properties desired of T_W.

If $e \in E_m$, then δe is a finite nonempty set of edges of X, so has finite stabilizer. Since $G_e \subseteq G_{\delta e}$ we see E_m has finite stabilizers, that is, T_m has finite edge stabilizers.

Let $s \in S_G$. The next step is to find a certain subset δ_s of E_m which *refines*

s in the sense that for all $v, v' \in VX$, if $v(e) = v'(e)$ for all $e \in \delta_s$ then $v(s) = v'(s)$. Deleting the edges of δ_s from T_m partitions VT_m, and hence partitions VX via the map $VX \to VT_m$; our requirement is precisely that the δ_s-partition of VX be finer than the *s-partition* determined by the function s on VX.

By a *component of* δ we shall mean the vertex set of a component of the graph $X - \delta$, viewed as a function on VX. Each component of δ has coboundary lying in δ, so has at most m elements. Thus each component of δ lies in $\mathscr{B}_m X$. Further δ has at most $m+1$ components, so there is a finite subset δ_m of E_m such that each component of δ lies in the ring generated by δ_m.

Let $S'_G = \{s' \in S_G \mid \delta s' \subseteq E_1\}$ and $S''_G = S_G - S'_G$, so S''_G is finite. Write $\delta'_s = \{e \mid VX : e \in \delta s \cap E_1\}$, and define $\delta_s = \delta'_s$ if $s \in S'_G$, and $\delta_s = \delta_m \cup \delta'_s$ if $s \in S''_G$.

Now suppose $v, v' \in VX$ and $v(e) = v'(e)$ for all $e \in \delta_s$. By choice of δ_m, we know that v, v' lie in the same component of $X - ((EX - E_1) \cap \delta s)$ so there exists a path p from v to v' having no repeated edges and which does not meet $(EX - E_1) \cap \delta s$. Also, if $e \in \delta s \cap E_1$ then $v(e) = v'(e)$, so v, v' lie in the same component of $X - \{e\}$, so e does not lie in p. Thus p does not meet $((EX - E_1) \cap \delta s) \cup (\delta s \cap E_1) = \delta s$, so s is constant on the vertices of p, and $v(s) = v'(s)$. We have proved the δ_s-partition refines the s-partition, as desired.

In particular, the partition of VX induced by deleting *all* the edges of T_m is finer than the gs-partition, for each $gs \in GS_G = E_G$, and hence finer than the partition determined by the map $VX \to VX \mid E_G$. But the latter is bijective by Corollary 4.4, so $VX \to VT_m$ is injective and we shall treat it as inclusion.

It remains to construct a G-map $VT_m \to V_G$. Let v be a vertex of T_m and write $K = G_v$. We want to find an element of V_G stabilized by K. Notice that for $e \in E_m$

$$\bigvee_{(g,s) \in G \times S_G} (\{e\} \cap g\delta_s) \subseteq \bigvee_{(g,s) \in G \times S''_G} (\{e\} \cap g\delta_m) \vee \bigvee_{(g,s) \in G \times S_G} (\{e\} \cap \delta gs),$$

and the latter is finite since e has finite stabilizer and S''_G, δ_m are finite. For each $f \in E_G$, set $\delta_f = \cup\{g\delta_s \mid (g,s) \in G \times S_G,\ gs = f\}$, so δ_f refines f, $g\delta_f = \delta_{gf}$ for all $g \in G$, and $\bigvee_{f \in E_G} (\{e\} \cap \delta_f)$ is finite for all $e \in E_m$.

Let S_K be a K-transversal in E_G. We shall examine the partition of Kv_0 induced by S_K.

Let $s \in S_K$. The s-partition of VX is refined by deleting δ_s from T_m. Hence, the s-partition of Kv_0 is refined by deleting $\delta_s \cap ET_K$ from the K-subtree T_K of T_m generated by Kv_0. Here T_K is K-finite, generated as K-set by the path joining v to v_0, so by Lemma 5.3, $\bigvee_{s\in S_K} \delta_s \cap ET_K$ is finite. Thus for almost all $s\in S_K$, $\delta_s \cap ET_K$ is empty, and for all $s\in S_K$, $\delta_s \cap ET_K$ is finite.

If any edge is deleted from T_K the component not containing v is a finite path, so deleting finitely many edges leaves the component containing v, and finitely many finite components. The edge stabilizers in T_K are finite, so one of the elements of the corresponding partition of Kv_0 is of the form $K'v_0$ for a cofinite subset K' of K.

Thus for almost all $s\in S_K$, $s|Kv_0$ is constant, and for all $s\in S_K$, $s|K'v_0$ is constant for a cofinite subset K' of K. Hence for almost all $s\in S_K$, $v_0|Ks$ is constant, and for all $s\in S_K$, $v_0|K'^{-1}s$ is constant for a cofinite subset K' of K. Thus we can construct a K-map w on $KS_K = E_G$ which is almost equal to v_0, and we have an element w of V_G stabilized by K. ∎

6.11 Theorem. *There exists a G-tree T with finite edge stabilizers having $Y \cup W$ as G-subgraph and having a G-map $VT \to V_G$.*

Proof. By Theorem 6.9, we have a map of G-graphs $\psi: Y\cup W \to T_Y$. Choose a G-transversal U in VT_Y, and form the G-forest $Z = \bigcup_{u\in U} G\otimes_{G_u} T_W$. By Lemma 3.4, $VY \cup W = \bigcup_{u\in U} G\otimes_{G_u}\psi^{-1}(u)$ so we can construct embeddings

$$VY\cup W \subseteq \bigcup_{u\in U} G\otimes_{G_u}(VY\cup W) \subseteq \bigcup_{u\in U} G\otimes_{G_u} VT_W = VZ.$$

There is then a G-map $\psi: VZ \to VT_Y$ which sends all of $g\otimes_{G_u} VT_W$ to gu; the restriction to the subset $VY\cup W$ agrees with the previous ψ. Let $\phi: ET_Y \to EY$ be inverse to the restriction $\psi: EY \to ET_Y$. Since ψ is a graph homomorphism, for each $e\in EY$ we have $\psi(\iota\phi e) = \iota(\psi\phi e) = \iota e$, and similarly $\psi(\tau\phi e) = \tau e$.

Form the fibred G-tree T with base T_Y, fibre T_W over each $u\in U$, and attaching maps $\iota\phi, \tau\phi: ET_Y \to VY \subseteq VZ$. If we identify $ET_Y = EY$ in T then $Y\cup W$ is a G-subgraph of T, and T has finite edge stabilizers. Also, we have a G-map

$$VT = \bigcup_{u\in U} G\otimes_{G_u} VT_W \to \bigcup_{u\in U} G\otimes_{G_u} V_G \to V_G,$$

as desired.

Let us record precisely what has been proved.

6.12 Theorem. *Suppose that G is finitely generated over H and that $gE_H \cap E_H = \varnothing$ for all $g\in G - H$. For any G-finite G-subset W of V_G and any H-tree T_H with vertex set V_H, the G-graph $W \cup GT_H$ embeds in a G-tree T with finite edge stabilizers for which there exists a G-map $VT \to V_G$.* ∎

7 Finitely generated extensions

The next project is to modify the tree just constructed to get vertex set precisely V_G. We shall need a number of new concepts.

7.1 Definitions. (i) Let V be a G-set and W a G-subset of V.

If V has the property that every G-map from $V - W$ to V is given by an automorphism of $V - W$, then V is said to be *G-incompressible over W*; otherwise V is *G-compressible over W*.

It is not difficult to see that V is G-incompressible over W if and only if for all $v \in V - W$, $v' \in V$, if $G_v \subseteq G_{v'}$ then $Gv = Gv'$ and $G_v = G_{v'}$.

Note that if V is G-incompressible over W then for every G-subset V' of V containing W, every G-map $V' - W \to V$ is injective and has image $V' - W$.

Let V_∞ denote the set of elements of V with infinite stabilizer. Suppose $G_{vv'}$ is finite for all distinct $v, v' \in V$; then, if $v, v' \in V_\infty$ and $G_v \subseteq G_{v'}$, then $G_{vv'} = G_v$ is infinite so $v = v'$. Thus, here V_∞ has a rigidity property even stronger than G-incompressibility over W.

(ii) Let T be a G-tree with finite edge stabilizers and Y a G-subgraph of T. If T' is a G-tree obtained from T by contracting edges, and VT' is a G-retract of VT containing VY, then we say that T' is obtained by *compressing T over Y*. If the only such tree T' is T itself we say that T is *incompressible over Y*; otherwise T is *compressible over Y*. An edge e of T is said to be compressible over Y if e has a vertex $v \in VT - VY$, and a vertex $v' \in VT$ and $Gv \neq Gv'$ and $G_e = G_v$; otherwise e is *incompressible over Y*.

By Remark 1.2(i), T can be compressed over Y to any G-subtree containing Y.

By omitting W and Y in all the above we have the concepts of *compressible* and *incompressible*. ∎

7.2 Lemma. *The following are equivalent for a G-tree T with finite edge stabilizers and a G-subgraph Y of T.*

(*a*) *VT is G-incompressible over VY.*

(*b*) *T is incompressible over Y.*

(c) *T has no compressible edges over Y.*

Proof. (*a*)$\Rightarrow$(*b*). Suppose T is compressible to a G-tree T' over Y. Then VT' is a proper subset of VT containing VY, and there is a G-map $VT - VY \to VT'$, so VT is G-compressible over VY.

(*b*)$\Rightarrow$(*c*). If T has a compressible edge e over Y, then contracting every edge in Ge compresses T over Y, so T is compressible over Y.

(*c*)$\Rightarrow$(*a*). Suppose that T has no incompressible edges over Y and consider any vertices $v' \in VT - VY$, $v \in VT$ with $G_{v'} \subseteq G_v$. We wish to show that $Gv = Gv'$ and $G_v = G_{v'}$, and we proceed by induction on the length of the geodesic p connecting v, v'. Clearly we may assume this length is positive. Let e be the first edge in p, and let v', v'' be the vertices of e. Then $G_e \subseteq G_{v'} = G_{vv'} = G_p \subseteq G_e$, so $G_{v'} = G_e \subseteq G_{v''}$ and $G_{v'}$ is finite. Since $v' \in VT - VY$, and e is not compressible over Y, $Gv' = Gv''$. Thus $v'' \in VT - VY$, and $G_{v'}$, $G_{v''}$ are conjugate finite groups with $G_{v'} \subseteq G_{v''}$, so $G_{v'} = G_{v''}$. Now, by induction on the length of p, we see that $Gv = Gv'$ and $G_v = G_{v'}$, as desired. ∎

7.3 Lemma. *If T, T' are G-trees and $VT \approx VT'$ as G-sets then $ET \approx ET'$ as G-sets.*

Proof. Let H be a subgroup of G, and N the normalizer of H in G.

For any G-set W let $n_W(H)$ denote the number of distinct G-subsets of W which are G-isomorphic to G/H; thus the function given by $H \mapsto n_W(H)$ completely determines the G-set structure of W. One way of calculating $n_W(H)$ is to take the N-subset $W(H)$ of elements of W with stabilizer precisely H, and collect together those which give the same G-orbit in W, that is, those which lie in the same N-orbit in $W(H)$. Therefore, $n_W(H) = |N \backslash W(H)|$. Notice also that $W(H) = W^H - \bigcup_{K>H} W^K$ where, as usual, W^H denotes the set of elements of W stabilized by H.

Write $V = VT$, $V' = VT'$, $E = ET$, $E' = ET'$, and let $\phi: V \to V'$ be an isomorphism of G-sets. This determines G-linear isomorphisms $\mathbb{Z}E \approx \omega\mathbb{Z}V \approx \omega\mathbb{Z}V' \approx \mathbb{Z}E'$, and we denote this composite by ψ. Since T^H, T'^H are subtrees of T, the isomorphisms restrict to isomorphisms $\mathbb{Z}E^H \approx \omega\mathbb{Z}V^H \approx \omega\mathbb{Z}V'^H \approx \mathbb{Z}E'^H$ so $\psi(\mathbb{Z}E^H) = \mathbb{Z}E'^H$. Similarly, $\psi\left(\sum_{K>H} \mathbb{Z}E^K\right) = \sum_{K>H} \mathbb{Z}E'^K$. These are N-submodules, so ψ induces an N-linear isomorphism on the cokernel $\mathbb{Z}[E(H)] \approx \mathbb{Z}[E'(H)]$. On applying $\mathbb{Z} \otimes_{\mathbb{Z}N}$, we find $\mathbb{Z}[N \backslash E(H)] \approx \mathbb{Z}[N \backslash E'(H)]$ as abelian groups, and thus $|N \backslash E(H)| = |N \backslash E'(H)|$. Hence, E and E' have the same G-set structure, as desired. ∎

7.4 Definition. If T is a G-finite G-tree with finite edge stabilizers we define the *size sequence* of T to be the eventually zero sequence of non-negative integers

$$\text{size}(T) = (|G \backslash E| - |G \backslash V|, \quad |G \backslash E_1|, \quad |G \backslash E_2|, \ldots),$$

where $E = ET$, $V = VT$, and for each $n \geq 1$, $E_n = \{e \in ET : |G_e| = n\}$.

Size sequences will be compared lexicographically, that is, $(m_0, m_1, m_2, \ldots) > (n_0, n_1, n_2, \ldots)$ if there exists some i such that $m_0 = n_0, \ldots, m_{i-1} = n_{i-1}, m_i > n_i$. ∎

The next result will be quite useful.

7.5 Lemma. *If T_1, T_2 are incompressible G-finite G-trees with finite edge stabilizers and there exists a G-map $VT_1 \to VT_2$ then $|G\backslash T_1| \geqslant |G\backslash VT_2|$. Further, size $(T_1) \geqslant$ size (T_2) and equality holds if and only if the map $VT_1 \to VT_2$ is an isomorphism.*

Proof. Let U_1, U_2 be G-transversals in VT_1, VT_2, respectively. Let $\tilde{T}$ be a fibred G-tree with base T_2 and fibre T_1 over u for each $u \in U_2$. As G-set

$$V\tilde{T} = \bigcup_{u \in U_2} G \otimes_{G_u} VT_1 \approx \bigcup_{u \in U_2} G/G_u \times VT_1 \approx VT_2 \times VT_1.$$

Since there exists a G-map $VT_1 \to VT_2$, we see that VT_1 is a G-retract of $V\tilde{T}$.

By construction, $\tilde{T}$ contains a copy of ET_2, and contracting all other edges yields T_2. The map $VT_1 \to V\tilde{T}$ carries each edge of T_1 to a path of $\tilde{T}_1$ to a path of $\tilde{T}$, and using these paths we can find a G-finite G-subtree of $\tilde{T}$. Adding finitely many G-orbits we can connect up ET_2. By Remark 1.2 (i), $\tilde{T}$ can be compressed to this G-finite subtree. By contracting G-orbits of compressible edges not in ET_2, we can compress $\tilde{T}$ to a G-finite G-tree T such that ET contains a copy of ET_2 and all compressible edges of T lie in ET_2, and contracting the edges of $ET - ET_2$ yields T_2.

Compress T to an incompressible G-tree T'. Then there exist G-maps $VT_1 \to V\tilde{T} \to VT \to VT'$, $VT' \to VT_1$. But VT_1 and VT' are incompressible G-sets, so the composites $VT_1 \to VT' \to VT_1$, $VT' \to VT_1 \to VT'$ must be bijective, and hence $VT_1 \approx VT'$ as G-sets. By Lemma 7.3, $ET_1 \approx ET'$, so for the purposes of this theorem we may take T_1 to be T'.

To summarize, we have a G-finite G-tree T with finite edge stabilizers such that T can be compressed to T_1, and T_2 can be obtained from T by contracting incompressible edges.

Let us treat ET_1 and ET_2 as subsets of ET, so $ET = ET_1 \cup ET_2$ and T_1 is obtained by contracting the edges of $ET - ET_1 = ET_2 - ET_1$ which are all compressible, and T_2 is obtained by contracting the edges of $ET - ET_2 = ET_1 - ET_2$, which are all incompressible.

We then think of the vertices of T_2 as the components of the graph $T - ET_2$, and partition VT_2 into two G-sets as follows. Let VT_{21} consist of those vertices of T_2 which contain only one vertex of T. There is then a natural identification of VT_{21} with a G-subset of VT, so we have an injective G-map $VT_{21} \to VT$. Consider the composite of the four G-maps $VT_{21} \to VT \to VT_1 \to VT \to VT_2$. Since VT_2 is incompressible, the image is VT_{21}, so the image of the tail $VT_1 \to VT \to VT_2$ contains VT_{21}.

Write $VT_{22} = VT_2 - VT_{21}$. Then VT_{22} consists of those vertices of T_2 which contain an edge of T, necessarily in $ET_1 - ET_2$, which means we have a surjective G-map $ET_1 - ET_2 \to VT_{22}$.

Combining the above yields a G-map $VT_1 \cup (ET_1 - ET_2) \to VT_2$ whose image contains $VT_{21} \cup VT_{22} = VT_2$. Thus the map is surjective, and $|G \backslash VT_2| \leqslant |G \backslash (VT_1 \cup (ET_1 - ET_2))| \leqslant |G \backslash T_1|$. This completes the first part.

We next consider the size sequences.

If $VT_1 \to VT_2$ is an isomorphism then by Lemma 7.3, $ET_1 \approx ET_2$ so $\text{size}(T_1) = \text{size}(T_2)$. Hence it remains to show that if $VT_1 \to VT_2$ is not an isomorphism then $\text{size}(T_1) > \text{size}(T_2)$.

Since T_1 is obtained from T by contracting compressible edges, and at each stage one edge orbit and one vertex orbit are lost, $|G \backslash ET_1| - |G \backslash VT_1| = |G \backslash ET| - |G \backslash VT|$. Since T_2 is obtained from T by contracting edges, and at each stage one edge orbit and at most one vertex orbit are lost, $|G \backslash ET_2| - |G \backslash VT_2| \leqslant |G \backslash ET| - |G \backslash VT| = |G \backslash ET_1| - |G \backslash VT_1|$. Thus the first co-ordinate of $\text{size}(T_1)$ dominates the first co-ordinate of $\text{size}(T_2)$. We may assume that equality holds. In particular, for any $f \in ET_1 - ET_2$, the vertices of f in T lie in different G-orbits. But such an f is incompressible in T, so for any vertex v of f in T we have $G_f \subset G_v$.

If $ET_2 - ET_1$ is empty then $ET = ET_1$ and $T = T_1$. Here it is clear that $\text{size}(T_1) > \text{size}(T_2)$ unless $T_1 = T_2$ in which case the map $VT_1 \to VT_2$ must be an isomorphism.

This leaves the case where $ET_2 - ET_1$ is nonempty. Choose $e \in ET_2 - ET_1$ with stabilizer of smallest possible order. Let u and v be the vertices of e, and let e be compressed to u in T_1, so $G_e = G_v$. Let $\bar{v}$ be the image of v in T_2 viewed as a subtree of T. If $\bar{v} = \{v\}$ then $G_{\bar{v}} = G_v = G_e$; but the image of e in T_2 is not compressible, so $\bar{v} = \{v\}$ and $\bar{u}$ are in the same G-orbit, and thus $\bar{u} = \{u\}$, and u, v are in the same G-orbit. This contradicts e being compressible in T. Hence, $\bar{v} \neq \{v\}$, so $\bar{v} \in VT_{22}$. Thus there is an edge f in $\bar{v}$ incident to v. Here, $f \in ET_1 - ET_2$, so $G_f \subset G_v = G_e$. By the minimality of $|G_e|$ for $ET_2 - ET_1$ we see that $\text{size}(T_1) > \text{size}(T_2)$. ∎

We have now arrived at the finitely generated case.

7.6 Theorem. *Suppose that G is finitely generated over H and that*

$gE_H \cap E_H = \emptyset$ for all $g \in G - H$. Then any H-tree T_H with vertex set V_H extends to a G-tree T_G with vertex set V_G.

Further, for any such T_G, the G-tree obtained by contracting T_H to a vertex can be compressed to an incompressible G-finite G-tree.

Here the trees involved have finite edge stabilizers by Lemma 5.1.

Proof. By Theorem 6.12, if W is any G-finite G-subset of $(V_G - GV_H)_\infty$ then there exists a G-tree T with finite edge stabilizers containing $W \cup GT_H$ as G-subgraph, and having a G-map $VT \to V_G$.

By hypothesis there exist $\{g_1, \ldots, g_m\} \subseteq G$ such that G is generated by $H \cup \{g_1, \ldots, g_m\}$. Choose a finite subtree X of T containing $\{v_0, g_1 v_0, \ldots, g_m v_0\}$. Contracting all the edges of T that lie in $GX \cup GT_H$ contracts $X \cup T_H$ to v_0, and makes v_0 stabilized by a generating set of G. Thus there is only one vertex left, so $GX \cup GT_H$ is connected. By Remark 1.2(i), T can be compressed to $GX \cup Y$. Since X is G-finite we can contract, one-by-one, G-orbits of edges compressible over GT_H until we arrive at a G-tree T_W which is incompressible over GT_H. Here $VT_W - GT_H$ is G-finite, and there exists a G-map $VT_W - GT_H \to V_G$. Since T_W is incompressible over GT_H the image cannot meet GT_H, so we have a G-map $VT_W - GT_H \to V_G - GT_H$. The compressing disturbs neither the vertices with infinite stabilizers nor the vertices of GT_H so there is an embedding $W \subseteq VT_W - GV_H$.

In summary, for an arbitrary G-finite G-subset W of $(V_G - GV_H)_\infty$, we have constructed a G-tree T_W with finite edge stabilizers incompressible over GT_H such that $VT_W - GV_H$ is G-finite and contains W, and there is a G-map $VT_W - GV_H \to V_G - GV_H$.

If we contract all the edges of GT_H then GT_H becomes a single orbit $G\bar{v}_0$. Let $\bar{T}_W$ denote the G-finite G-tree obtained by contracting all the edges in T_W which lie in GT_H, so $V\bar{T}_W = (VT_W - GV_H) \cup G\bar{v}_0$. This does not disturb W so there is an embedding $W \to V\bar{T}_W$, and hence $|G \backslash W| \leqslant |G \backslash (V\bar{T}_W)_\infty|$.

In particular, such a tree T_W is defined for the case $W = \emptyset$.

Take W to be any G-finite G-subset of $(V_G - GV_H)_\infty$ such that W contains the image of $(VT_\emptyset - GV_H)_\infty \to (V_G - GV_H)_\infty$. Then there is a G-map $(VT_\emptyset)_\infty \to W \cup GV_H \subseteq VT_W$. Since every finite subgroup of G stabilizes a vertex of VT_W we have a G-map $VT_\emptyset \to VT_W$. It follows easily that we have a G-map $V\bar{T}_\emptyset \to V\bar{T}_W$. These trees will be G-compressible if H is infinite, but in any event compressing them does not increase $|G \backslash \bar{T}_\emptyset|$ and does not change $|G \backslash (V\bar{T}_W)_\infty|$, so, by Lemma 7.5, $|G \backslash \bar{T}_\emptyset| \geqslant |G \backslash (V\bar{T}_W)_\infty| \geqslant |G \backslash W|$. Hence the number of G-orbits in

$(V_G - GV_H)_\infty$ is at most $|G\backslash \bar{T}_\emptyset|$, which is finite. Thus we can take $W = (V_G - GV_H)_\infty$ and set $T_G = T_W$.

We now have a G-tree T_G with finite edge stabilizers having GT_H as subgraph, such that T_G is incompressible over GT_H, and there exists an embedding $(V_G)_\infty \subseteq VT_G - GV_H$, and a G-map $VT_G - GV_H \to V_G - GV_H$. Since every finite subgroup of G stabilizes a vertex of T, we get G-maps $V_G \to VT_G$, $VT_G \to V_G$, which are the identity on GV_H. The composite $VT_G \to V_G \to VT_G$ is the identity on GV_H, and is bijective on $VT - GV_H$, since T_G is incompressible over GT_H. Thus VT_G is a G-retract of V_G. By Remark 1.2(ii), we can extend T_G to a G-tree with vertex set V_G.

The second paragraph of the proof confirms the final statement of the theorem. ∎

8 The general case

8.1 Lemma. *Let T be a G-tree with finite edge stabilizers, and let H stabilize a vertex v_0 of T. If G is finitely generated over H then G_{v_0} is finitely generated over H, and for each $v \in VT - Gv_0$, G_v is finitely generated.*

Proof. Let $H \cup \{g_1, \ldots, g_n\}$ generate G.

Let $U = G/H$ as G-set, and write $u_0 = 1H$. There is then a G-map $\phi: U \to VT$ sending u_0 to v_0. By Lemma 3.4, $\phi^{-1}(v_0) = G_{v_0}u_0$. We now construct a G-graph using the following setting. Draw each vertex v of T as a circle, and for each edge e in star(v), add one vertex to the perimeter of the circle and attach e there. Inside the circle add one vertex for each element of $\phi^{-1}(v)$. This gives a diagram on which G acts in an obvous way.

Let $i \in [1, n]$, and suppose that the T-geodesic from v_0 to g_iv_0 is $v_0, e_1^{\varepsilon_1}, v_1, \ldots, v_{m-1}, e_m^{\varepsilon_m}, v_m = g_iv_0$. We now add to our diagram the $m+1$ edges obtained by joining u_0 to e_1 inside v_0, e_1 to e_2 inside $v_1, \ldots, e_{m-1}$ to e_m inside v_{m-1}, e_m to g_iu_0 inside v_m, giving us a path of length $2m+1$ as in Fig. III.2. We do this for each $i \in [1, n]$, and then we take X to be the G-graph made up of the G-translates of these n paths. Since the paths are finite, X is G-finite. Notice also that the vertices not in U and the edges all have finite stabilizers.

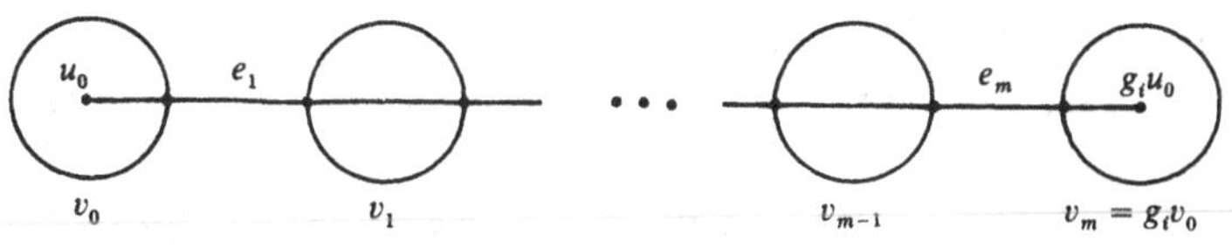

Fig. III.2

Contracting all the edges of X identifies all the vertices in the n paths with u_0, and the image is stabilized by $H \cup \{g_1, \ldots, g_n\}$ so there is only vertex. Thus X is connected.

Extend ϕ to a G-map $X \to T$ in the obvious way. Let $v \in VT$. If v is not in the image of ϕ then v is not in the subtree T' of T generated by Gv_0, and it follows that G_v is finite, so G_v is finitely generated. We may now assume that v is in the image of ϕ, so the graph $X_v = \phi^{-1}(v)$ is nonempty. By Lemma 3.4, $G \otimes_{G_v} X_v$ embeds in X, so X_v is G_v-finite. It is clear that if X_v is not connected then X is not connected, so X_v *is* connected. By the Structure Theorem I.9.2, G_v is generated by finitely many elements together with finitely many vertex stabilizers from different G_v-orbits, one vertex being chosen arbitrarily. But each vertex not in U has finite stabilizer. Thus, if $v \in VT - Gv_0$ then G_v is finitely generated, and G_{v_0} is finitely generated over $G_{u_0} = H$. ∎

8.2 Theorem. *Suppose that G is finitely generated over H and that $gE_H \cap E_H = \emptyset$ for all $g \in G - H$; suppose further that $H \leqslant K \leqslant G$ and whenever $K \leqslant L \leqslant G$ and L is finitely generated over H and $LE_K = E_L$ then $gE_K \cap E_K = \emptyset$ for all $g \in L - K$.*

If an H-tree T_H with vertex set V_H extends to a K-tree T_K with vertex set V_K, then T_K extends to a G-tree T_G with vertex set V_G.

Proof. Recall from Notation 4.2, that for any subset E' of E, $V(E') = \{v \in V \mid v \nabla v_0 \subseteq E'\}$.

Put $E_0 = E_G$, $V_0 = V(E_0) = V_G$, $G_0 = G_{V_0} = G$, and for $n \geqslant 1$, put $E_n = G_{n-1}E_K$, $V_n = V(E_n)$ and $G_n = G_{V_n} \cap G_{n-1}$. Notice that $E_n \subseteq E_{n-1}$ and hence $V_n \subseteq V_{n-1}$.

It is helpful to picture E_K as the edge set of a K-subtree of a G-tree, so GE_K is the edge set of a G-subforest, and then V_1 is the vertex set of the component T_1 containing v_0, and G_1 is its stabilizer, so T_1 is a G_1-tree containing E_K. Repeating this procedure over and over gives a descending sequence of trees. The content of the proof is to realize precisely this situation and to show that the procedure terminates.

Let $n \geqslant 0$, and inductively assume that G_n contains K and is finitely generated over H. We shall show that G_{n+1} contains K and is finitely generated over H, and at the same time we shall construct a G_n-tree T_n with vertex set V_n such that V_{n+1} is the vertex set of a G_{n+1}-subtree T'_{n+1} containing T_H.

By Lemma 4.3, for each $n \geqslant 0$ we have G_n-isomorphisms $V_n \to V_n | E_n \to V_n | (E_n - E_{n+1}) \times V_n | E_{n+1}$. The first factor is a restriction

G_n-map $\phi: V_n \to V_n|(E_n - E_{n+1})$. Write $\bar{v}_0 = v_0|(E_n - E_{n+1})$ and let U be a G_n-transversal for $V_n|(E_n - E_{n+1})$ containing $\bar{v}_0$. For each $u \in U$, $\phi^{-1}(u) \approx \{u\} \times V_n|E_{n+1} \approx V_n|E_{n+1}$ as G_{nu}-set. Notice that $\phi^{-1}(\bar{v}_0) = V_{n+1}$, so $\bar{G}_{nv_0} = G_{n+1}$. Thus $V_n \approx \bigvee_{u \in U} G_n \otimes_{G_{nu}} (V_n|E_{n+1})$. We shall construct T_n as a fibred G_n-tree with base having vertex set $V_n|(E_n - E_{n+1})$ and fibre over $u \in U$ having vertex set $V_n|E_{n+1}$; note that here all edge stabilizers are finite by Lemma 5.1.

Since $E_K \subseteq G_n E_K = E_{n+1}$, K stabilizes $\bar{v}_0 = v_0|(E_n - E_{n+1})$ so $K \subseteq G_{n+1}$. By our induction hypothesis, G_n is finitely generated over H which stabilizes $\bar{v}_0$, so by Theorem 7.6 there exists a G_n-tree with vertex set $V_n|(E_n - E_{n+1})$. This is our base. By Lemma 8.1, $G_{n\bar{v}_0} = G_{n+1}$ is finitely generated over H, and for $u \in U - \{v_0\}$, G_{nu} is finitely generated.

As G_{n+1}-set, $\phi^{-1}(\bar{v}_0) = V_{n+1} \approx V_n|E_{n+1}$, so by Theorem 7.6 there exists a G_{n+1}-tree T'_{n+1} with vertex set V_{n+1} containing $G_{n+1}T_H$ as G_{n+1}-subforest. This is our fibre over $\bar{v}_0$.

For $u \in U - \{v_0\}$, as G_{nu}-set $\phi^{-1}(\bar{v}_0) \approx V_n|E_{n+1}$, so by Theorem 7.6 there exists a G_{nu}-tree T_u with vertex set $V_n|E_{n+1}$. This is our fibre over u.

By Definition 3.5, we have a fibred G_n-tree T_n with base and fibres as above. There is a natural identification $VT_n = V_n$, and V_{n+1} is the vertex set of a G_{n+1}-subtree T'_{n+1} containing T_H as desired.

Since V_{n+1} is a fibre, $gV_{n+1} \cap V_{n+1} = \emptyset$ for all $g \in G_n - G_{n+1}$, and contracting the G_n-orbits of edges which lie in T'_{n+1} contracts T'_{n+1} to a vertex with stabilizer G_{n+1}.

It follows by induction that $gV_n \cap V_n = \emptyset$ for all $g \in G - G_n$, so in fact $G_n = G_{V_n}$.

We now proceed to construct a sequence of G-trees all having vertex set V_G. The first term is $T^{(0)} = T_0$, and here V_1 is the vertex set of a subtree T'_1 containing T_H. The next term $T^{(1)}$ is obtained by replacing T'_1 with T_1; formally, we contract all the edges in the G-orbit of T'_1 to get a tree to use as a base, and use T_1 as the v_0-fibre, with fibres over points in other G-orbits being trivial. Here the attaching maps are already specified, and all we are changing is G-orbits of edges which lie in T'_1.

We continue this way introducing $T_0, T_1, T_2, \ldots$ to get a sequence of G-trees $T^{(0)}, T^{(1)}, T^{(2)}, \ldots$ all with vertex set V_G and containing T_H, such that in $T^{(n)}$, V_n is the vertex set of the G_n-subtree T_n constructed above. Here $gT_n \cap T_n = \emptyset$ for all $g \in G - G_n$.

We claim that $G_{n-1} = G_n$ for some n. This is obvious if some G_n is finite, so we may assume all the G_n are infinite. By the second part of Theorem 7.6 we can form an incompressible G-finite G-tree $\bar{T}^{(\infty)}$ from

$T^{(0)}$ by contracting T_H to a vertex and compressing the resulting tree. Similarly, for each $n \geqslant 0$, we can form an incompressible G-finite G-tree $\bar{T}^{(n)}$ from $T^{(n)}$ by contracting T_n to a vertex and compressing the resulting tree; notice that, as it is infinite, G_n is the stabilizer of a vertex of $\bar{T}^{(n)}$. Since $T_0 \supseteq T_1 \supseteq \cdots \supseteq T_H$ there exist G-maps $V\bar{T}^{(0)} \leftarrow V\bar{T}^{(1)} \leftarrow \cdots \leftarrow V\bar{T}^{(\infty)}$, so $\operatorname{size}(\bar{T}^{(0)}) \leqslant \operatorname{size}(\bar{T}^{(1)}) \leqslant \cdots \leqslant \operatorname{size}(\bar{T}^{(\infty)})$ by Lemma 7.5. But $\operatorname{size}(\bar{T}^{(\infty)})$ is an eventually zero sequence, so the inequalities cannot all be strict and there must exist some $n \geqslant 1$ with $\operatorname{size}(\bar{T}^{(n+1)}) = \operatorname{size}(\bar{T}^{(n)})$. By Lemma 7.5, the map $V\bar{T}^{(n)} \to V\bar{T}^{(n-1)}$ is an isomorphism; since the vertex of $V\bar{T}^{(n)}$ with stabilizer G_n is mapped to the vertex of $V\bar{T}^{(n-1)}$ with stabilizer G_{n-1}, we have $G_{n-1} = G_n$ as claimed.

Now $G_{n-1}v_0 = G_n v_0 \subseteq V_n$ so $E(G_{n-1}v_0) \subseteq E(V_n) = EV(G_{n-1}E_K) = G_{n-1}E_K$, that is, $E_{G_{n-1}} \subseteq G_{n-1}E_K$; the reverse inclusion always holds, so $E_{G_{n-1}} = G_{n-1}E_K$. By the hypotheses of the theorem, $gE_K \cap E_K = \varnothing$ for all $g \in G_{n-1} - K$. Since $V_{G_{n-1}} = V(E_{G_{n-1}}) = V(G_{n-1}E_K) = V_n$ we can apply Theorem 7.6 to extend T_K to a G_{n-1}-tree with vertex set V_n. To avoid extra notation, and without loss of generality, we take T_n to be this tree. Then $T^{(n)}$ has the desired properties. ∎

8.3 Lemma. *If G is finitely generated over H then $E_G - GE_H$ is G-finite.*

Proof. Let S be a finite subset of G such that $S \cup H$ generates G. Let $E' = GE_H \cup G\left(\bigcup_{s \in S} v_0 \nabla sv_0\right)$. Then E' is a G-subset of E_G, and each $g \in S \cup H$ stabilizes $v_0 | (E - E')$, so G stabilizes $v_0 | (E - E')$. Hence $E_G \subseteq E'$, so $E_G = E'$ and $E_G - GE_H \subseteq G\left(\bigcup_{s \in S} v_0 \nabla sv_0\right)$ is G-finite. ∎

8.4 Theorem. *Suppose that G is countably generated over H and that $gE_H \cap E_H = \varnothing$ for all $g \in G - H$. Then any H-tree T_H with vertex set V_H extends to a G-tree T_G with vertex set V_G.*

Proof. Let $g_1, g_2, \ldots$ be a countable sequence in G such that G is generated by $H \cup \{g_1, g_2, \ldots\}$.

We now construct an ascending sequence of subgroups $H = G_0 \subseteq G_1 \subseteq \cdots$ such that, for each $n \geqslant 0$,

(1) G_n contains g_n,

(2) G_n is finitely generated over H,

(3) whenever $G_n \leqslant L \leqslant G$, and L is finitely generated over H, if $LE_{G_n} = E_L$

then $gE_{G_n} \cap E_{G_n} = \emptyset$ for all $g \in L - G_n$; that is, if $L \otimes_{G_n} EG_n \to E_L$ is surjective then it is bijective.

Suppose that $n \geq 1$ and that we have constructed G_{n-1}. Consider any subgroup K of G that is finitely generated over H, and contains $G_{n-1} \cup \{g_n\}$; by Lemma 8.3, $K\backslash(E_K - KE_H)$ is finite. Hence among all such subgroups K, there is one for which $|K\backslash(E_K - KE_H)|$ achieves a minimum; let G_n be such a subgroup, so G_n satisfies (1), (2). If it does not satisfy (3) then there exists a subgroup L of G such that L contains G_n and is finitely generated over H, and $L \otimes_{G_n} E_{G_n} \to E_L$ is surjective but not bijective. Deleting $L \otimes_{G_n} G_n E_H = LE_H$ from both sides we still have a properly surjective map $L \otimes_{G_n} (E_{G_n} - G_n E_H) \to E_L - LE_H$. Hence $|L\backslash(E_L - LE_H)| < |L\backslash(L \otimes_{G_n} (E_{G_n} - G_n E_H))| = |G_n\backslash(E_{G_n} - G_n E_H)|$ which contradicts the minimality of $G_n\backslash(E_{G_n} - G_n E_H)$. Therefore G_n satisfies (3), as claimed.

Let $T_0 = T_H$, and take as an induction hypothesis that we have a G_n-tree T_n with vertex set V_{G_n}. Theorem 8.2 shows that T_n extends to G_{n+1}-tree T_{n+1} with vertex set $V_{G_{n+1}}$. The tree $T = \bigcup_{n \geq 0} T_n$ then has all the desired properties. ∎

We can now achieve our objective.

8.5 The Almost Stability Theorem. *If E is a G-set with finite stabilizers, A a nonempty set, and V a G-stable almost equality class in (E, A), then there exists a G-tree with finite edge stabilizers and vertex set V.*

Proof. Index the elements of G with some ordinal γ so $G = \{g_\beta \mid \beta \in [0, \gamma)\}$. We shall construct an ascending chain of subgroups G_β, $\beta \in [0, \gamma]$, such that $G_0 = 1$ and for each $\beta \in [0, \gamma]$, writing E_β, V_β for E_{G_β}, V_{G_β}, respectively, we have

$$G_\beta \supseteq \{g_\alpha \mid \alpha \in [0, \beta)\},$$

$$gE_\beta \cap E_\beta = \emptyset \text{ for all } g \in G - G_\beta,$$

$$G_\beta \text{ is countably generated over } G_{\beta-1} \text{ if } \beta \text{ is a succesor ordinal,}$$

$$G_\beta = \bigcup_{\alpha < \beta} G_\alpha \text{ if } \beta \text{ is a limit ordinal.}$$

Suppose that $\beta \in [1, \gamma]$, and we have constructed G_α, $\alpha \in [0, \beta)$.

If β is a limit ordinal we define $G_\beta = \bigcup_{\alpha < \beta} G_\alpha$ and it is clear that the conditions are satisfied.

If β is a successor ordinal we construct G_β as follows. Let K_0 be the subgroup of G generated by $G_{\beta-1} \cup \{g_\beta\}$. Suppose that $n \geq 0$, and that we have constructed a group K_n which is finitely generated over $G_{\beta-1}$. Let K_{n+1} be the subgroup of G generated by $K_n \cup \{g \in G \mid gE_{K_n} \cap E_{K_n} \neq \varnothing\}$. We claim that $\{g \in G \mid gE_{K_n} \cap E_{K_n} \neq \varnothing\}$ is a finite union of double K_n-cosets. Let $S_{\beta-1}$, S be K_n-transversals for $K_n E_{\beta-1}$, $E_{K_n} - K_n E_{B-1}$, respectively. By Lemma 8.3, S is finite. Since $S_{\beta-1}$ is a G-transversal for $GE_{\beta-1}$ and stabilizers are finite, there are only finitely many $g \in G$ such that $gS \cap (S \cup S_{\beta-1}) \neq \varnothing$. Since $G_{\beta-1}$ contains all $g \in G$ with $gS_{\beta-1} \cap S_{\beta-1} \neq \varnothing$ the claim now follows easily. Thus K_{n+1} is finitely generated over K_n. Let $G_\beta = \bigcup_{n \geq 0} K_n$, so $G_\beta \supseteq G_{\beta-1} \cup \{g_\beta\}$, and G_β is countably generated over $G_{\beta-1}$. If $gE_\beta \cap E_\beta \neq \varnothing$ for some $g \in G$ then, since $E_\beta = \bigcup_{n \geq 0} E_{K_n}$, there is some n such that $gE_{K_n} \cap E_{K_n} \neq \varnothing$, so $g \in K_{n+1} \subseteq G_\beta$. Thus the conditions are satisfied.

By transfinite induction the chain exists and clearly $G_\gamma = G$.

We next construct an ascending chain T_β, $\beta \in [0, \gamma]$, of subtrees of the complete graph on V, such that each T_β is a G_β-tree with vertex set V_β.

To start the construction we take $T_0 = \{v_0\}$. Suppose now that $\beta \in [1, \gamma]$, and we have constructed T_α, $\alpha \in [0, \beta)$.

If β is a limit ordinal we define $T_\beta = \bigcup_{\alpha < \beta} T_\alpha$ which is a G_β-tree with vertex set $\bigcup_{\alpha < \beta} V_\alpha = V_\beta$.

If β is successor ordinal then by Theorem 8.4, $T_{\beta-1}$ extends to a G_β-tree T_β with vertex set V_β.

By transfinite induction we arrive at a G-tree T_γ with vertex set V_G. But V_G is a G-retract of V since the composite $V_G \to V \to V \mid E_G$ is bijective, so by Remark 1.2(ii) T_γ extends to a G-tree T with vertex set V. By Lemma 5.1, T has finite edge stabilizers. ∎

In fact with no extra work we could have proved the relative version.

8.6 Theorem *If $gE_H \cap E_H = \varnothing$ for all $g \in H - H$ then any H-tree T_H with vertex set V_H extends to a G-tree T_G with vertex set V_G.*

Notes and comments

The Almost Stability Theorem 8.5 is original. Many of the ideas involved in the proof can be traced back to various authors. The starting point of the theory is the work of Stallings (1968, 1971). His techniques are largely subsumed in the proof of Theorem II.2.20, but for historical interest his approach, as modified by Dunwoody (1979), is singled out for description in Section 2. Section 3 is based on Cohen (1973). The $H = 1$ case of Section 6 is from Dunwoody (1979). Lemma 7.3 is from Dicks (1980). The arguments in Section 8 generalize techniques developed by Swan (1969) and refined by Cohen (1973).

IV

Applications of the Almost Stability Theorem

In this chapter we see the strength of the Almost Stability Theorem through its applications. In Section 1 it is shown that action on a tree with finite edge stabilizers lifts from a subgroup of finite index; in particular, a group with a free subgroup of finite index acts on a tree with stabilizers of bounded order. Section 2 gives the fundamental application, showing how to manufacture a tree out of a derivation to a projective module. Section 3 determines the groups of cohomological dimension one over an arbitrary ring. In Section 4 this is generalized to a characterization of projective augmentation modules. Section 5 gives a complete description of the situation where an augmentation ideal of a subgroup generates a summand in the augmentation ideal of the group. Section 6 gives a characterization of arbitrary groups which have more than one end. Section 7 briefly discusses accessibility, and gives a cohomological criterion for a finitely generated group to be accessible.

The reader is assumed to be familiar with elementary module theory, but no deep background knowledge of cohomology theory is needed. All that is used are two items of terminology, namely the first cohomology group and cohomological dimension, and the definitions are recalled as needed.

1 Subgroups of finite index

We want to show that action on a tree with finite edge stabilizers can be lifted from a subgroup of finite index. The idea of the proof is to encode the tree for the subgroup into an amost equality class, lift the almost equality class to the larger group, and then decode the new almost equality class into a new tree.

1.1 Lemma. *Let H be a subgroup of G of finite index, and E an H-set. For each H-stable almost equality class V in $(E, \mathbb{Z}_2)$ there exists a G-stable almost equality class W in $(G \otimes_H E, \mathbb{Z}_2)$ such that $W|E = V$.*

Proof. Let $g_1 = 1, \ldots, g_n$ be a right H-transversal in G, and identify $E = 1 \otimes E$ in $G \otimes_H E$, so $G \otimes_H E = g_1 E \vee \cdots \vee g_n E$. Choose an element $v \in V$ and view $v \subseteq E$. Set $w = g_1 v \cup \cdots \cup g_n v = g_1 v \vee \cdots \vee g_n v \subseteq G \otimes_H E$, and let W be the almost equality class of w in $(G \otimes_H E, \mathbb{Z}_2)$. Since $g_1 = 1$ we see $w|E = v$ and $W|E = V$. Finally, if $g \in G$ then there are unique $h_1, \ldots, h_n \in H$ such that $\{gg_1, \ldots, gg_n\} = \{g_1 h_1, \ldots, g_n h_n\}$, so $gw = gg_1 v \cup \cdots \cup gg_n v = g_1 h_1 v \cup \cdots \cup g_n h_n v =_a g_1 v \cup \cdots \cup g_n v = w$. Thus w is almost G-stable, so W is G-stable. ∎

1.2 Corollary. *If H is a subgroup of finite index in G and T an H-tree then there exists a G-stable almost equality class W in $(G \otimes_H ET, \mathbb{Z}_2)$ and an H-map $W \to VT$.*

Proof. For each $v \in VT$, let $\bar{v}: ET \to \mathbb{Z}_2$ be the function corresponding to the set of all edges of T which point to v. By Example III.1.3, the map $VT \to (ET, \mathbb{Z}_2)$, $v \mapsto \bar{v}$, carries VT to an H-stable almost equality class $\tilde{V}$ having VT as an H-retract.

By Lemma 1.1 there is a G-stable almost equality class W in $(G \otimes_H ET, \mathbb{Z}_2)$ such that $W|ET = \tilde{V}$, so we have H-maps $W \to W|ET = \tilde{V} \to VT$. ∎

1.3 Theorem. *If H is a subgroup of finite index in G and T an H-tree with finite edge stabilizers then there exists a G-tree $\tilde{T}$ with finite edge stabilizers having an H-map $V\tilde{T} \to VT$.*

Proof. By Corollary 1.2, there exists a G-stable almost equality class W in $(G \otimes_H ET, \mathbb{Z}_2)$ and an H-map $W \to VT$. Here $G \otimes_H ET$ has finite stabilizers, so by the Almost Stability Theorem III.8.5, there exists a G-tree $\tilde{T}$ with finite edge stabilizers and vertex set W, so there is an H-map $V\tilde{T} = W \to VT$. ∎

1.4 Example. Let $G = \langle a, b, c \mid a^p = b^q = c^r = abc = 1 \rangle$, where $p, q, r \geqslant 2$ and $1/p + 1/q + 1/r < 1$.

It is a classical fact that G has a subgroup H of finite index which is an infinite surface group; see, for example, Zieschang, Vogt and Coldewey (1980). In particular, $H = B \underset{C}{*} t$ with C infinite cyclic; see, for example, the proof of Theorem V.4.9. By Theorem I.7.6 there exists an H-tree T with edge set H/C, and no vertex stabilized by H. By Corollary 1.2 there is a G-stable almost equality class W in $(G/C, \mathbb{Z}_2)$ with no element stabilized by G.

If G acts on a tree then each of a, b, c must stabilize vertices, and it is an easy exercise to show that two of them must stabilize the same vertex, and hence G stabilizes that vertex. But H stabilizes no element of VT, so G stabilizes no element of W, and therefore W is not the vertex set of a G-tree.

In the smallest case, where $(p, q, r) = (2, 3, 7)$, G has as a homomorphic image the simple group $\mathrm{PSL}_2(\mathbf{F}_7)$ of order 168, with kernel $H = \langle x_1, x_2, x_3, y_1, y_2, y_3 \mid [x_1, y_1][x_2, y_2][x_3, y_3] \rangle = \langle x_1, x_2, y_1, y_2, x_3 \rangle \underset{C}{*} y_3$, where $C = \langle x_3 \rangle$ and $x_3^{y_3} = x_3[y_2, x_2][y_1, x_1]$. A history of this particular example is given in Gray (1982). ∎

1.5 Corollary. *If G has a subgroup H of finite index which acts on a tree T with finite stabilizers then G itself acts on a tree with finite stabilizers.*

Proof (*see also Remark 3.19*). By Theorem 1.3 there is a G-tree $\tilde{T}$ and an H-map $V\tilde{T} \to VT$, so $V\tilde{T}$ has finite H-stabilizers, and thus has finite G-stabilizers. ∎

This enables us to characterize groups with a free subgroup of finite index.

1.6 Theorem. *The following are equivalent:*

(*a*) *There exists a G-tree such that the orders of the vertex stabilizers are bounded by some integer.*

(*b*) *G is the fundamental group of a graph of groups such that the orders of the vertex groups are bounded by some integer.*

(*c*) *G has a free subgroup of finite index.*

Proof. $(a) \Rightarrow (b)$ by the Structure Theorem I.4.1.

$(b) \Rightarrow (c)$. Suppose $G = \pi(G(-), Y)$ where the orders of the $G(v)$, $v \in VY$, are bounded by some integer, and let n be the least common multiple of these orders. By Theorem I.7.4 there exists a homomorphism $G \to \mathrm{Sym}\, n$ which is injective on each $G(v)$. By Proposition I.7.10, the kernel is a free subgroup of finite index.

$(c) \Rightarrow (a)$. Suppose that G has a free subgroup F of finite index. By Theorem I.8.3, there exists an F-free F-tree. It follows from Corollary 1.5 that there exists a G-tree with F-free vertex set, and here, for each vertex v, G_v acts freely on $F\backslash G$, so the order of G_v divides $(G:F)$. ∎

1.7 Remark. For any finite group A there is an interesting correspondence between connected A-graphs and extensions of free groups by A, as follows.

By Theorem I.9.2 any connected A-graph X gives rise to an extension of the free group $\pi(X)$ by A.

Conversely, every extension $1 \to F \to G \to A \to 1$, with F free, arises in this way, for G acts with finite stabilizers on a tree T, by Theorem 1.6, and $X = F\backslash T$ is a connected A-graph with fundamental group F, by Corollary I.4.2. Of course, there is some flexibility allowed in the choice of T, and hence X. ∎

Let us make more precise statements of Theorem 1.6 in the cases where G is torsion-free or finitely generated.

1.8 Theorem. *Every torsion-free group with a free subgroup of finite index is free.*

Proof. If G is torsion-free and has a free subgroup of finite index, then by Theorem 1.6 $(c) \Rightarrow (a)$ there exists a G-free G-tree, so G is free by Theorem I.8.3. ∎

1.9 Corollary. *The following are equivalent:*

(*a*) *There exists a G-finite G-tree with finite stabilizers.*

(*b*) *G is the fundamental group of a finite graph of finite groups.*

(*c*) *G has a finitely generated free subgroup of finite index.*

Proof. The equivalence of $(a), (b)$ and (c) is just the case of Theorem 1.6 where G is finitely generated. For $(b) \Rightarrow (c)$, one uses the fact that a subgroup of finite index in G is again finitely generated, by Theorem I.8.5. For $(c) \Rightarrow (a)$, one uses the fact that any G-tree contains a G-finite subtree if G is finitely generated. ∎

1.10 Definition. Suppose that G satisfies the equivalent conditions of Corollary 1.9, so G has a finitely generated free subgroup F of finite index, and $G = \pi(G(-), Y)$ where Y is finite and $G(y)$ is finite for each $y \in Y$.

Let $T = T(G(-), Y)$. Since F is torsion-free, F acts freely on T. By Corollary I.4.2, $F \approx \pi(F\backslash T)$ and $\chi(F) = \chi(F\backslash T)$; that is,

$$1 - \operatorname{rank} F = \sum_{v \in VY} |F\backslash G/G(v)| - \sum_{e \in EY} |F\backslash G/G(e)|.$$

But any finite subgroup H of G acts freely on $F\backslash G$, so $|F\backslash G/H| = (G:F)/|H|$, and we deduce

$$(1) \qquad \frac{1 - \operatorname{rank} F}{(G:F)} = \sum_{v \in VY} \frac{1}{|G(v)|} - \sum_{e \in EY} \frac{1}{|G(e)|}.$$

The rational number occurring in both sides of (1) is called the *Euler characteristic* of G, denoted $\chi(G)$; clearly it is independent of the choice of F and $(G(-), Y)$. ∎

1.11 Remarks. Let the notation be as above.

(i) If G is finite, $\chi(G) = 1/|G|$.

(ii) If G is free then $\chi(G) = 1 - \operatorname{rank} G$ as in Definition I.8.1, since here we can take $F = G$.

(iii) Condition (*c*) of Corollary 1.9 is preserved by going up by finite index to a larger group and down by finite index to a smaller group. If H is any subgroup of G of finite index then $\chi(H) = (G:H)\chi(G)$, for by replacing F with $F \cap H$ we may assume that F lies in H, so by definition, $\chi(H) = \chi(F)/(H:F)$ and $\chi(G) = \chi(F)/(G:F) = \chi(H)(H:F)/(G:F) = \chi(H)/(G:H)$.

(iv) Consider an extension of groups $1 \to A \to B \to C \to 1$.

By the preceding remark, if C is finite then A satisfies condition (*c*) of Corollary 1.9 if and only if B does, and then $\chi(B) = \chi(A)/|C| = \chi(A)\chi(C)$.

If A is finite then it is not difficult to show that B satisfies condition (*c*) of Corollary 1.9 if and only if C does, and then $\chi(B) = \chi(C)/|A| = \chi(A)\chi(C)$.

(v) If n is the lowest common multiple of the orders of the finite subgroups of G then, from the right hand side of (1), $\chi(G) \in n^{-1}\mathbb{Z}$.

Hence by the Sylow theorems, if q is a prime power dividing the denominator of $\chi(G)$ then G has a subgroup of order q.

(vi) For any prime p, if A is a p-group and F is a finitely generated free group with $\operatorname{rank} F \not\equiv 1 (\operatorname{mod} p)$ then any extension of groups $1 \to F \to G \to A \to 1$ must split for the following reasons. Here $\chi(G) = \chi(F)/|A|$ in lowest terms. By the preceding remark, G has a subgroup B of the same order as A. Since F is torsion-free, B maps injectively to A.

(vii) If we choose F to be normal and set $A = G/F$ then $X = F\backslash T$ is a finite connected A-graph and $F = \pi(X)$. Each $G(v)$ embeds in A, and

$$\begin{aligned}\chi(F) &= \chi(X) = |VX| - |EX| \\ &= \sum_{v \in VY} |Av| - \sum_{e \in EY} |Ae| = \sum_{v \in VY} (A:G(v)) - \sum_{e \in EY} (A:G(e)).\end{aligned}$$

Conversely, by Theorem I.9.2, a finite group A acting on a finite connected graph X gives an extension G of the free group $F = \pi(X)$ by A.

In Example I.9.6, where X is the one-skeleton of the cube and A its automorphism group, we have $G = D_3 \underset{D_1}{*} D_2$ and rank $F = 5$. Here $(A:D_3) + (A:D_2) - (A:D_1) = 8 + 12 - 24 = 20 - 24 = \chi(X) = \chi(F)$. ∎

We now mention several natural groups with normal free subgroups of finite index and their expressions as fundamental groups of graphs of finite groups. By the preceding remark, one way to code the information is to describe a finite quotient group acting on a finite graph, and we leave the construction of the appropriate finite graphs as exercises.

1.12 Examples. (i) $\mathrm{GL}_2(\mathbb{Z}) \approx D_6 \underset{D_2}{*} D_4$ and there is an extension $1 \to F_2 \to \mathrm{GL}_2(\mathbb{Z}) \to D_{12} \to 1$; here (1) takes the form

$$\chi(\mathrm{GL}_2(\mathbb{Z})) = -\tfrac{1}{24} = \tfrac{1}{12} + \tfrac{1}{8} - \tfrac{1}{4}.$$

The isomorphism comes from Example I.5.2(i), and the extension arises from a choice of surjection $D_6 \underset{D_2}{*} D_4 \to D_{12}$, which is injective on D_6 and D_4 so the kernel is free, and the rank must be two by (1).

Similarly one can show the following.

(ii) $\mathrm{PGL}_2(\mathbb{Z}) \approx D_3 \underset{D_1}{*} D_2$, $\chi(\mathrm{PGL}_2(\mathbb{Z})) = -1/12$, and there is an extension $1 \to F_2 \to \mathrm{PGL}_2(\mathbb{Z}) \to D_6 \to 1$.

(iii) $\mathrm{SL}_2(\mathbb{Z}) \approx C_6 \underset{C_2}{*} C_4$, $\chi(\mathrm{SL}_2(\mathbb{Z})) = -1/12$, and there is an extension $1 \to F_2 \to \mathrm{SL}_2(\mathbb{Z}) \to C_{12} \to 1$.

(iv) $\mathrm{PSL}_2(\mathbb{Z}) \approx C_3 * C_2$, $\chi(\mathrm{PSL}_2(\mathbb{Z})) = -1/6$, and there is an extension $1 \to F_2 \to \mathrm{PSL}_2(\mathbb{Z}) \to C_6 \to 1$. ∎

1.13 Example. In Example I.5.2(v) it was observed that $\mathrm{Aut}(F_2) = A \underset{C}{*} B$ where A, B, C are extensions of F_2 by certain finite subgroups D_4, D_6 and D_2 of $\mathrm{GL}_2(\mathbb{Z})$, respectively. Thus the Euler characteristics of A, B, C are $-1/8$, $-1/12$, $-1/4$, respectively.

By Theorem 1.6, A, B, C must be fundamental groups of finite graphs of finite groups; they are actually expressible as free products with amalgamation of dihedral groups as follows.

$A = AD_4 \underset{AD_1}{*} AD_2$, where

$$AD_4 = \{(x, y), (x^{-1}, y^{-1}), (y, x^{-1}), (y^{-1}, x), (x, y^{-1}), (x^{-1}, y), (y, x), (y^{-1}, x^{-1})\},$$

$$AD_2 = \{(x, y), (x^{-1}, y), (y^{-1}x^{-1}y, y^{-1}), (y^{-1}xy, y^{-1})\},$$

$$AD_1 = \{(x, y), (x^{-1}, y)\}, \quad \text{noncentral in } AD_4;$$

$B = BD_3 \underset{BD_1}{*} BD_2$, where

$$BD_3 = \{(x, y), (y, x), (x^{-1}y, x^{-1}), (y^{-1}, y^{-1}x), (y^{-1}x, y^{-1}), (x^{-1}, x^{-1}y)\},$$

$$BD_2 = \{(x, y), (y, x), (x^{-1}, y^{-1}), (y^{-1}, x^{-1})\},$$

$$BD_1 = \{(x, y), (y, x)\};$$

$C = CD_2 * CD_1$, where

$$CD_2 = \{(x,y),(y,x),(x^{-1},y^{-1}),(y^{-1},x^{-1})\},$$
$$CD_1 = \{(x,y),(y^{-1}x^{-1}y,y^{-1})\}.$$

The map $C \to A$ arises from the inclusions $CD_2 \subseteq AD_4$, $CD_1 \subseteq AD_2$, while the map $C \to B$ arises from identifying $CD_2 = BD_2$ and $(y^{-1}x^{-1}y, y^{-1}) = bcb^{-1}$, where $b = (x^{-1}y, x^{-1}) \in BD_3, c = (x^{-1}, y^{-1}) \in BD_2$.

To prove that A, B, C have the claimed descriptions, one first verifies that $b^{-1}c^{-1}bc = x$, $bcb^{-1}c^{-1} = y$ and that the Euler characteristics are correct. The details are left as an exercise. ∎

2 Derivations to projective modules

In this section we prove the consequence of the Almost Stability Theorem which will be used in all subsequent applications.

2.1 Definitions. Let M be a G-module.

We write $\mathrm{Der}(G, M)$ for the set of all derivations $d: G \to M$. This is an abelian group under pointwise addition. We write $\mathrm{Inn}(G, M)$ for the subgroup consisting of all inner derivations $G \to M$. The *first cohomology group with coefficients in M* is the quotient $H^1(G, M) = \mathrm{Der}(G, M)/\mathrm{Inn}(G, M)$.

If $d: G \to M$ is a derivation then we write M_d for the G-set with underlying set M and G-action given by $g \cdot m = gm + dg$ for all $g \in G, m \in M$; it is readily verified that this is actually a G-set. ∎

2.2 Definitions. Let A be an abelian group.

The G-set (G, A) is an abelian group under pointwise addition, and is easily seen to be a G-module; it is called the G-module *co-induced* from A.

We denote by AG the G-module $\mathbb{Z}G \otimes_{\mathbb{Z}} A$; it is called the G-module *induced* from A.

As abelian group AG is a direct sum of copies of G indexed by A, and the elements are expressed as formal sums $\sum_{g \in G} a_g g$ with $a_g \in A$ being 0 for almost all $g \in G$. The function $G \to A, g \mapsto a_g$, is then almost equal to 0. In this way we can identify AG with the almost equality class of 0 in (G, A), and so view AG as G-submodule of AG.

We view A with trivial G-action as the G-submodule of (G, A) consisting of the constant functions.

Thus $A + AG$ is the set of almost constant functions, and $A \cap AG$ is either A or 0, depending if G is finite or infinite, respectively.

Let $\mathscr{A}(G, A)$ be the G-submodule of (G, A) consisting of the almost G-stable functions $v: G \to A$, that is, $gv =_a v$ for all $g \in G$.

For $v \in \mathscr{A}(G, A)$, the derivation ad $v: G \to (G, A)$, $g \mapsto gv - v =_a 0$, has image in AG, and so determines a derivation $d_v: G \to AG$. Hence there is a natural additive map $\mathscr{A}(G, A) \to \mathrm{Der}(G, AG)$, $v \mapsto d_v$. ∎

2.3 Proposition. *For any abelian group A there is an exact sequence of abelian groups* $0 \to A \to \mathscr{A}(G, A) \to \mathrm{Der}(G, AG) \to 0$. ∎

Proof. The kernel of $\mathscr{A}(G, A) \to \mathrm{Der}(G, AG)$ is the set of maps stabilized by G, that is, the constant maps, and it remains to show that $\mathscr{A}(G, A) \to \mathrm{Der}(G, AG)$ is surjective.

Let $d \in \mathrm{Der}(G, AG)$, so for every $x \in G$, $dx \in AG \subseteq (G, A)$. Define $v \in (G, A)$ to be the map which sends $x \in G$ to $-(dx)(x) \in A$. We shall show that $d(g) = gv - v$ for all $g \in G$.

Let $g, x \in G$. Then $dx = d(g(g^{-1}x)) = g(d(g^{-1}x)) + dg$, so $dg - dx = -g(d(g^{-1}x))$. Thus $(dg)(x) + v(x) = (dg)(x) - (dx)(x) = -(g(d(g^{-1}x)))(x) = -(d(g^{-1}x))(g^{-1}x) = v(g^{-1}x) = (gv)(x)$. Rearranging, we have $(dg)(x) = (gv)(x) - v(x) = (gv - v)(x)$. Since this holds for all $x \in G$, we have $dg = gv - v$, as claimed.

Also $gv - v = dg =_a 0$, so $v \in \mathscr{A}(G, A)$. ∎

2.4 Corollary. *If A is an abelian group then* $H^1(G, AG) \approx \mathscr{A}(G, A)/(AG + A)$. ∎

2.5 Theorem. *For any abelian group A and derivation $d: G \to AG$, there exists a G-tree T with finite edge stabilizers such that $VT = (AG)_d$.*

Proof. By Proposition 2.3 there exists $v \in \mathscr{A}(G, A)$ such that $dg = gv - v$ for all $g \in G$. There is a natural bijective map from AG to the almost equality class $V = v + AG$ of v given by $p \mapsto v + p$, and this is an isomorphism of G-sets $(AG)_d \approx V$. By the Almost Stability Theorem III.8.5, there exists a G-tree T with finite edge stabilizers and vertex set $(AG)_d$. ∎

There is another form of this which will also be useful; we first require the following.

2.6 Proposition. *A G-module P is a G-summand of an induced module if and only if there exists an additive function* $\mathrm{tr}: P \to P$ *such that, if $p \in P$ then* $\mathrm{tr}(gp) = 0$ *for almost all $g \in G$, and* $p = \sum_{g \in G} g\,\mathrm{tr}(g^{-1}p)$.

Proof. Suppose P is a G-summand of AG for some abelian group A. Let π be the G-linear retraction onto P. Every element of P can be written uniquely in the form $p = \sum_{x \in G} a_x x$. Define $\operatorname{tr}(p) = \pi(a_1)$; clearly $\operatorname{tr}: P \to P$ is additive. For all $g \in G$, $\operatorname{tr}(g^{-1}p) = \operatorname{tr}\left(\sum_{x \in G} a_{gx} x\right) = \pi(a_g)$. Thus $\sum_{g \in G} g \operatorname{tr}(g^{-1}p) = \sum_{g \in G} g\pi(a_g) = \pi\left(\sum_{g \in G} g a_g\right) = \pi(p) = p$.

Conversely, suppose we are given such a function $\operatorname{tr}: P \to P$. It is straightforward to show that the additive map $P \to \mathbb{Z}G \otimes_{\mathbb{Z}} P$, $p \mapsto \sum_{g \in G} g \otimes \operatorname{tr}(g^{-1}p)$, is G-linear, and is a left inverse of the multiplication map $\mathbb{Z}G \otimes_{\mathbb{Z}} P \to P$, $g \otimes p \mapsto gp$. Hence P is isomorphic to a G-summand of the induced module $\mathbb{Z}G \otimes_{\mathbb{Z}} P$. ∎

2.7 Definition. When the equivalent conditions of Proposition 2.6 hold then we say that P is *G-projective* and call tr a *G-trace map* for P. ∎

2.8 Corollary. *If P is a G-projective G-module and $d: G \to P$ a derivation then there exists a G-tree T with finite edge stabilizers such that VT contains P_d as a G-retract.*

Proof. Here there is an abelian group A, and an idempotent G-linear endomorphism π of AG with image P. In particular, we have a derivation $G \xrightarrow{d} P \subseteq AG$.

By Theorem 2.5 there exists a G-tree T with finite edge stabilizers and vertex set $(AG)_d$.

Since $\pi(g \cdot r) = \pi(gr + dg) = g\pi r + dg = g \cdot \pi r$ for all $g \in G$, $r \in AG$, we see that π is an idempotent G-map on $(AG)_d = VT$ with image P_d. Thus P_d is a G-retract of VT. ∎

2.9 Remarks. In the situation of Corollary 2.8 there are maps $\phi: P \to VT$, $\psi: VT \to P$ such that $\psi\phi$ is the identity on P and if $g \in G$, $v \in VT$, $p \in P$, then $g(\phi p) = \phi(gp + dg)$, $\psi(gv) = g(\psi v) + dg$.

Consequently, ϕ is injective, so $g\phi(p) = \phi(p)$ if and only if $gp + dg = p$. Hence $G_{\phi(p)} = \operatorname{Ker}(d + \operatorname{ad} p)$.

If $g \in G_v$ then $(d + \operatorname{ad} \psi v)(g) = dg + g\psi(v) - \psi(v) = \psi(gv) - \psi(v) = \psi(v) - \psi(v) = 0$. Thus $G_v \subseteq \operatorname{Ker}(d + \operatorname{ad} \psi v)$.

In particular, the subgroups of G on which d is inner are precisely the subgroups of G which stabilize a vertex of T.

The map ψ induces a G-linear map $\psi: \omega\mathbb{Z}[VT] \to P$ with $\psi(v - w) =$

$\psi(v)-\psi(w)$. For $v_0=\psi(0)$ we then have $\psi(gv_0-v_0)=dg$, so d factors through the derivation $G\to\omega\mathbb{Z}[VT]$, $g\mapsto gv_0-v_0$, or equivalently the derivation $T[v_0,\_v_0]:G\to\mathbb{Z}[ET]$. ∎

2.10 Corollary. *If M is a G-module having a G-projective G-submodule P and an element m such that $m+P$ is a G-subset of M then there exists a G-tree T with finite edge stabilizers such that VT contains $m+P$ as a G-retract.*

Proof. We have a derivation $d:G\to P$ by $d(g)=gm-m\in P$, and $m+P\approx P_d$. Now the result follows from Corollary 2.8. ∎

3 Cohomological dimension one

Throughout this section, let R be a nonzero ring.

To fix notation, we quickly review the concept of cohomological dimension; this will be repeated in slightly more detail in Definition V.2.3.

For later reference we state the following without proof.

3.1 Proposition. *For any R-module P the following are equivalent:*

(a) P is an R-summand of a free R-module; that is, there exists a free R-module F and an R-module Q and an R-linear isomorphism $F\approx P\oplus Q$.

(b) Every surjective R-linear map to P is R-split, that is, if $M\to P$ is any surjective R-linear map then there exists an R-linear map $P\to M$ such that $P\to M\to P$ is the identity. ∎

3.2 Definitions. An R-module P satisfying the equivalent conditions of Proposition 3.1 is said to be *R-projective* or simply *projective* if the context is clear.

An *R-projective resolution* of a left R-module M is an exact sequence of R-linear maps $\cdots\to P_n\to\cdots\to P_0\to M\to 0$ where all the P_n are projective left R-modules. If there is some integer $n\geqslant -1$ such that $P_m=0$ for all $m>n$ we write $0\to P_n\to\cdots\to P_0\to M\to 0$; the *length* of the resolution is the smallest such integer n, or ∞ if no such n exists.

The least possible length of a projective resolution of M is called the *projective dimension* of M as R-module, denoted $\mathrm{pd}_R\,M$. Thus $\mathrm{pd}_R\,M=-1$ if and only if $M=0$.

We make R into a left RG-module with trivial G-action, and define $\mathrm{cd}_R\,G$ to be $\mathrm{pd}_{RG}\,R$ called the *cohomological dimension of G over R*. For $R=\mathbb{Z}$ we write $\mathrm{cd}\,G$ in place of $\mathrm{cd}_{\mathbb{Z}}\,G$. ∎

The next result shows that any two projective resolutions are closely related.

3.3 Schanuel's Lemma. *If* $0 \to K \to P \to M \to 0$, $0 \to K' \to P' \to M \to 0$ *are exact sequences of left R-modules with P, P' projective then* $K \oplus P' \approx K' \oplus P$ *as R-modules.*

Proof. Denote the two maps by $\alpha: P \to M$, $\alpha': P' \to M$, and let $A = \{(p, p') \in P \oplus P' \mid \alpha p = \alpha' p'\}$. Projection onto the first co-ordinate gives a surjective R-linear map $\pi: A \to P$ with the same kernel as α', that is, K'. But P is projective so π is R-split and thus $A \approx P \oplus K'$. By symmetry $A \approx P' \oplus K$. ∎

3.4 Notation. Let RG denote the group ring. The map $\varepsilon: RG \to R$, $\sum_{g \in G} r_g g \mapsto \sum_{g \in G} r_g$ is called the *augmentation homomorphism*. It is a homomorphism of rings as well as of RG-modules. The kernel, ωRG, is called the *augmentation ideal* of RG. ∎

It is a simple matter to characterize the groups of cohomological dimension 0.

3.5 Proposition. *If R is a nonzero ring then* $\mathrm{cd}_R G = 0$ *if and only if G is finite with order invertible in R. In this event* $R \approx RGe$, *where* $e = |G|^{-1} \sum_{g \in G} g$.

Proof. The augmentation homomorphism $\varepsilon: RG \to R$ has an RG-linear right inverse if and only if there exists $e = \sum_{g \in G} r_g g \in RG$ such that $ge = e$ for all $g \in G$, and $\sum_{g \in G} r_g = 1$. But this is equivalent to $r_g = r_1$ for all $g \in G$ and $|G| r_1 = 1$. The result follows easily. ∎

We need a number of preliminary results before we can determine the groups of cohomological dimension one.

3.6 Lemma. *Suppose that P is a projective left RG-module.*

(i) *P is G-projective with an R-linear G-trace map.*

(ii) *If H is any subgroup of G then P is RH-projective.*

Proof. (i) Here P is an RG-summand of AG for some free R-module A, and we now argue exactly as in the proof of Proposition 2.6.

(ii) The hypothesis and conclusion are preserved under taking direct summands and direct sums of RG-modules so we may assume $P = RG$. Any H-transversal for the left H-action on G gives an RH-basis of RG, so RG is RH-free. ∎

3.7 Proposition. *If H is a subgroup of G then $\operatorname{cd}_R H \leqslant \operatorname{cd}_R G$.*

Proof. Any RG-projective resolution of R is an RH-projective resolution of R by Lemma 3.6(ii), so the result follows. ∎

3.8 Proposition. *If $\operatorname{cd}_R G$ is finite then the order of each finite subgroup of G is invertible in R.*

Proof. By Proposition 3.7 we may assume that G is finite. Let $\operatorname{cd}_R G = n$. If $n = 0$ we are finished by Proposition 3.5. If $n > 0$ we get a contradiction as follows. Let $0 \to P_n \xrightarrow{\alpha} P_{n-1} \to \cdots \to P_0 \to R \to 0$ be an RG-projective resolution. Since this sequence is R-split, α has an R-linear left inverse $\beta: P_{n-1} \to P_n$. By Lemma 3.6(i), P_n has an R-linear G-trace map $\operatorname{tr}: P_n \to P_n$. Define $\tilde{\beta}: P_{n-1} \to P_n$ by $\tilde{\beta}(p) = \sum_{g \in G} g \operatorname{tr}(\beta(g^{-1}p))$. It is readily verified that $\tilde{\beta}$ is an RG-linear left inverse of α. Thus $\operatorname{cd}_R G < n$, the desired contradiction. ∎

3.9 Notation. For any G-set U we write RU or $R[U]$ for the free R-module on U. The elements are expressed as formal sums $\sum_{u \in U} r_u u$ with $r_u \in R$ being 0 for almost all $u \in U$. ∎

3.10 Lemma. *If U is a G-set and U_0 a G-transversal in U, then there is an RG-linear isomorphism $RU \approx \bigoplus_{u \in U_0} RG \otimes_{RG_u} R$, with gu corresponding to $g \otimes_{RG_u} 1$ for all $g \in G$, $u \in U_0$.* ∎

We leave the converse of the next result as an exercise.

3.11 Lemma. *If U is a G-set such that the order of each stabilizer is finite and invertible in R then RU is RG-projective.*

Proof. By hypothesis and Proposition 3.5, R is projective as RG_u-module, for each $u \in U_0$, so RU is RG-projective by Lemma 3.10. ∎

3.12 Proposition. *If T is a G-tree then there is an exact sequence of left RG-modules $0 \to RET \xrightarrow{\partial} RVT \xrightarrow{\varepsilon} R \to 0$.*

Proof. The sequence in Theorem I.6.6 is split as a sequence of $\mathbb{Z}$-modules, so remains exact under $R \otimes_{\mathbb{Z}}$-; alternatively, one can use the proof of Theorem I.6.6. ∎

We can now prove the main theorem of the section.

3.13 Theorem. *If R is a nonzero ring then the following are equivalent:*

(*a*) $\mathrm{cd}_R\, G \leqslant 1$;

(*b*) *G acts on a tree T with stabilizers being finite and having order invertible in R;*

(*c*) *G is the fundamental group of a graph of finite groups having order invertible in R.*

Proof. (*a*)⇒(*b*). Suppose that $\mathrm{cd}_R\, G \leqslant 1$. By Schanuel's Lemma 3.3, ωRG is RG-projective. By Lemma 3.6(i), ωRG is G-projective, and $1 + \omega RG$ is a G-subset of RG, so by Corollary 2.10, there is a G-tree T such that VT has $1 + \omega RG$ as a G-retract. But $1 + \omega RG \subseteq RG - \{0\}$, and it is easy to see that $RG - \{0\}$ has finite stabilizers, so VT has finite stabilizers. The orders are invertible in R by Proposition 3.8.

(*b*)⇒(*a*). Suppose (*b*) holds. By Lemma 3.11, RET and RVT are RG-projective, so by Proposition 3.12, $0 \to RET \to RVT \to R \to 0$ is a projective RG-resolution of R, so $\mathrm{cd}_R\, G \leqslant 1$.

(*b*)⇒(*c*) by the Structure Theorem I.4.1.

(*c*)⇒(*b*) by Theorem I.7.6. ∎

Let us specialize Theorem 3.13 to the case where $R = \mathbb{Z}$ or G is torsion-free.

3.14 Corollary. *The following are equivalent:*

(*a*) *G acts freely on a tree.*

(*b*) *G is free.*

(*c*) *ωRG is RG-free and G is torsion-free.*

(*d*) *ωRG is RG-projective, that is, $\mathrm{cd}_R\, G \leqslant 1$, and G is torsion-free.*

(*e*) *$\omega \mathbb{Z}G$ is $\mathbb{Z}G$-free.*

(*f*) *$\omega \mathbb{Z}G$ is $\mathbb{Z}G$-projective, that is, $\mathrm{cd}\, G \leqslant 1$.*

Proof. (*a*)⇒(*b*) by Theorem I.8.3.

(*b*)⇒(*c*). If G is freely generated by a set S then by Theorem I.8.2, the

Cayley graph is a tree T. The exact sequence for T is $0 \to R[G \times S] \to RG \xrightarrow{\varepsilon} R \to 0$ so $\omega RG \approx R[G \times S]$ which is RG-free.

$(c) \Rightarrow (d)$ is clear.

$(d) \Rightarrow (a)$ by Theorem 3.13.

On taking $R = \mathbb{Z}$ and using Proposition 3.8, we see $(c), (d)$ become $(e), (f)$. ∎

For emphasis we restate one part.

3.15 The Stallings–Swan Theorem. $\operatorname{cd} G \leqslant 1$ *if and only if G is free.* ∎

The finitely generated case also deserves mention, and here we can use Corollary 1.9.

3.16 Corollary. *If G is finitely generated then the following are equivalent.*

(a) G has a free subgroup of finite index, and every finite subgroup of G has order invertible in R.

(b) G acts on a tree with stabilizers having order invertible in R.

(c) $\operatorname{cd}_R G \leqslant 1$. ∎

Using Corollary 1.5 and Theorem 3.13 we have the following.

3.17 Corollary. *If H is a subgroup of G of finite index, then* $\operatorname{cd}_R G = 1$ *if and only if* $\operatorname{cd}_R H = 1$ *and all finite subgroups of G have order invertible in R.* ∎

The analogue of Corollary 3.17 for dimensions higher than one is not known, except where R is commutative; see Serre's Extension Theorem V.5.3. We conclude this section by discussing special cases.

3.18 Lemma. *Suppose H is a subgroup of G of finite index.*

(i) *If* $\operatorname{cd}_R G$ *is finite then* $\operatorname{cd}_R G = \operatorname{cd}_R H$.

(ii) *If* $(G:H)$ *is invertible in R then* $\operatorname{cd}_R G = \operatorname{cd}_R H$.

Proof. By Proposition 3.7, $\operatorname{cd}_R H \leqslant \operatorname{cd}_R G$.

(i) Let $n = \operatorname{cd}_R G$ and choose an RG-projective resolution $0 \to P_n \xrightarrow{\alpha} P_{n-1} \to \cdots \to P_0 \to R \to 0$. In the expectation of getting a contradiction, assume that $\operatorname{cd}_R H < n$. Then $n \geqslant 1$ and α has an RH-linear left inverse $\beta: P_{n-1} \to P_n$. By Lemma 3.6(i), P_n has an R-linear G-trace map $\operatorname{tr}: P_n \to P_n$. Since β is RH-linear, and $(G:H) < \infty$ it follows that for any $p \in P_{n-1}$, $\mathrm{tr}(\beta(gp)) = 0$ for almost all $g \in G$. Thus there is a well-defined map $\tilde{\beta}: P_{n-1} \to P_n$ such that $\tilde{\beta}(p) = \sum_{g \in G} g^{-1}\,\mathrm{tr}(\beta(gp))$ for all $p \in P_{n-1}$. It is readily verified that $\tilde{\beta}$ is an RG-linear left inverse of α, so $\mathrm{cd}_R\, G < n$, the desired contradiction.

(ii) Suppose that $\mathrm{cd}_R\, H$ is finite, and choose a projective RG-resolution $\cdots \to P_0 \to R \to 0$. Let $n = \mathrm{cd}_R\, H$ so we have an exact sequence $0 \to K \xrightarrow{\alpha} P_n \to \cdots \to P_0 \to R \to 0$ of RG-modules, and α has an RH-linear left inverse $\beta: P_n \to K$. Define $\tilde{\beta}: P_n \to K$ by $\tilde{\beta}(p) = (G:H)^{-1} \sum_{gH \in G/H} g\beta(g^{-1}p)$ for all $p \in P_n$. Then $\tilde{\beta}$ is an RG-linear left inverse for α, so $\mathrm{cd}_R G \leqslant n$, as desired. ∎

3.19 Remark. Theorem 3.13 can be used to give another proof of Corollary 1.5 as follows. Suppose that H is a subgroup of G of finite index acting on a tree with finite stabilizers. By Theorem 3.13(b)$\Rightarrow$(a), $\mathrm{cd}_{\mathbb{Q}}\, H \leqslant 1$. By Lemma 3.18(ii), $\mathrm{cd}_{\mathbb{Q}}\, G \leqslant 1$, so by Theorem 3.1($a$)$\Rightarrow$($b$), G acts on a tree with finite stabilizers. ∎

4 Projective augmentation modules

Throughout this section let A be an abelian group, and U a G-set.

4.1 Definitions. We write AU, or sometimes $A[U]$, for the G-module $\mathbb{Z}U \otimes_{\mathbb{Z}} A$. As abelian group, AU is a direct sum of copies of A indexed by the elements of U. The elements of AU will be written as formal sums $\sum_{u \in U} a_u u$, with $a_u \in A$ being 0 for all but finitely many $u \in U$. The G-action is given by $g\left(\sum_{u \in U} a_u u\right) = \sum_{u \in U} a_u gu$.

If U_0 is a G-transversal in U then $U = \bigvee_{u \in U_0} Gu \approx \bigvee_{u \in U_0} G/G_u$ and $AU = \bigoplus_{u \in U_0} A[Gu] \approx \bigoplus_{u \in U_0} A[G/G_u] \approx \bigoplus_{u \in U_0} \mathbb{Z}G \otimes_{\mathbb{Z}G_u} A$.

The map $\varepsilon: AU \to A$, $\sum_{u \in U} a_u u \mapsto \sum_{u \in U} a_u$, is *called the augmentation map*, and the kernel, denoted ωAU, is called the *augmentation module* of U with coefficients in A. Since ε is G-linear, ωAU is a G-submodule of AU.

Let n be a cardinal. We say n is *invertible on* A, or $n \in \mathrm{Aut}\, A$, if n is finite, and the map $A \to A$, $a \mapsto na$, is bijective, or equivalently has an additive one-sided inverse, which is then two-sided. ∎

Theorem 4.8 will describe the situation where ωAU is G-projective, the case where A is zero or U is empty requiring no comment.

We start with the most important step.

4.2 Theorem. *If A is a nonzero abelian group and U a nonempty G-set such that ωAU is G-projective then there exists a G-tree T such that U is a G-subset of VT and $T-U$ has finite stabilizers.*

Proof. Choose a nonzero $a_0 \in A$, and $u_0 \in U$. Then $a_0 u_0 + \omega AU$ is a G-subset of AU, so by Corollary 2.10 there is a G-tree with finite edge stabilizers such that VT contains $a_0 u_0 + \omega AU$ as a G-retract. In particular, $a_0 U$ is a G-subset of VT isomorphic to U. It is easy to see that the retracted part, $VT-(a_0 u_0 + \omega AU)$, has finite stabilizers since edge stabilizers are finite. Thus it remains to show that each $v \in (a_0 u_0 + \omega AU) - a_0 U$ has finite stabilizer. Write $v = a_1 u_1 + \cdots + a_n u_n$ with nonzero $a_i \in A$; here $n \geqslant 1$ since $a_1 + \cdots + a_n = \varepsilon(v) = a_0 \neq 0$ and $n \geqslant 2$ since $v \notin a_0 U$. Clearly G_v permutes $u_1, \ldots, u_n$ so permutes the vertices $a_0 u_1, \ldots, a_0 u_n$. In particular, $G_v a_0 u_1$ is finite. Also, $v \neq a_0 u_1$ since $v \notin a_0 U$, but edge stabilizers are finite so $G_{v, a_0 u_1}$ is finite. Thus G_v is finite. ∎

We now need several lemmas on G-projective modules.

4.3 Lemma. *If P is a G-projective G-module then P is H-projective for each subgroup H of G.*

Proof. Choose an H-transversal S in G for the left H-action, so as H-set, $G = H \times S$. Then for any abelian group A, there is an H-module isomorphism $AG \approx BH$, where $B = AS$. The result now follows. ∎

4.4 Lemma. *AU is G-projective if and only if $|G_u| \in \operatorname{Aut} A$ for each $u \in U$.*

Proof. For any G-transversal S in U, $AU = \bigoplus_{s \in S} AGs$, so it suffices to consider the case $U = Gu$ for some $u \in U$.

Suppose $|G_u| \in \operatorname{Aut} A$, and let $p = \sum_{gu \in Gu} a_{gu} gu$ denote an arbitrary element of AGu. Let $\operatorname{tr}: AGu \to AGu$ be the map such that $\operatorname{tr}(p) = |G_u|^{-1} a_u u$. Clearly tr is additive. For any $g \in G$, $\operatorname{tr}(g^{-1}p) = |G_u|^{-1} a_{gu} u$, and it follows easily that $\sum_{g \in G} g \operatorname{tr}(g^{-1}p) = p$.

Conversely, suppose that AGu is G-projective, and hence G_u-projective, by Lemma 4.3. Let $\operatorname{tr}: AGu \to AGu$ be a G_u-trace map, and $\varepsilon: AGu \to A$ the augmentation map. For any $a \in A$, $a = \varepsilon(au) = \varepsilon\left(\sum_{h \in G_u} h \operatorname{tr}(h^{-1} au)\right) = \sum_{h \in G_u} \varepsilon(h \operatorname{tr}(au)) = |G_u| \varepsilon(\operatorname{tr}(au))$. Thus $A \to A$, $a \mapsto \varepsilon(\operatorname{tr}(au))$, is an inverse of multiplication by $|G_u|$. ∎

4.5 Corollary. *ωAU is G_u-projective for all $u \in U$ if and only if $|G_{uu'}| \in \operatorname{Aut} A$ for all distinct $u, u' \in U$.*

Proof. Let u_0 be an element of U. Then there is a G_{u_0}-linear isomorphism $\omega AU \to A[U - \{u_0\}]$, $\sum_{u \in U} a_u u \mapsto \sum_{u \in U - \{u_0\}} a_u u$, with inverse $A[U - \{u_0\}] \to \omega AU$ determined by $u \mapsto u - u_0$. Thus by Lemma 4.4, ωAU is G_{u_0}-projective if and only if $|G_{u_0 u}| \in \operatorname{Aut} A$, for all $u \in U - \{u_0\}$. The result now follows. ∎

4.6 Lemma. *A G-module summand P of AU is G-projective if and only if P is G_u-projective for each $u \in U$.*

Proof. One direction is clear from Lemma 4.3.

Now suppose that $\pi: AU \to P$ is a G-linear retraction, and that for each $u \in U$ we have a G_u-trace map $\operatorname{tr}_u: P \to P$. Let p be an arbitrary element of AU. Choose a G-transversal S in U so there is a unique expression $p = \sum_{s \in S} \sum_{gs \in Gs} a_{gs} gs$ with $a_{gs} \in A$ almost all zero. Let $\operatorname{tr}: AU \to P$ be the map such that $\operatorname{tr} p = \sum_{s \in S} \operatorname{tr}_s(\pi(a_s s))$. Clearly tr is additive, and for each $g \in G$, $\operatorname{tr}(g^{-1}p) = \sum_{s \in S} \operatorname{tr}_s(\pi(a_{gs} s))$. For each $s \in S$, let T_s be a transversal in G for the right G_s-action. Then

$$\begin{aligned}
\sum_{g \in G} g \operatorname{tr}(g^{-1}p) &= \sum_{g \in G} g \sum_{s \in S} \operatorname{tr}_s(\pi(a_{gs} s)) = \sum_{s \in S} \sum_{g \in G} g \operatorname{tr}_s(\pi(a_{gs} s)) \\
&= \sum_{s \in S} \sum_{t \in T_s} \sum_{x \in G_s} tx \operatorname{tr}_s(\pi(a_{ts} x^{-1} s)) \quad \text{since } s = x^{-1}s, \\
&= \sum_{s \in S} \sum_{t \in T_s} t \sum_{x \in G_s} x \operatorname{tr}_s(x^{-1} \pi(a_{ts} s)) \\
&= \sum_{s \in S} \sum_{t \in T_s} t \pi(a_{ts} s) = \pi\left(\sum_{s \in S} \sum_{t \in T_s} a_{ts} ts \right) \\
&= \pi(p).
\end{aligned}$$

Hence, for any $p \in P$, $\sum_{g \in G} g \operatorname{tr}(g^{-1}p) = \pi(p) = p$, and the restriction of tr to P is a G-trace map on P. ∎

4.7 Lemma. *If G is finite then ωAU is G-projective if and only if $\mathrm{HCF}_{u \in U}(G:G_u) \in \operatorname{Aut} A$, and for all distinct $u, u' \in U$, $|G_{uu'}| \in \operatorname{Aut} A$.*

In this event the augmentation map $\varepsilon: AU \to A$ is G-split, that is, there is a G-linear map $\phi: A \to AU$ such that $\varepsilon\phi(a) = a$ for all $a \in A$.

Proof. Let $n = \mathrm{HCF}_{u \in U}(G:G_u)$, and let S be a G-transversal in U, so there exists a map $\psi: S \to \mathbb{Z}$, with $\psi(s) = 0$ for almost all $s \in S$ such that

$\sum_{s \in S} |Gs|\psi(s) = n$. This extends to a G-map $\psi: U \to \mathbb{Z}$ by $\psi(gs) = \psi(s)$ for $(g, s) \in G \times S$. Notice that $\psi(u) = 0$ for almost all $u \in U$, and $\sum_{u \in U} \psi(u) = \sum_{s \in S} |Gs|\psi(s) = n$.

Suppose that $n \in \text{Aut}\, A$. It is easy to see that $\phi: A \to AU, a \mapsto \sum_{u \in U} \psi(u)n^{-1}au$, is G-linear, and $\varepsilon\phi a = \sum_{u \in U} \psi(u)n^{-1}a = nn^{-1}a = a$. Thus ωAU is a G-summand of AU. Now assume further that $|G_{uu'}| \in \text{Aut}\, A$ for all distinct $u, u' \in U$. By Corollary 4.5, ωAU is G_u-projective for all $u \in U$. Now by Lemma 4.6, ωAU is G-projective.

Now suppose conversely that ωAU is G-projective. By Lemma 4.3 and Corollary 4.5, $|G_{uu'}| \in \text{Aut}\, A$ for all distinct $u, u' \in U$. Let tr be a G-trace on ωAU and u_0 any element of U. Consider the pairing $A \times U \to A$, $(a, u) \mapsto \langle a, u \rangle$, such that for $a \in A$, $au_0 - \sum_{g \in G} g^{-1}\text{tr}(agu_0 - au_0) = \sum_{u \in U} \langle a, u \rangle u$. It is straightforward to show that for all $x \in G$,

$$(x-1)\left(\sum_{g \in G} g^{-1}\,\text{tr}\,(agu_0 - au_0)\right) = axu_0 - au_0.$$

Thus the element $\sum_{u \in U} \langle a, u \rangle u = au_0 - \sum_{g \in G} g^{-1}\,\text{tr}\,(agu_0 - au_0)$ of AU is stabilized by G, so $\langle a, - \rangle$ is constant on Gs, for each $s \in S$; applying ε we see that $\sum_{s \in S} \langle a, s \rangle |G_s| = a$. The map $A \to A, a \mapsto \sum_{s \in S} \langle a, s \rangle (|G_s|/n)$, is additive, and is inverse to multiplication by n. Thus $n \in \text{Aut}\, A$. ∎

We now come to the main theorem of this section.

4.8 Theorem. *Let A be a nonzero abelian group and U a nonempty G-set. Then ωAU is G-projective if and only if for all distinct $u, u' \in U$, $|G_{uu'}| \in \text{Aut}\, A$, and there exists a G-tree T with finite edge stabilizers having U as a G-subset of VT such that for every $v \in VT - U$, $\text{HCF}_{u \in U}(G_v : G_{vu}) \in \text{Aut}\, A$ (and in particular G_v is finite).*

Proof. Suppose that for all distinct $u, u' \in U$, $|G_{uu'}| \in \text{Aut}\, A$, and there exists a G-tree T with finite edge stabilizers having U as a G-subset of VT such that for every $v \in VT - U$, $\text{HCF}_{u \in U}(G_v : G_{vu}) \in \text{Aut}\, A$. For each $u \in U$, ωAU is G_u-projective by Corollary 4.5, and for each $v \in VT - U$, G_v is finite, so by Lemma 4.7, ωAU is G_v-projective and there exists a G_v-linear mapping $\phi_v: A \to AU$ which splits the augmentation map. Choose any G-transversal S in $VT - U$ and construct a G-linear map $\phi: A[VT - U] \to AU$ by

$ags \mapsto g\phi_s(a)$ for all $(a, g, s) \in A \times G \times S$. Since ϕ respects the augmentations, it follows that ωAU is a G-summand of $\omega A[VT] \approx A[ET]$. Now ωAU is G_v-projective for all $v \in VT$, and hence G_e-projective for all $e \in ET$, so by Lemma 4.6, ωAU is G-projective.

Conversely, suppose that ωAU is G-projective. By Lemma 4.3 and Corollary 4.5, for all distinct $u, u' \in U$, $|G_{uu'}| \in \text{Aut}\, A$. By Theorem 4.2 there exists a G-tree T with finite edge stabilizers having U as a G-subset of VT such that for every $v \in VT - U$, G_v is finite. For any $v \in VT - U$, ωAU is G_v-projective by Lemma 4.3, so $\text{HCF}_{u \in U}(G_v : G_{vu}) \in \text{Aut}\, A$ by Lemma 4.7. ∎

The above proof also shows the following.

4.9 Corollary. *Let A be a nonzero abelian group and U a nonempty G-set. Then ωAU is G-projective if and only if there exists a G-tree T such that U is a G-subset of VT and $T - U$ has finite stabilizers, and ωAU is G_v-projective for each vertex v of T.* ∎

4.10 Remark. It can be deduced from Theorem 4.8 that ωAG is G-projective if and only if there exists a G-tree T such that $|G_v| \in \text{Aut}\, A$ for all $v \in VT$. ∎

For $A = \mathbb{Z}$ we can make a stronger statement.

4.11 Theorem. *For a nonempty G-set V, the following are equivalent:*

(*a*) $G = F * \left(\underset{v \in V_0}{*} G_v \right)$ *for some G-transversal V_0 in V, and some free subgroup F of G.*

(*b*) *There exists a G-tree with G-free edge set and vertex set V.*

(*c*) *$\omega \mathbb{Z} V$ is $\mathbb{Z}G$-free.*

(*d*) *$\omega \mathbb{Z} V$ is $\mathbb{Z}G$-projective.*

(*e*) *$\omega \mathbb{Z} V$ is G-projective.*

Proof. (*a*)⇔(*b*) by Theorems I.4.1 and I.7.6.

(*b*)⇒(*c*). Suppose there exists a G-tree T with $VT = V$ and $ET = G \times S$ for some set S. Then $\omega \mathbb{Z} V = \omega \mathbb{Z}[VT] \approx \mathbb{Z}[ET] = \mathbb{Z}[G \times S]$ is $\mathbb{Z}G$-free with basis S.

(*c*)⇒(*d*)⇒(*e*) clearly.

(*e*)⇒(*b*). Suppose that $\omega \mathbb{Z} V$ is G-projective. By Theorem 4.8 there is a G-tree T with finite edge stabilizers, having V as a G-subset of VT such that $VT - V$ has finite stabilizers, and

(1) for all distinct $u, u' \in V$, $G_{uu'} = 1$,

(2) for each $v \in VT - V$, $\text{HCF}_{u \in V}(G_v : G_{uv}) = 1$.

Let $v \in VT - V$ and write $H = G_v$. We shall show that H stabilizes an element of V. Let S be an H-transversal in V. We have

$$|H| > \left|\bigcup_{u \in U}(H_u - \{1\})\right| = \sum_{u \in U}(|H_u| - 1) \quad \text{by (1)}$$
$$= \sum_{s \in S}|Hs|(|H_s| - 1) = \sum_{s \in S}(|H| - |Hs|).$$

Hence there is at most one $s \in S$ such that $|H| - |Hs| \geqslant |H|/2$, or equivalently $|H| > |Hs|$. In any event, we can choose an $s \in S$ such that $|H| = |Hu|$ for all $u \in V - Hs$. Then $\text{HCF}_{u \in V}|Hu| = |Hs|$, and by (2), $|Hs| = 1$. That is, H stabilizes s, as desired.

For each $u \in V$, let $T_u = \{u\} \cup \{t \in T | 1 \neq G_t \subseteq G_u\}$. Clearly T_u is a G_u-subtree of T containing u. By (1), the $T_u, u \in V$, are pairwise disjoint, so we can form a new G-tree by contracting T_u to u for each $u \in U$, without increasing the stabilizers. Thus we may assume that $T_u = \{u\}$ for each $u \in V$, so by the preceding paragraph, G acts freely on $T - V$. If $VT \neq V$ then some element u of V is joined by an edge e to an element v of $VT - V$. Contracting ge for each $g \in G$ gives a new G-tree. Repeating this procedure transfinitely often gives a G-tree T with vertex set V and G-free edge set. ∎

5 Splitting augmentation ideals

Throughout this section, let R be a nonzero ring and H a subgroup of G.

5.1 Theorem. *The following are equivalent:*

(a) The exact sequence $0 \to RG\omega RH \to \omega RG \to \omega R[G/H] \to 0$ *is* RG*-split.*

(b) There exists a G*-tree* T *with finite edge stabilizers and a vertex* v_0 *such that* $G_{v_0} = H$ *and for every vertex* v *if* $v \neq v_0$ *and either* G_v *is infinite or* $v \in Gv_0$ *then* $|H_v| \in UR$.

Proof. Let $\beta: RG \to R[G/H]$ be the natural surjection. It is straightforward to show that (a) is equivalent to

(a') There exists an RG-linear map $\gamma: \omega R[G/H] \to RG$ such that $\beta\gamma: \omega R[G/H] \to R[G/H]$ is the inclusion map.

We shall show that (b) is equivalent to (a').

$(a') \Rightarrow (b)$. Suppose that (a') holds. Composing the derivation $G \to \omega R[G/H]$, $g \mapsto gH - 1H$, with γ gives a derivation $d: G \to RG$, $g \mapsto \gamma(gH - 1H)$. Notice that $\operatorname{Ker} d = H$, since γ is injective, and that $\beta d: G \to R[G/H]$ is given by $\operatorname{ad}(1H)$.

By Theorem 2.5, there exists a G-tree T with finite edge stabilizers, and vertex set $(RG)_d$. Thus there is a bijective map $\phi: RG \to VT$ such that if $p \in RG$, $g \in G$ then $g(\phi p) = \phi(gp + dg)$, so $G_{\phi p} = \operatorname{Ker}(d + \operatorname{ad} p)$.

Let $v_0 = \phi(0)$, so $G_{v_0} = \operatorname{Ker}(d + \operatorname{ad} 0) = H$.

Notice ϕ is an H-map, so for any $p \in RG$, $H_p \subseteq H_{\phi p}$. Thus the coefficients of p are constant on $H_{\phi p}$-orbits. Passing mod H, we find that the coefficient of $1H$ in $\beta(p) \in R[G/H]$ lies in $R|H_{\phi p}|$. Thus if the coefficient of $1H$ in βp is -1 then $|H_{\phi p}| \in UR$.

Now suppose $v \in Gv_0 - \{v_0\}$. Then for some $g \in G - H$, $v = gv_0 = g\phi(0) = \phi(g0 + dg) = \phi(dg)$. The coefficient of $1H$ in $\beta(dg) = gH - 1H$ is -1, so $|H_{\phi(dg)}| \in UR$, that is, $|H_v| \in UR$.

Consider any $v \in VT - Gv_0$ with G_v infinite. Since T has finite edge stabilizers, G_v acts on $Gv_0 = G/H$ with finite stabilizers, so with infinite orbits; hence the only element of $R[G/H]$ stabilized by G_v is 0. Let $p = \phi^{-1}(v)$. Then $G_v = G_{\phi p} = \operatorname{Ker}(d + \operatorname{ad} p) \subseteq \operatorname{Ker} \beta(d + \operatorname{ad} p) = \operatorname{Ker}(\beta d + \operatorname{ad} \beta p) = \operatorname{Ker} \operatorname{ad}(1H + \beta p) = G_{1H + \beta p}$. Thus G_v stabilizes $1H + \beta p \in R[G/H]$ so $1H + \beta p = 0$. Hence the coefficient of $1H$ in βp is -1 and $|H_{\phi p}| \in UR$, that is $|H_v| \in UR$.

$(b) \Rightarrow (a')$. Suppose that (b) holds. Let

$$T_H = \{v_0\} \cup \{t \in T \mid |H_t| \notin UR\}.$$

Since H stabilizes v_0 it follows that T_H is an H-subtree of T. By the hypothesis, $VT_H \subseteq \{v_0\} \cup \{v \in VT - Gv_0 \mid G_v \text{ is finite}\}$.

We claim that for all $v \in VT$,

(1) $\quad Gv \cap T_H$ is H-finite.

We may assume that $Gv \cap T_H$ is nonempty, and that $v \in VT_H$. If $v_0 \in Gv$ then $Gv \cap T_H = \{v_0\}$; thus we may assume that $v_0 \notin Gv$ and that G_v is finite. Let W be an H-transversal in $Gv \cap T_H$ and consider an element w of W. Choose $g_w \in G$ such that $g_w v = w$. Since T has finite edge stabilizers, $H_w = G_{v_0 w}$ is finite, and by definition of T_H, $|H_w| \notin UR$, so we can choose $h_w \in H_w$ whose order does not belong to UR. Now $g_w^{-1} h_w g_w$ stabilizes $g_w^{-1} w = v$. In this way we construct a function $W \to G_v$, $w \mapsto g_w^{-1} h_w g_w$. Suppose $u, w \in W$ and $g_w^{-1} h_w g_w = g_u^{-1} h_u g_u$. Let $g = g_w g_u^{-1}$, so $h_w = g h_u g^{-1}$ stabilizes gv_0. Thus $|H_{gv_0}| \notin UR$, so by hypothesis $gv_0 = v_0$. Thus $g \in H$, $Hg_u = Hg_w$ and $Hu = Hg_u v = Hg_w v = Hw$. But W is an H-transversal so $u = w$. Thus $W \to G_v$ is injective, and since G_v is finite, W is finite. This proves (1).

Let

$$c_\iota = \{e \in ET \mid \iota e \in VT_H, \tau e \notin VT_H\}$$
$$c_\tau = \{e \in ET \mid \tau e \in VT_H, \iota e \notin VT_H\}.$$

These are H-subsets of ET. We claim that for each $e \in ET$, $Ge \cap c_\iota$ is H-finite. Clearly we may assume that $Ge \cap c_\iota$ is nonempty and that $\iota e \in VT_H$. If $\iota e = v_0$ then for all $g \in G$, if $ge \in c_\iota$ then $gv_0 = \iota ge \in VT_H$, so $g \in H$; here $Ge \cap c_\iota = He$, which is H-finite. Thus we may assume that $\iota e \neq v_0$, so $G_{\iota e}$ is finite. Hence the natural map $Ge \cap c_\iota \to G\iota e \cap VT'$ has finite fibres, and the image is H-finite by (1), so $Ge \cap c_\iota$ is H-finite, as claimed.

Similarly $Ge \cap c_\tau$ is H-finite.

Let b_ι, b_τ be H-transversals in c_ι, c_τ, respectively, so for each $e \in ET$, $Ge \cap (b_\iota \cup b_\tau)$ is finite. Define a function $p: ET \to R$, by

$$p(e) = \begin{cases} -|H_e|^{-1} & \text{if } e \in b_\iota \\ |H_e|^{-1} & \text{if } e \in b_\tau \\ 0 & \text{if } e \in ET - (b_\iota \cup b_\tau). \end{cases}$$

Since T is a tree, $\omega R[VT]$ has an R-basis consisting of $\tau e - \iota e$, $e \in ET$, so there is an R-linear map $\gamma: \omega R[VT] \to R[G]$ such that $\gamma(\tau e - \iota e) = \sum_{g \in G} p(g^{-1}e)g$ for all $e \in ET$. Clearly γ is $R[G]$-linear.

We now apply the map $\beta: RG \to R[G/H]$. For $e \in ET$, $\beta(\gamma(\tau e - \iota e)) = \sum_{Hg \in G/H} \left(\sum_{h \in H} p(h^{-1}g^{-1}e) \right) gH$. Thus the coefficient of gH in $\beta\gamma(\tau e - \iota e)$ is

$$\sum_{h \in H} p(h^{-1}g^{-1}e) = \sum_{f \in Hg^{-1}e} |H_f| p(f)$$
$$= \begin{cases} -1 & \text{if } g^{-1}e \in c_\iota \\ 1 & \text{if } g^{-1}e \in c_\tau \\ 0 & \text{if } g^{-1}e \in ET - (c_\iota \cup c_\tau). \end{cases}$$

For any vertices u, v of T, the expression of $v - u$ in terms of our basis is given by the T-geodesic, and it follows that the coefficient of gH in $\beta\gamma(v - u)$ is

$$\begin{cases} -1 & \text{if } g^{-1}u \in VT_H \text{ and } g^{-1}v \notin VT_H, \\ 1 & \text{if } g^{-1}v \in VT_H \text{ and } g^{-1}u \notin VT_H, \\ 0 & \text{in all other events.} \end{cases}$$

Hence for $x \in G - H$, the coefficient of gH in $\beta\gamma(xv_0 - v_0)$ is

$$\begin{cases} -1 & \text{if } g^{-1} \in H, \\ 1 & \text{if } g^{-1}x \in H, \\ 0 & \text{in all other events.} \end{cases}$$

Thus $\beta\gamma(xv_0 - v_0) = xH - 1H$ for all $x \in G$.

On identifying $Gv_0 = G/H$ we obtain an $R[G]$-linear map $\gamma: \omega R[G/H] \to R[G]$ such that $\beta\gamma: \omega R[G/H] \to R[G/H]$ is the inclusion map, so (a') holds. ∎

We note some special cases.

5.2 Corollary. *If H is a torsion-free subgroup of G, then the exact sequence $0 \to RG\omega RH \to \omega RG \to \omega R[G/H] \to 0$ is RG-split if and only H is a free factor of G, that is, $G = H * K$ for some subgroup K of G.*

Proof. If $G = H * K$ then the corresponding tree T satisfies condition (b) of Theorem 5.1, since here $H_v = 1$ for all $v \in VT - \{v_0\}$.

Conversely, if the sequence is split then by Theorem 5.1, there exists a G-tree T with finite edge stabilizers and a vertex v_0 of T with $G_{v_0} = H$. Choose a fundamental G-transversal Y in T containing v_0, and then contract toward v_0 all edges in Y with nontrivial stabilizer. No edge incident to v_0 gets contracted, since H is torsion-free.

We therefore construct a G-tree with trivial edge stabilizers and a vertex having stabilizer H. By the Structure Theorem I.4.1, G is the free product of a free group and certain vertex groups, one of which is H. Thus H is a free factor of G. ∎

5.3 Corollary. *An element x of G generates an infinite cyclic free factor of G if and only if there exists a derivation $d: G \to RG$ with $d(x) = 1$.*

Proof. Suppose that x generates an infinite cyclic free factor of G. Then there is an HNN-decomposition $G = H * x$ for some subgroup H of G, and we let T be the corresponding tree. The isomorphism $\omega R[VT] \approx R[ET]$ here takes the form $\omega R[G/H] \approx RG$, with $xH - H$ mapping to 1. Hence we have a derivation $G \xrightarrow{\text{ad}\,1H} \omega R[G/H] \approx RG$ taking x to 1.

Conversely, suppose that $d: G \to RG$ has $d(x) = 1$. If x has order $n \geqslant 1$ then $0 = d(1) = d(x^n) = x^{n-1} + \cdots + x + 1 \neq 0$, a contradiction. Thus the subgroup H generated by x is infinite cyclic. Since ωRG has R-basis $(g - 1 \mid g \in G - \{1\})$ there is an R-linear map $\gamma: \omega RG \to RG(x-1)$ such that $\gamma(g-1) = d(g)(x-1)$ for all $g \in G$. It is readily verified that γ is RG-linear, and fixes $x - 1$, so is the identity on $RG(x-1) = RG\omega RH$. Thus $RG\omega RH$ splits off ωRG, and by Corollary 5.2, H is a free factor of G. ∎

5.4 Corollary. *If G is finite, and H a subgroup of G then*

$$0 \to RG\omega RH \to \omega RG \to \omega R[G/H] \to 0$$

is RG-split if and only if $|H \cap H^g| \in UR$ for all $g \in G - H$.

Proof. By Theorem 5.1, the sequence splits if and only there exists a G-tree T with finite edge stabilizers and a vertex v_0 of T with $G_{v_0} = H$ such that $|H_v| \in U(R)$ whenever $v \in Gv_0 - \{v_0\}$. This is equivalent to $|H \cap H^g| \in UR$ for all $g \in G - H$ and there exists a G-tree T having finite edge stabilizers and a vertex v_0 with $G_{v_0} = H$. The tree condition is automatic, since we can take the tree corresponding to the expression $G = G \underset{H}{*} H$. ∎

6 Ends of groups

We now quickly review M.H. Stone's classical Representation Theorem. Details and proofs can be found in Simmons (1963), Appendix 3.

6.1 Review. Recall that a *Boolean space* is a totally disconnected compact Hausdorff topological space.

The *Boolean space* $\mathscr{X}B$ of a Boolean ring B is the subspace of $(B, \mathbb{Z}_2)$ consisting of all ring homomorphisms $B \to \mathbb{Z}_2$, where $\mathbb{Z}_2$ has the discrete topology, and $(B, \mathbb{Z}_2)$ the corresponding product topology. In terms of subsets of B, the elements of $\mathscr{X}B$ are the ultrafilters on B, that is, the complements of the prime ideals of B.

The *Boolean ring* $\mathscr{B}X$ of a Boolean space is the subring of $(X, \mathbb{Z}_2)$ consisting of all continuous maps $X \to \mathbb{Z}_2$. In terms of subsets of X, the elements of $\mathscr{B}X$ are the clopen sets, that is, subsets of X which are both closed and open.

Stone's Representation Theorem states that the natural double-dual maps $B \to \mathscr{B}\mathscr{X}B$, $X \to \mathscr{X}\mathscr{B}X$ are isomorphisms of Boolean rings and spaces, respectively. Thus we have an anti-equivalence between the category of Boolean rings and the category of Boolean spaces. ∎

Let us consider the Boolean rings of Chapter II.

6.2 Definitions. Let X be a connected graph.

Recall that the *Boolean ring* $\mathscr{B}X$ of X is the subring of $(VX, \mathbb{Z}_2)$ consisting of all b such that $\delta b = \{e \in EX \mid b(\iota e) \neq b(\tau e)\}$ is finite.

The *Boolean space* $\mathscr{X}X$ of X is then defined as $\mathscr{X}\mathscr{B}X$. There is a natural double-dual map $VX \to \mathscr{X}X$, $v \mapsto \bar{v}$, and we shall denote the image by $\bar{V}X$. The subspace $\mathscr{X}X - \bar{V}X$ of $\mathscr{X}X$ is called the *space of ends* of X, denoted $\mathscr{E}X$, and the elements are called *ends* of X.

Let $\mathscr{P}X$ denote the set of infinite paths in X with no repeated vertices. There is a map from $\mathscr{P}X$ to $\mathscr{X}X$, defined as follows. Let $p, p' \in \mathscr{P}$, and $b \in \mathscr{B}X$. Since δb is finite, the restriction of b to the vertices in p almost agrees with a constant function $c=0$ or 1; we define $\bar{p}(b)=c$. It is not difficult to see that $\bar{p}:\mathscr{B}X \to \mathbb{Z}_2$ is a ring homomorphism, so $\bar{p}\in\mathscr{X}X$. Further $\bar{p}=\bar{p}'$ if and only if there is no finite subset δ of EX such that the tails of p, p' lie in different components of $X-\delta$.

The theory is most satisfactory for trees and for locally finite graphs; in these cases, the ends can be thought of as 'vertices at infinity'.

In general, the topology which VX inherits from $\mathscr{X}X$ is discrete if and only if X is locally finite. In this event $\mathscr{X}X$ is a compactification of the discrete space VX, and $\mathscr{E}X$ consists of those elements in $\mathscr{X}X$ of the form $\bar{p}$ for some $p\in\mathscr{P}$. Here there is another description of $\mathscr{E}X$. Treat $\mathbb{Z}_2[VX]$ as an ideal of $\mathscr{B}X$ and form the Boolean ring $\bar{\mathscr{B}}X=\mathscr{B}X/\mathbb{Z}_2[VX]$; its elements can be thought of as certain almost equality classes. The surjection $\mathscr{B}X \to \bar{\mathscr{B}}X$ induces an embedding $\mathscr{X}\bar{\mathscr{B}}X \to \mathscr{X}\mathscr{B}X=\mathscr{X}X$ and the image is precisely $\mathscr{E}X$. Here the elements of $\mathscr{E}X$ are the ring homomorphisms $\mathscr{B}X \to \mathbb{Z}_2$ which vanish on $\mathbb{Z}_2[VX]$. ∎

Let us give a more precise description of the Boolean ring of a tree.

6.3 Discussion. Let T be a tree.

For $e\in ET$ recall that $e|VT$ denotes the vertex set of the component of $T-\{e\}$ containing τe. Thus $\delta(e|VT)=\{e\}$, which is finite, so $e|VT\in\mathscr{B}T$. Thus the image of the structure map $ET\to(VT,\mathbb{Z}_2)$ lies in $\mathscr{B}T$. We shall treat ET as a subset of $\mathscr{B}T$, so we have $ET\subseteq\mathscr{B}T\subseteq(VT,\mathbb{Z}_2)$.

Let $s\in\mathscr{B}T$. For any $v, v'\in VT$, if $s(v)\neq s(v')$ then the T-geodesic from v to v' passes through some $e\in\delta s$, which means that $e(v)\neq e(v')$. It follows that s can be expressed in terms of Boolean operations using the finitely many elements $e|VT$, $e\in\delta s$. Hence ET is a generating set for $\mathscr{B}T$, so every ring homomorphism $\mathscr{B}T\to\mathbb{Z}_2$ is completely specified by its values on ET, and we have a natural embedding $\mathscr{X}T\subseteq(ET,\mathbb{Z}_2)$. This composes with the double-dual map $VT\to\mathscr{X}T$ to give the costructure map for T, and we have $VT\subseteq\mathscr{X}T\subseteq(ET,\mathbb{Z}_2)$.

We have seen that ET is a generating set for $\mathscr{B}T$. For each edge e of T let e^1, e^* denote e and its inverse, respectively. Let $X\subseteq(ET,\mathbb{Z}_2)$ be the set of functions $x:ET\to\mathbb{Z}_2$ which respect the nesting, that is, for all $e, e'\in ET$, $\varepsilon,\varepsilon'\in\{1,*\}$ if $e^\varepsilon, e'^{\varepsilon'}$ point away from each other in T then $x(e)^\varepsilon x(e')^{\varepsilon'}=0$ in $\mathbb{Z}_2$.

Since the elements of X respect relations which hold in $\mathscr{B}T$, we see $\mathscr{X}T\subseteq X$.

We leave it as an exercise to use the proof of Theorem II.1.5 to show that the elements of X are the vertices of T together with the ends of T, which we can identify with infinite reduced paths starting at a given vertex. Hence $X \subseteq \mathscr{X}T$, so $X = \mathscr{X}T$.

But X can also be viewed as the Boolean space of the Boolean ring presented with generating set ET and the nesting relations. Now by the Stone Representation Theorem, the latter ring is $\mathscr{B}T$. Explicitly, $\mathscr{B}T$ is the Boolean ring presented with generating set ET and relations $e^{\varepsilon}e'^{\varepsilon'} = 0$ for all $e, e' \in ET$ and $\varepsilon, \varepsilon' \in \{1, *\}$ such that $e^{\varepsilon}, e'^{\varepsilon'}$ point away from each other in T. ∎

There is an analogous construction for groups.

6.4 Definition. The *Boolean ring of* $G, \mathscr{B}G$, is the subring of $(G, \mathbb{Z}_2)$ consisting of those $b \in (G, \mathbb{Z}_2)$ that are almost right G-stable, that is, $bg =_a b$ for all $g \in G$, where $bg(x) = b(xg^{-1})$ for all $x \in G$.

In the presence of a finite generating set, there is a connected locally finite Cayley graph X, and Proposition 6.7 will show that the two Boolean rings $\mathscr{B}G, \mathscr{B}X$ coincide.

The set $\mathscr{A}(G, \mathbb{Z}_2)$ of almost G-stable functions in $(G, \mathbb{Z}_2)$ is a Boolean ring, and there is a Boolean ring isomorphism $\mathscr{B}G \approx \mathscr{A}(G, \mathbb{Z}_2)$, $b \mapsto b^{-1}$, where $b^{-1}(g) = b(g^{-1})$ for each $g \in G$.

The *Boolean space* of $G, \mathscr{X}G$, is defined to be the space $\mathscr{X}\mathscr{B}G$. There is an injective double-dual map $G \to \mathscr{X}G$ and the complement $\mathscr{X}G - G$ is denoted $\mathscr{E}G$, called the *space of ends* of G. The number of elements of $\mathscr{E}G$ is denoted $e(G)$, called the *number of ends* of G.

View $\mathbb{Z}_2 G$ as an ideal of $\mathscr{B}G$ and form the Boolean ring $\bar{\mathscr{B}}G = \mathscr{B}G/\mathbb{Z}_2 G$. The surjection $\mathscr{B}G \to \bar{\mathscr{B}}G$ determines an embedding $\mathscr{X}\bar{\mathscr{B}}G \to \mathscr{X}\mathscr{B}G$ which gives an identification $\mathscr{X}\bar{\mathscr{B}}G = \mathscr{E}G$. The elements of $\mathscr{E}G$ are the ring homomorphisms $\mathscr{B}G \to \mathbb{Z}_2$ which vanish on $\mathbb{Z}_2 G$. In particular, G is an open discrete subset of a compact space, $\mathscr{X}G$, and the complement is $\mathscr{E}G$. Notice that $e(G)$ is finite if and only if $\bar{\mathscr{B}}G$ is finite; in this event $e(G) = \dim_{\mathbb{Z}_2} \bar{\mathscr{B}}G$.

By Corollary 2.4, we have abelian group isomorphisms $H^1(G, \mathbb{Z}_2 G) \approx \mathscr{A}(G, \mathbb{Z}_2)/(\mathbb{Z}_2 G + \mathbb{Z}_2) \approx \bar{\mathscr{B}}G/*$, where in the latter each b is identified with $b^* = 1 + b$. ∎

6.5 Proposition. *The number of ends of G is zero if and only if G is finite.*

Proof. If G is finite then $\mathscr{B}G = \mathbb{Z}_2 G$, so $\bar{\mathscr{B}}G = 0$. If $\bar{\mathscr{B}}G = 0$ then $\mathscr{B}G = \mathbb{Z}_2 G$, but $1 \in \mathscr{B}G$, so G is finite. ∎

6.6 Proposition. *If H is a subgroup of finite index in G then restriction gives a ring isomorphism $\bar{\mathscr{B}}G \to \bar{\mathscr{B}}H$, and $e(H) = e(G)$.*

Proof. The restriction map is clearly a ring homomorphism, and it is surjective by Lemma 1.1. If $b \in \mathscr{B}G$ such that $b|H =_a 0$ then for each $g \in G$, $bg|H =_a b|H =_a 0$ so $b|Hg^{-1} =_a 0$. Since $(G:H)$ is finite it follows that $b =_a 0$, so $\bar{\mathscr{B}}G \to \bar{\mathscr{B}}H$ is injective. ∎

6.7 Proposition. *Suppose G is finitely generated.*

Let X be the Cayley graph with respect to a finite generating set S of G. Then X is locally finite, and $\mathscr{B}G = \mathscr{B}X$ in $(G, \mathbb{Z}_2) = (VX, \mathbb{Z}_2)$, so $\mathscr{E}G = \mathscr{E}X$.

In particular, $\mathscr{E}G$ is metrizable.

Proof. Let $b \in (G, \mathbb{Z}_2) = (VX, \mathbb{Z}_2)$. Then $\delta b = \{(g, s) \in G \times S \mid b(gs) \neq b(g)\} = \{(g, s) \in G \times S \mid bs^{-1}(g) \neq b(g)\}$.

If δb is finite then $bs^{-1} =_a b$ for all $s \in S$, and since S generates G, b is almost right G-stable.

Conversely, if b is almost right G-stable, then δb is finite since S is finite. Hence $\mathscr{B}G = \mathscr{B}X$.

The number of finite subsets of EX is countable, and each finite subset of EX gives rise to only finitely many elements of $\mathscr{B}X$, so $\mathscr{B}X$ is countable. Hence $\bar{\mathscr{B}}G$ is countable. Since the elements of $\bar{\mathscr{B}}G$ give a base for the open topology of $\mathscr{E}G = \mathscr{X}\bar{\mathscr{B}}G$, it follows from Urysohn's Lemma that $\mathscr{E}G$ is metrizable. ∎

6.8 Example. Let n be a natural number, and $G = F_n$ a free group of rank n. Form the Cayley graph X with respect to a free generating set of G, so $\mathscr{B}X = \mathscr{B}G$. Here X is a G-tree. As mentioned in Discussion 6.3, $\mathscr{E}X$ can be identified with the set of all reduced infinite paths starting at $v = 1$. This has a natural metric with the distance between p, p' being the inverse of the number of vertices they have in common.

Notice that $e(F_0) = 0$, $e(F_1) = 2$, and for $n \geqslant 2$, $e(F_n) = 2^{\aleph_0}$. ∎

6.9 Example. Let n be a natural number, and $G = \mathbb{Z}^n$ a free abelian group of rank n. Form the Cayley graph X with respect to a free abelian generating set of G, so $\mathscr{B}X = \mathscr{B}G$ by Proposition 6.7. Here X is a lattice in $\mathbb{R}^n$. As noted in the previous example, $e(\mathbb{Z}^0) = 0$, $e(\mathbb{Z}^1) = e(F_1) = 2$. For $n \geqslant 2$, it is easy to see that for each finite set δ of edges of X, $X - \delta$ has exactly one infinite component, so X has one end. Thus for $n \geqslant 2$, $e(\mathbb{Z}^n) = 1$. ∎

The main result concerning ends of groups gives equivalent conditions for G to have more than one end.

6.10 Theorem. *For any nonzero abelian group A, the following are equivalent:*

(*a*) $e(G) > 1$.

(*b*) $H^1(G, AG) \neq 0$; *that is, there exists an outer derivation* $d: G \to AG$.

(*c*) *There exists a G-tree with finite edge stabilizers such that no vertex is stabilized by G.*

(*d*) *One of the following holds:*

(*d*1) $G = B \underset{C}{*} D$, *where* $B \neq C \neq D$ *and C is finite;*

(*d*2) $G = B \underset{C}{*} x$, *where C is finite;*

(*d*3) *G is countably-infinite locally-finite.*

(*e*) $e(G)$ *is* 2 *or* ∞.

If G is torsion-free, (*d*) is equivalent to G being either infinite cyclic or a nontrivial free product.

Proof. (*a*)$\Rightarrow$(*b*). Suppose (*a*) holds. Then there exists an almost right G-stable function $G \to \mathbb{Z}_2$ which is not almost constant. By inverting, we get an almost G-stable function $G \to \mathbb{Z}_2$ which is not almost constant. By choosing any injective set map $\mathbb{Z}_2 \to A$, we obtain an almost G-stable function $v: G \to A$ which is not almost constant. By Proposition 2.3 we have an outer derivation $d_v: G \to AG$.

(*b*)$\Rightarrow$(*c*). Suppose we have an outer derivation $d: G \to AG$. By Theorem 2.5 and Remark 2.9, there exists a G-tree T with finite edge stabilizers such that for each vertex v, d is inner on G_v, so $G_v \neq G$.

(*c*)$\Rightarrow$(*d*) by Theorem I.4.12.

(*d*1), (*d*2)$\Rightarrow$(*e*). If (*d*1) or (*d*2) holds then there exists a G-tree T with finite edge stabilizers, one edge orbit, and no vertex stabilized by G. It follows that no vertex has valency one.

Consider first the case where every vertex of T has valency two, so T is homeomorphic to $\mathbb{R}$. Here the edge groups have index at most two in the vertex groups. By Theorem I.7.4, G has a normal subgroup N of finite index which acts freely on T, so is free. It follows that N must be infinite cyclic. By Example 6.8, $e(N) = 2$, and, by Proposition 6.6, $e(G) = e(N) = 2$. We remark that this case corresponds to $(B:C) = (D:C) = 2$ for (*d*1) and $B = C$ for (*d*2), so there is a finite normal subgroup C such that $G/C \approx C_2 * C_2$ or C_∞.

This leaves the case where some vertex of T has valency at least three. By replacing G with a subgroup of index two, if necessary, we may assume that every vertex has valency at least three. Here, for any edges $e \neq f$ of $T, T - \{e, f\}$ has three infinite components.

Let v be a vertex of T. The map $G \to VT$, $g \mapsto gv$, determines a ring homomorphism $(VT, \mathbb{Z}_2) \to (G, \mathbb{Z}_2)$, and it is not difficult to show that the image of $\mathscr{B}T$ lies in $\mathscr{B}G$. Further, the composite $ET \to \mathscr{B}T \to \mathscr{B}G \to \bar{\mathscr{B}}G$ is injective, since for any distinct edge e, f of T, each of the three components of $T - \{e, f\}$ meets Gv in an infinite set. Hence $\bar{\mathscr{B}}G$ is infinite, and $e(G) = \infty$.

$(d3) \Rightarrow (e)$. Suppose $G = \bigcup_{n \geq 1} G_n$, where $G_1 \subset G_2 \subset \cdots$ are finite subgroups of G. Partition G into finite sets by setting $v_1 = G_1$, and $v_n = G_n - G_{n-1}$, for each $n \geq 2$. For each $S \subseteq \mathbb{N}$, let $v_s = \bigcup_{n \in S} v_n$. For any $g \in G$, there is some N such that $g \in G_N$, and then $v_n = v_n g$ for all $n > N$. Hence for each subset S of $\mathbb{N}$, v_S is almost right G-stable. It follows that there exists an injective ring homomorphism $(\mathbb{N}, \mathbb{Z}_2) \to \mathscr{B}G$; moreover the inverse image of $\mathbb{Z}_2 G$ corresponds to the set of finite subsets of $\mathbb{N}$. It follows that $\bar{\mathscr{B}}G$ is infinite, so G has infinitely many ends.

$(e) \Rightarrow (a)$ is clear. ∎

6.11 Remarks. Notice $e(G) = 0, 1, 2$ or ∞.

If $e(G) = 0$ then $\mathscr{E}G$ is empty, $\bar{\mathscr{B}}G = (\mathscr{E}G, \mathbb{Z}_2) = 0$ and $H^1(G, \mathbb{Z}_2 G) \approx \bar{\mathscr{B}}G/* = 0$.

If $e(G) = 1$ then $\mathscr{E}G$ is the one-element space, $\bar{\mathscr{B}}G = (\mathscr{E}G, \mathbb{Z}_2) = \mathbb{Z}_2$ and $H^1(G, \mathbb{Z}_2 G) \approx \bar{\mathscr{B}}G/* = 0$.

If $e(G) = 2$ then $\mathscr{E}G$ is the discrete two-element space, $\bar{\mathscr{B}}G = (\mathscr{E}G, \mathbb{Z}_2) \approx \mathbb{Z}_2 \times \mathbb{Z}_2$, and $H^1(G, \mathbb{Z}_2 G) \approx \bar{\mathscr{B}}G/* \approx \mathbb{Z}_2$.

If $e(G) = \infty$ then $\mathscr{E}G$ is an infinite Boolean space, $\bar{\mathscr{B}}G$ is an infinite Boolean ring and $H^1(G, \mathbb{Z}_2 G) \approx \bar{\mathscr{B}}G/*$ is an infinite vector space over $\mathbb{Z}_2$. ∎

We leave it to the reader to use the proof of Theorem 6.10 to verify the next result.

6.12 Theorem. *The following are equivalent.*

(*a*) $e(G) = 2$.

(*b*) *G has an infinite cyclic subgroup of finite index.*

(*c*) *G has a finite normal subgroup N such that G/N is C_∞ or $C_2 * C_2$; equivalently, G is $N \rtimes C_\infty$ or $A \underset{N}{*} B$ with $(A:N) = (B:N) = 2$.*

(d) There is an action of G on a tree T homeomorphic to $\mathbb{R}$ with finite stabilizers and one edge orbit. ∎

6.13 Irrelevant remark. There is an intuitive notion of a 'reasonable' G-tree T with finite edge stabilizers, and it is amusing to consider the number of ends of such a T and see how the integer 1 fails to fit into the pattern. There are five possibilities:

$e(G)=0$, $e(T)=0$, with T consisting of a single vertex;
$e(G)=1$, $e(T)=0$, with T consisting of a single vertex;
$e(G)=2$, $e(T)=2$, with T homeomorphic to $\mathbb{R}$;
$e(G)=\infty$, $e(T)=1$, with G countably-infinite locally-finite;
$e(G)=\infty$, $e(T)=\infty$. ∎

7 Accessibility

7.1 Definition. A G-tree T is said to be *terminal* if each edge stabilizer is finite and each vertex stabilizer has at most one end.

If T is a G-tree with finite edge stabilizers but the stabilizer of some vertex v has more than one end, then G_v acts on a tree T_v with finite edge stabilizers without stabilizing a vertex, and blowing up v to T_v gives a G-tree T' such that there exists a G-map $VT' \to VT$, but no G-map $VT \to VT'$. We think of T' as being strictly larger than T.

The group G is said to be *accessible* if there exists a terminal G-tree, and otherwise G is *inaccessible*. For example, any free group is accessible since it acts freely on some tree.

If G is inaccessible then any G-tree with finite edge stabilizers is not terminal so can be blown up to a strictly larger G-tree with finite edge stabilizers; since this can then be repeated any finite number of times, we get an infinite sequence. This happens also if G has a terminal G-tree which is not G-finite. However, if G has a G-finite terminal G-tree then it follows from Lemma III.7.5 that that there is no infinite sequence of blowing up G-trees with finite edge stabilizers, where each term is strictly larger than its predecessor. ∎

7.2 Example. The group

$$G = \langle a_1, b_1, a_2, b_2, a_3, b_3, \ldots \mid b_1 = [a_2, b_2], b_2 = [a_3, b_3], \ldots \rangle$$

is inaccessible. To see this, we suppose there exists a terminal G-tree T and get a contradiction.

Notice that G can be expressed as an infinite free product with amalgamations $\langle a_1, b_1\rangle \underset{B_1}{*} \langle a_2, b_2\rangle \underset{B_3}{*} \langle a_3, b_3\rangle * \ldots$, where B_n is the infinite cyclic group with generator b_n mapped to $[a_{n+1}, b_{n+1}]$ in $\langle a_{n+1}, b_{n+1}\rangle$. It follows that for each $n \geqslant 1$, $a_1, a_2, \ldots, a_n, b_n$ freely generate a free subgroup of G. The union over all n gives all of G, so G is locally free, and in particular, torsion-free. The edge stabilizers of T, being finite, must be trivial.

Let v be a vertex of T.

We show next that T has a G-finite subtree. Let $\bar{T}$ be the G-tree formed by contracting every edge in T whose orbit meets the T-geodesic from v to $b_0 v$, where $b_0 = [a_1, b_1]$. Write $\bar{v}$ for the image of v in $\bar{T}$, so b_0 stabilizes $\bar{v}$. In the action of $H = \langle a_1, b_1\rangle$ on $\bar{T}$, the edge stabilizers are trivial, so by the Structure Theorem I.4.1, $H_{\bar{v}}$ is a free factor of H, containing b_0. If $H_{\bar{v}}$ is free of rank one then there is a retraction homomorphism from H to $H_{\bar{v}}$ collapsing the other free factor; but here $H_{\bar{v}}$ is abelian so the retraction collapses the commutator b_0, which is a contradiction. This forces $H_{\bar{v}} = H$, because H is free of rank two. Thus a_1, b_1 stabilize $\bar{v}$. Since $b_1 = [a_2, b_2]$, it follows similarly that a_2, b_2 stabilize $\bar{v}$, and so on. By induction, all the generators of G stabilize $\bar{v}$, so G stabilizes $\bar{v}$. This means that all the edges contracted in T are connected together, constituting the edges of a G-finite G-subtree, which is again a terminal G-tree.

Thus we may assume that T itself is G-finite.

Let N be the subgroup of G generated by the vertex stabilizers of T. Then N is a normal subgroup and G/N is free of finite rank. Let n be an integer greater than the rank of G/N and express G as $\langle a_1, a_2, \ldots, a_n\rangle * K$, where $K = \langle a_{n+1}, b_{n+1}, a_{n+2}, b_{n+2}, \ldots \mid b_{n+1} = [a_{n+2}, b_{n+2}], \ldots\rangle$. By viewing this as an n-fold HNN extension, we can form a G-tree T' with trivial edge stabilizers and vertex set G/K. For each vertex v of T, G_v has at most one end and therefore stabilizes a vertex of T' so lies in a conjugate of K. Thus N lies in the normal closure of K, and G/N maps onto $\langle a_1, \ldots, a_n\rangle$, contradicting the choice of n. ∎

7.3 Conjecture. Every finitely generated group is accessible.

Discussion. It follows from the Grushko–Neumann Theorem I.10.6 that any torsion-free finitely generated group is accessible. More generally, Linnell (1983) showed that any finitely generated group having a finite subgroup of largest order is accessible. Theorem VI.6.3 will show that finitely *presented* groups are accessible. ∎

In this section we discuss two aspects of finitely generated accessible groups.

7.4 Proposition. *If G is finitely generated and accessible then there exists an incompressible terminal G-finite G-tree T.*

Further if T' is any other such G-tree then $VT \approx VT'$, $ET \approx ET'$ as G-sets.

Proof. Since G is accessible, there exists a terminal G-tree T. Since G is finitely generated T has a G-finite subtree which is again terminal so we may assume that T itself is G-finite. We can then repeatedly contract compressible edges until we arrive at an incompressible G-finite G-tree T', and VT' is a G-subset of VT, so T' is terminal.

Now suppose T, T' are incompressible terminal G-trees. For each vertex v of T, G_v has at most one end, so stabilizes a vertex of T'. This means that there exists a G-map $\phi: VT \to VT'$. Similarly, there exists a G-map $\psi: VT' \to VT$. Since VT is G-incompressible by Lemma III.7.2, the composite $\psi\phi: VT \to VT$ is bijective, so ϕ is injective and ψ is surjective. By symmetry ϕ and ψ are bijective, so ϕ is an isomorphism of G-sets. By Lemma III.7.3, $ET \approx ET'$ as G-sets. ∎

We conclude this section with a homological criterion for accessibility of finitely generated groups.

7.5 Theorem. *Let R be a nonzero ring and G a finitely generated group. Then G is accessible if and only if $H^1(G, RG)$ is finitely generated as right RG-module.*

Proof. Suppose that G is accessible. By Proposition 7.4 there is a G-finite terminal G-tree T. Let v be a vertex of T. The derivation $T[v,v]: G \to RET$ induces a map $\operatorname{Hom}_R(RET, RG) \to \operatorname{Der}(G, RG) \to H^1(G, RG)$. Further, this map is independent of the choice of v since, for any vertex v', $T[v,\_v] - T[v',\_v'] = \operatorname{ad} T[v',v]$. Now $\operatorname{Hom}_R(RET, RG)$ is finitely generated as right RG-module, so it remains to show that $\operatorname{Hom}_R(RET, RG) \to H^1(G, RG)$ surjective, as it is clearly right RG-linear. By Remark 2.9, for any derivation $d: G \to RG$, there exists a G-tree T' with finite edge stabilizers and a vertex v' such that d factors through $T'[v',\_v']: G \to RET'$. But T is terminal so there exists a G-map $VT \to VT'$. By altering d by an inner derivation, we may replace v' with the image of v. Then under the left RG-linear maps $RET \approx \omega RVT \to \omega RVT' \approx RET'$, $T[v, gv] \mapsto gv - v \mapsto gv' - v' \mapsto T'[v', gv']$. Hence the class of d in $H^1(G, RG)$ is in the image of the above map, as desired.

Conversely, suppose that $H^1(G, RG)$ is finitely generated as right RG-module. Choose a finite set of derivations $d_1, \ldots, d_n: G \to RG$ which give rise to an RG-generating set of $H^1(G, RG)$. Let $d = (d_1, \ldots, d_n): G \to RG^n$, so d is a derivation. By Remark 2.9, there exists a G-tree T with finite edge stabilizers such that d is inner on G_v for each $v \in VT$. Since G is finitely generated it follows that for any set E', any derivation $G \to R[G \times E']$ is inner on G_v for every $v \in VT$. It suffices to show that for any G-tree T' with finite edge stabilizers and any vertex v of T, G_v stabilizes a vertex of T'. Without loss of generality we assume G_v is infinite. Let E' be a G-transversal in ET'. The map $E' \to R[G \times E']$, $e \mapsto \sum_{g \in G_e} (g, e)$, induces a left RG-linear embedding $RET' \to R[G \times E']$. Any vertex v' of T' gives a derivation $T'[v', \_v']: G \to RET' \to R[G \times E']$ which, by the above, is inner on G_v. Since G_v is infinite one can show that $T'[v', \_v']: G \to RET'$ must also be inner. When composed with $RET' \approx \omega RVT' \subseteq RVT'$ the derivation becomes $\operatorname{ad} v'$, so G_v stabilizes an element of $v' + \omega RVT'$, necessarily nonzero. Hence G_v has a finite orbit in VT'; by Corollary I.4.8, G_v stabilizes a vertex of T' as desired. ∎

7.6 Corollary. *If G is finitely generated then G is accessible if and only if $\mathscr{B}G$ is finitely generated as a G-module.*

Proof. Recall from Definition 6.4 that $H^1(G, \mathbb{Z}_2 G) \approx \bar{\mathscr{B}}(G)/*$. The result now follows from the $R = \mathbb{Z}_2$ case of Theorem 7.5. ∎

7.7 Remark. Suppose that G is finitely generated and $\mathscr{B}G$ is finitely generated as G-module. Let X be the Cayley graph of G with respect to some finite generating set of G. Since $\mathscr{B}G$ is finitely generated as G-module, $\mathscr{B}G = \mathscr{B}_m X$ for some m. By Theorem II.2.20 there is a G-finite tree G-subset E of $\mathscr{B}G$ which generates $\mathscr{B}G$ as a Boolean ring. By Theorem II.1.5 there is a G-tree $T = T(E)$, with finite edge stabilizers, and we claim that T is a terminal G-tree. Let $\{d_1, d_2, \ldots, d_n\} \subseteq \operatorname{Der}(G, \mathbb{Z}_2 G)$ be the image under the natural map $\mathscr{B}(G) \to \mathscr{B}(G)/* \approx \operatorname{Der}(G, \mathbb{Z}_2 G)$ of a G-transversal in E. Arguing as in the proof of Theorem 7.5, we see that T is G-terminal. ∎

Notes and comments

Theorem 1.3 is new. Example 1.4 was pointed out to us by G.P. Scott. Theorem 1.6(b)⇒(c) was proved by Karrass, Pietrowski and Solitar (1973). They conjectured (c)⇒(b) and proved it for G finitely generated; the countable case was done by Cohen (1973); and Scott (1974) finished off the general case. Theorem 1.8 was proved by Stallings (1968) for G finitely generated, and Swan (1969) completed the result. The formula in Definition 1.10 is taken from Serre (1977). Far more general results than those of Remarks 1.11(i)–(vi) can be stated and proved; the interested reader should consult Brown (1982) and the references cited there. Example 1.12 is taken from Serre (1977), and Example 1.13 seems to be new.

For the essential case $A = \mathbb{Z}_2$, Proposition 2.3 goes back to Specker (1949, 1950) and Stallings (1968). The G-projective G-modules are the projective $(\mathbb{Z}G, \mathbb{Z})$-modules of Hochschild (1956). Theorem 2.5 is new and generalizes results of Dunwoody (1979).

Theorem 3.13 is from Dunwoody (1979); one application of it was to characterize left hereditary group rings in Dicks (1979).

Theorem 3.15 was proved by Stallings (1968) for G finitely generated and then by Swan (1969) for arbitrary G.

Theorem 4.8 is new, although the case where U consists of a single G-orbit was done in Dicks (1980).

Theorem 4.11 is new. It was conjectured by Wall (1971) and proved by him in the case where G is finitely generated accessible. It was proved by Dunwoody (1979) in the case where G is finitely generated, and in Dicks (1980) in the case where V consists of a single G-orbit.

Theorem 5.1 was proved by Dicks (1981) for G finitely generated over H, and by P.A. Linnell for arbitrary G; his proof, which is unpublished, uses cohomology and Corollary 2.8.

The 'if' direction of Corollary 5.2 is from Lewin (1970). The 'only if' direction was done by Dunwoody (1979′) for G finitely generated accessible over H, by Dicks (1981) for G finitely generated over H, and the general case is new.

Corollary 5.3 was proved by D. Piollet in the case where G is free; the general case is new.

The Boolean ring of a group and its relation to the ends was first mentioned by Specker (1950). The crucial part of Theorem 6.10, namely $(b) \Rightarrow (d)$, was proved by Stallings (1968) in the case where G is finitely generated, and by Holt (1981) for G locally finite; the general case is new. The other interesting implication is $(a) \Rightarrow (e)$, which was proved by Hopf (1943) for G finitely generated, by Holt (1981) for G locally finite, and the general case is new.

Theorem 6.12 combines results of Hopf (1943) and Wall (1967).

Example 7.2 goes back to Kurosh (1937). Conjecture 7.3 was made by Wall (1971) to give one possible approach to prove what is now our Theorem 4.12. Proposition 7.4 combines results of Scott and Wall (1979) and Dicks (1980). Theorem 7.5 is from Dunwoody (1979).

V

Poincaré duality

This chapter presents some of the main results on Poincaré duality groups, and culminates in the proof that, in dimension two, the groups involved are precisely the infinite surface groups.

The argument is algebraic, and a knowledge of homological algebra will be useful since the proofs of many standard results will be omitted.

A familiarity with manifolds will be needed to appreciate the motivation, especially in Section 1, where the reader is expected to believe local-coefficient Poincaré–Lefschetz duality for compact manifolds.

Throughout this chapter R denotes a nonzero ring and K a nonzero commutative ring.

1 Introduction

Let $n \geqslant 0$ be an integer, and M a G-module.

Recall that if $P \to \mathbb{Z}$ is a projective $\mathbb{Z}G$-resolution of $\mathbb{Z}$ then the homology of the complex $M \otimes_{\mathbb{Z}G} P$ is the *homology* of G with *coefficients in* M, denoted $H_*(G, M) = \bigoplus_{i \in \mathbb{Z}} H_i(G, M)$; similarly, the homology of the cocomplex $\mathrm{Hom}_{\mathbb{Z}G}(P, M)$ is the *cohomology* of G with *coefficients in* M, denoted $H^*(G, M) = \bigoplus_{i \in \mathbb{Z}} H^i(G, M)$. These concepts will be reviewed in Sections 2 and 3.

Recall also that a manifold Y is *closed* if it is compact, without boundary, and it is *aspherical* if it is path-connected and the homotopy group $\pi_i Y$ is trivial for each $i \geqslant 2$.

1.1 Theorem. *If G is the fundamental group of a closed aspherical n-manifold Y, then there is a G-module Ω whose underlying abelian group is infinite* *cyclic, and for each G-module M and integer i, there is a natural isomorphism* $H^i(G, M) \approx H_{n-i}(G, \Omega \otimes_{\mathbb{Z}} M)$ *where G acts diagonally on* $\Omega \otimes_{\mathbb{Z}} M$.

Proof. Let $\tilde{Y}$ be the universal covering space of Y, so G acts on $\tilde{Y}$ with quotient Y. Moreover G acts properly and freely on $\tilde{Y}$, where G acting *properly* means that for each compact subset c of $\tilde{Y}$, $\{g \in G \mid gc \cap c \neq \emptyset\}$ is finite.

Let $P \to \mathbb{Z}$ be the augmented singular chain complex of $\tilde{Y}$, so P is a $\mathbb{Z}G$-complex in a natural way.

Since P acts properly, the homology of the complex $M \otimes_{\mathbb{Z}G} P$ is the homology of Y with local coefficients in M, $H^*(Y, M) = \bigoplus_{i \in \mathbb{Z}} H^i(Y, M)$; similarly, the homology of the cocomplex $\mathrm{Hom}_{\mathbb{Z}G}(P, M)$ is the cohomology of Y with local coefficients in M, $H_*(Y, M) = \bigoplus_{i \in \mathbb{Z}} H_i(Y, M)$.

Since G acts freely on $\tilde{Y}$, P is free as $\mathbb{Z}G$-module. By the asphericity of Y, $\tilde{Y}$ is contractible, so $P \to \mathbb{Z}$ is exact, and is therefore a free $\mathbb{Z}G$-resolution of $\mathbb{Z}$. Hence $H^*(Y, M) = H^*(G, M)$ and $H_*(Y, M) = H_*(G, M)$.

Since $\tilde{Y}$ has exactly two possible orientations, there is an *orientation map* $\varepsilon^\circ : G \to \{\pm 1\}$ defined by

$$\varepsilon^\circ(g) = \begin{cases} +1 \text{ if } g \text{ preserves the orientations of } \tilde{Y} \\ -1 \text{ if } g \text{ reverses the orientations of } \tilde{Y}. \end{cases}$$

For example, ε° is trivial if and only if Y is orientable. In any event ε° is a group homomorphism to the automorphism group $\{\pm 1\}$ of the infinite cyclic group; this gives us the G-module Ω.

By Poincaré duality (with local coefficients) for compact manifolds there is a natural isomorphism $H^i(Y, M) \approx H_{n-i}(Y, \Omega \otimes_{\mathbb{Z}} M)$ given by a cap product, and hence there are natural isomorphisms $H^i(G, M) \approx H^i(Y, M) \approx H_{n-i}(Y, \Omega \otimes_{\mathbb{Z}} M) \approx H_{n-i}(G, \Omega \otimes_{\mathbb{Z}} M)$. ∎

The groups satisfying the hypothesis of the theorem are called *geometric duality groups* of *dimension n*, or simply GD-groups of dimension n, or GD^n-groups.

The groups satisfying the conclusion of the theorem are called *Poincaré duality groups* of *dimension n*, or simply PD-groups of dimension n, or PD^n-groups.

In each case, if the G-module Ω has trivial action, we say that G is *orientable*.

Thus GD^n-groups are PD^n-groups, and the object of this chapter is to prove the converse for $n \leqslant 2$. What happens for $n \geqslant 3$ is not known.

1.2 Examples. Let Y be a closed aspherical n-manifold, $\tilde{Y}$ its universal covering space, $G = \pi(Y)$, the fundamental group, and ε° the orientation map.

(i) If Y is the n-torus, that is, the Cartesian product of n circles, then $\tilde{Y} = \mathbb{R}^n$ and $G = \mathbb{Z}^n$ acting by addition. Thus $\mathbb{Z}^n$ is an orientable GD^n-group.

(ii) The only path-connected zero-manifold is a point, so if $n = 0$ then $Y = \tilde{Y} = \mathbb{R}^0, G = 1$. Thus the only GD^0-group is trivial, and orientable.

(iii) The only path-connected closed one-manifold is the circle, S^1, so if $n = 1$ then $Y = S^1$, $\tilde{Y} = \mathbb{R}$, $G = \mathbb{Z}$ acting by addition. Thus the only GD^1-group is infinite cyclic, and orientable.

(iv) A path-connected closed two-manifold, called a *surface*, is uniquely described as being obtained by adding h handles to a sphere, projective plane, or Klein bottle for some $h \geqslant 0$; the first gives orientable surfaces, and the latter two give unorientable surfaces. The corresponding fundamental groups have presentation

$$\langle x_1, y_1, \ldots, x_h, y_h \mid [x_1, y_1] \cdots [x_h, y_h] = 1 \rangle,$$
$$\langle w, x_1, y_1, \ldots, x_h, y_h \mid w^2 [x_1, y_1] \cdots [x_h, y_h] = 1 \rangle,$$
$$\langle w, z, x_1, y_1, \ldots, x_h, y_h \mid w^2 z^2 [x_1, y_1] \cdots [x_h, y_h] = 1 \rangle,$$

respectively; these are the *surface groups.*

The sphere and the projective plane each have universal covering space the sphere, S^2, so are not aspherical; the remaining surfaces all have covering space homeomorphic to the plane, $\mathbb{R}^2$, so are aspherical. Thus the GD^2-groups are precisely the infinite surface groups. They can be presented as

$$\langle x_1, \ldots, x_m \mid [x_1, x_2] \cdots [x_{m-1}, x_m] = 1 \rangle, \quad m \text{ even}, m \geqslant 2,$$
$$\langle x_1, \ldots, x_m \mid x_1^2 \cdots x_m^2 = 1 \rangle, \quad m \geqslant 2.$$

The former are orientable; for the latter $\varepsilon^\circ(x_i) = -1$, $i \in [1, m]$. ∎

Hence the trivial group, the infinite cyclic group and the infinite surface groups are PD-groups of dimensions zero, one and two, respectively. It will be shown that these are the only ones, in Theorems 4.1, 4.4 and 12.4, respectively.

There is a generalization of Theorem 1.1 suggested by Poincaré–Lefschetz duality.

Let W be a nonempty G-set. View the augmentation module $\omega \mathbb{Z} W$ as a graded G-module concentrated in degree one, and suppose that $P \to \omega \mathbb{Z} W$ is a projective $\mathbb{Z}G$-resolution. The homology of the complex $M \otimes_{\mathbb{Z}G} P$

is the *homology* of the pair (G, W) with *coefficients in* M, denoted $H_*((G, W), M) = \bigoplus_{i\in\mathbb{Z}} H_i((G, W), M)$; similarly, the homology of the cocomplex $\mathrm{Hom}_{\mathbb{Z}G}(P, M)$ is the *cohomology* of (G, W) with *coefficients in* M, denoted $H^*((G, W), M) = \bigoplus_{i\in\mathbb{Z}} H^i(G, M)$.

To extend the notation to include the case where W is empty, we define the homology and cohomology of the pair $(G, \varnothing)$ to be the homology and cohomology of G.

1.3 Theorem. *Let Y be a compact n-manifold, $\tilde{Y}$ its universal covering space, and suppose that $\tilde{Y}$ and the components of the boundary $\partial\tilde{Y}$ are all contractible; let G be the fundamental group of Y, and W the G-set of components of $\partial\tilde{Y}$. Then there is a G-module Ω whose underlying abelian group is infinite cyclic, and for each G-module M and integer i, there is a natural isomorphism $H^i(G, M) \approx H_{n-i}((G, W), \Omega \otimes_{\mathbb{Z}} M)$, where G acts diagonally on $\Omega \otimes_{\mathbb{Z}} M$.*

Proof. The case where W is empty is covered by Theorem 1.1, so we may assume that W is nonempty.

Let $P \to \mathbb{Z}$ be the augmented singular chain complex of $\tilde{Y}$, and Q the singular chain complex of $\partial\tilde{Y}$. Then P, Q are $\mathbb{Z}G$-complexes and, in a natural way, P contains Q as a $\mathbb{Z}G$-subcomplex, so P/Q is a $\mathbb{Z}G$-complex.

As in the proof of Theorem 1.1, the homology of the complex $M \otimes_{\mathbb{Z}G} (P/Q)$ is the homology of the pair $(Y, \partial Y)$ with coefficients in M, $H_*((Y, \partial Y), M) = \bigoplus_{i\in\mathbb{Z}} H_i((Y, \partial Y), M)$, and the homology of the cocomplex $\mathrm{Hom}_{\mathbb{Z}G}(P, M)$ is $H^*(Y, M)$.

Also $P \to \mathbb{Z}$ is a free $\mathbb{Z}G$-resolution and $H^*(Y, M) \approx H^*(G, M)$.

Since each component of $\partial\tilde{Y}$ is contractible, the homology of the $\mathbb{Z}G$-complex Q is $\mathbb{Z}W$ concentrated in degree zero. From the exact triangle for homology corresponding to $0 \to Q \to P \to P/Q \to 0$, we see that the homology of P/Q is $\omega\mathbb{Z}W$ concentrated in degree one. Thus $H^*((Y, \partial Y), M) \approx H^*((G, W), M)$.

Also as in the proof of Theorem 1.1, $\tilde{Y}$ has two orientations, and there is an orientation map $\varepsilon^\circ : G \to \{1, -1\}$ and an orientation module Ω.

By Poincaré–Lefschetz duality (with local coefficients) for a compact manifold, for each integer i there is a natural isomorphism $H^i(Y, M) \approx H_{n-i}((Y, \partial Y), \Omega \otimes_{\mathbb{Z}} M)$ given by a cap product. Hence there are natural isomorphisms $H^i(G, M) \approx H^i(Y, M) \approx H_{n-i}((Y, \partial Y), \Omega \otimes_{\mathbb{Z}} M) \approx H_{n-i}((G, W), \Omega \otimes_{\mathbb{Z}} M)$. ∎

If (G, W) arises as in the hypotheses of Theorem 1.3, we say that (G, W) is a *geometric duality pair* of *dimension n*, or simply a GD-pair of dimension n, or a GD^n-pair.

If (G, W) satisfies the conclusion of Theorem 1.3, we say that (G, W) is a *Poincaré duality pair* of *dimension n*, or simply a PD-pair of dimension n, or a PD^n-pair.

In either case if the module Ω has trivial G-action, then (G, W) is said to be *orientable.*

Thus every GD^n-pair is a PD^n-pair. Over the course of Theorems 4.1, 4.4, 7.7, 10.3 and 12.4 we shall see that the converse is true for $n \leqslant 2$; it is not known if the converse holds for $n \geqslant 3$.

1.4 Examples. Let Y be a compact n-manifold-with-boundary, and let $\tilde{Y}$ be the universal covering space. Suppose that $\tilde{Y}$ and all components of $\partial\tilde{Y}$ are contractible. Let $G = \pi Y$, and let W be the G-set ∂Y, so W is nonempty.

(i) There is no zero-manifold-with-boundary, so there is no GD^0-pair (G, W) with W nonempty.

(ii) The only connected one-manifold-with-boundary is a line segment, with boundary consisting of the two endpoints. This is its own universal covering space. Thus in the only GD^1-pair (G, W) with W nonempty, G is trivial and W has two elements. It is orientable.

(iii) All connected two-manifolds-with-boundary arise by taking points out of the closed surfaces described in Example 1.2(iv), so are spheres, projective planes or Klein bottles with $h \geqslant 0$ handles and $p + 1 \geqslant 1$ punctures. In all cases the universal covering space is contractible; further, the components of the boundary are contractible except in the case of the sphere with one puncture, that is, the disc, where the boundary is a circle. All other two-manifolds-with-boundary give GD^2-pairs. They can be described as follows.

In the orientable case, Y is a disc with $h \geqslant 0$ handles and $p \geqslant 0$ holes such that $h + p \geqslant 1$. Here G is free of rank $2h + p$ on generators $t_1, \ldots, t_p$, $x_1, \ldots, x_h$, $y_1, \ldots, y_h$, and $W = \bigvee_{i=0}^{p} G/\langle t_i \rangle$, where $t_0 = t_1 \cdots t_p [x_1, y_1] \cdots [x_h, y_h]$.

In the unorientable case, Y is a Möbius strip with $m \geqslant 0$ cross-caps and $p \geqslant 0$ holes. Then G is free of rank $m + p + 1$ on generators $t_1, \ldots, t_p$, $x_0, \ldots, x_m$ and $W = \bigvee_{i=0}^{p} G/\langle t_i \rangle$, where $t_0 = t_1 \cdots t_p x_0^2 \cdots x_m^2$. Here $\varepsilon^\circ(x_i) = -1$, $i \in [0, m]$, $\varepsilon^\circ(t_i) = 1$, $i \in [1, p]$.

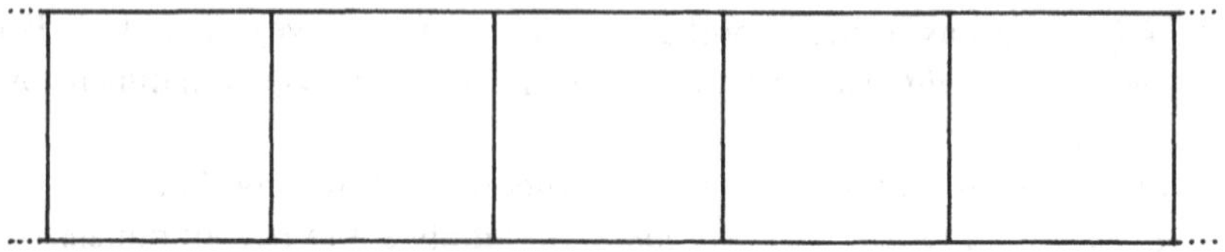

Fig. V.1

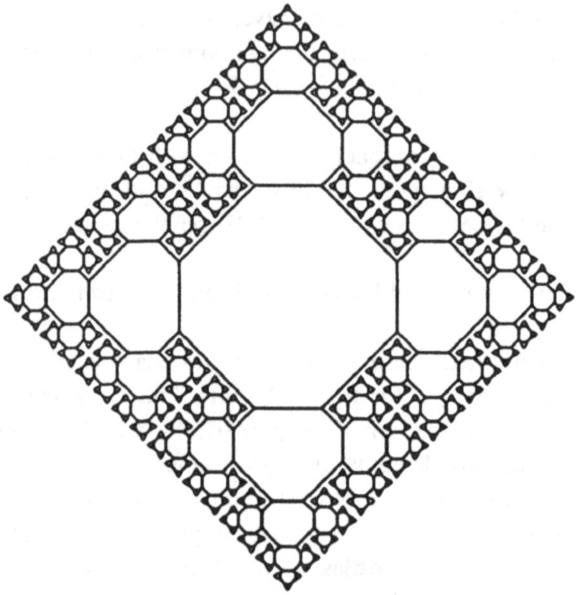

Fig. V.2

In all cases, G is given with a free generating set, and $\tilde{Y}$ can be thought of as a thickened version of the G-tree arising as the Cayley graph.

The case where G has rank one corresponds to Y being an annulus or a Möbius strip. We get $\tilde{Y}$ by thickening the usual tree homeomorphic to $\mathbb{R}$, and this can be depicted as in Fig. V.1. Here the elements of W are the (two) sides of $\tilde{Y}$; G acts trivially on W if Y is the annulus, and G acts transitively on W if Y is the Möbius strip.

The case where G has rank two corresponds to Y being one of the following: a disc with two holes, a torus with one hole, a Klein bottle with one hole, or a Möbius strip with one hole. We get $\tilde{Y}$ by thickening the tree of Fig. I.1, page 7, to get Fig. V.2. Here W consists of he infinitely many components of the boundary. ∎

2 PD-modules

We now review the theory of homology and cohomology of R-modules, omitting proofs of standard results. The main object is to develop the notion of a Poincaré duality module.

Throughout this section, let A be a right R-module, and B, C left R-modules, and $n \geqslant 0$ an integer.

2.1 Definitions. By a *graded* abelian group P we mean an abelian group given with specified subgroups P_i, $i \geqslant 0$, such that $P = \bigoplus_{i \geqslant 0} P_i$; we make the convention that $P_i = 0$ for all $i < 0$. Each $p \in P_i$ is said to be *homogeneous* of *degree* i, denoted $\deg p = i$; notice $\deg 0 = i$ for all $i \in \mathbb{Z}$. We say that P is *concentrated* in some degree i if $P_j = 0$ for all $j \neq i$.

By a *graded map* $\alpha: P \to Q$, we mean an additive map of graded abelian groups, such that for some $r \in \mathbb{Z}$, $\alpha(P_i) \subseteq Q_{i+r}$ for all $i \in \mathbb{Z}$; then α is said to have *degree* r, denoted $\deg \alpha = r$. Again $\deg 0 = r$ for all $r \in \mathbb{Z}$.

By a *differential* graded abelian group P we mean a graded abelian group given with a graded map $\partial = \partial_P: P \to P$ such that $\partial^2 = 0$, that is, $\operatorname{Im} \partial \subseteq \operatorname{Ker} \partial$. We then define $Z(P) = \operatorname{Ker} \partial$, $B(P) = \operatorname{Im} \partial$, and $H(P) = Z(P)/B(P) = (\operatorname{Ker} \partial)/(\operatorname{Im} \partial)$; these are all graded abelian groups. The elements of $Z(P)$ are called the *cycles* of P, the elements of $B(P)$ the *boundaries* of P, and $H(P)$ is called the *homology group* of P. If the degree of ∂ is -1 or $+1$ then P is called a *complex* or *cocomplex*, respectively. If $H(P) = 0$ then P is said to be *acyclic*.

By a *chain map* $\alpha: P \to Q$ we mean a graded map of differential graded abelian groups which commutes with the differentials in the obvious sense. There is then induced a map $H(\alpha): H(P) \to H(Q)$.

If $\alpha, \beta: P \to Q$ are chain maps of degree zero then by a *homotopy* $s: \alpha \approx \beta$ we mean a function $s: P \to Q$ such that $s\partial_P + \partial_Q s = \alpha - \beta: P \to Q$. In this event we say that α, β are *homotopic*, and one can show easily that $H(\alpha) = H(\beta): H(P) \to H(Q)$.

If $\alpha: P \to Q$ and $\beta: Q \to P$ are chain maps of degree zero such that $\alpha\beta: Q \to Q$ is homotopic to 1_Q, and $\beta\alpha: P \to P$ is homotopic to 1_P then β is said to be a *chain inverse* of α. Here $H(\alpha)$ is the inverse of $H(\beta)$. A chain map with a chain inverse is called a *chain equivalence.* ∎

For later reference, we state the fundamental theorem of homological algebra.

2.2 Theorem. *Suppose* $0 \to P' \xrightarrow{\alpha} P \xrightarrow{\beta} P'' \to 0$ *is an exact sequence of* *differential graded abelian groups. Then there exists a connecting map* $\delta: H(P'') \to H(P')$ *characterized by the property that* $\delta(z'' + B(P'')) = z' + B(P')$ *if* $z' \in Z(P')$, $z'' \in Z(P'')$ *and there is some* $p \in P$ *with* $\beta p = z''$, $\alpha z' = \partial p$. *The triangle of graded abelian groups*

$$\begin{array}{ccc} H(P') & \xleftarrow{\ \delta\ } & H(P'') \\ & {\scriptstyle H(\alpha)}\searrow \qquad \nearrow{\scriptstyle H(\beta)} & \\ & H(P) & \end{array}$$

is exact. ∎

It is sometimes helpful to think of δ as $\alpha^{-1}\partial_p\beta^{-1}$; in particular, $\deg \delta = \deg \partial_P - \deg \alpha - \deg \beta$.

2.3 Definitions. All of the above definitions extend from abelian groups to an arbitrary abelian category, and, in particular, to left and right R-modules. Thus, for example, we can speak of left R-module complexes and right R-module complexes.

An R-module complex $P = \left(\bigoplus_{i \geq 0} P_i, \partial\right)$ is said to be *projective* if P is projective as R-module.

Suppose P is a projective right R-module complex, and $\alpha: P \to A$ a chain map of degree zero where A is viewed as a graded right module concentrated in degree zero. If $H(\alpha): H(P) \to H(A) = A$ is an isomorphism, then we say that $\alpha: P \to A$ is a *projective right R-resolution* of A. Such a resolution always exists, and in fact P can be taken to be free.

Similarly, there exists a projective left R-resolution $\beta: Q \to B$.

If $\beta': Q' \to B'$ is a projective left R-resolution and $\phi: B \to B'$ is an R-linear map then it can be shown that there exists a chain map $\Phi: P \to P'$ of degree zero such that $\beta'\Phi = \phi\beta: P \to B'$, and Φ is called a *lift* of ϕ. It can be shown further that if $\Phi': P \to P'$ is another lift of ϕ then there exists an R-linear homotopy $s: \Phi \approx \Phi'$, that is, s is an R-linear map $P \to P'$ which is a homotopy.

In a natural way $\mathrm{Hom}_R(Q, C)$ is a cocomplex, and its homology is denoted $\mathrm{Ext}_R(Q, C) = \bigoplus_{i \geq 0} \mathrm{Ext}^i_R(Q, C)$, a graded abelian group. If $\beta': Q' \to B$ is another projective left R-resolution of B then the identity map has lifts $\Phi: Q \to Q'$, $\Phi': Q' \to Q$ and then $\Phi'\Phi: Q \to Q$, $\Phi\Phi': Q' \to Q'$ are lifts of the identity maps so there are R-linear homotopies $s: \Phi'\Phi \approx 1_Q$, $s': \Phi\Phi' \approx 1_{Q'}$. It follows that the cocomplexes $\mathrm{Hom}_R(Q, C)$ and $\mathrm{Hom}_R(Q', C)$ are chain equivalent, so have isomorphic homology groups. Thus $\mathrm{Ext}_R(B, C)$ is independent of the choice of projective left R-resolution of B.

In a natural way $\mathrm{Ext}_R(-,-)$ is a bifunctor.

It is easy to show that $\mathrm{Ext}_R(B, C)$ converts direct sums to direct products in the first variable, and commutes with finite direct sums in the second variable.

Notice that if B is projective then $\mathrm{Ext}_R(B, C)$ is $\mathrm{Hom}_R(B, C)$ concentrated in degree zero.

If B is such that $\underset{i\in I}{\mathrm{dir.lim.}}(\mathrm{Ext}_R(B, C_i)) = \mathrm{Ext}_R(B, \underset{i\in I}{\mathrm{dir.lim.}}\, C)$ for every directed system $(C_i | i \in I)$ of left R-modules then $\mathrm{Ext}_R(B, -)$ is said to *commute with direct limits.* ∎

2.4 Lemma. *If* $0 \to B' \to B \to B'' \to 0$ *is an exact sequence of left R-modules and* $P' \to B'$, $P'' \to B''$ *are projective left R-resolutions then there exists a projective left R-resolution* $P \to B$ *and an exact sequence of left R-complexes* $0 \to P' \to P \to P'' \to 0$ *which lifts the original exact sequence.* ∎

2.5 Theorem. *If* $0 \to B' \to B \to B'' \to 0$, $0 \to C' \to C \to C'' \to 0$ *are exact sequences of left R-modules then there exist exact triangles*

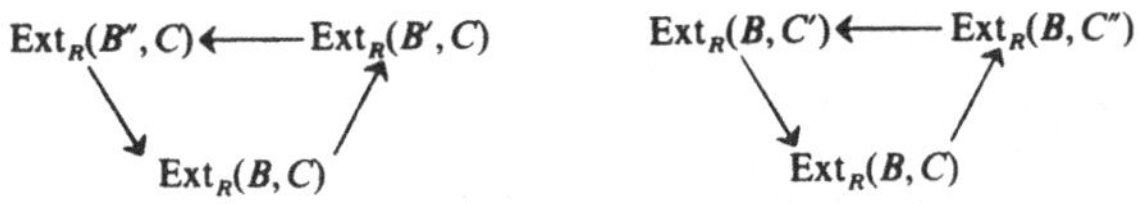

of graded abelian groups.

Proof. We get the right triangle from the definition and Theorem 2.2; we get the left triangle by using Lemma 2.4 and Theorem 2.2. ∎

We now examine how Ext gives information about finiteness conditions on modules.

2.6 Lemma. *If* $0 \to B' \to Q_{n-1} \to \cdots \to Q_0 \to B \to 0$ *is an exact sequence of left R-modules with each* Q_m *projective then there is an exact sequence*

$$\mathrm{Hom}_R(Q_{n-1}, C) \to \mathrm{Hom}_R(B', C) \to \mathrm{Ext}^n_R(B, C) \to 0$$

of abelian groups.

Proof. The result is clear for $n = 0$, where $Q_{-1} = 0$, $B' = B$, so we need consider only $n \geqslant 1$. Define $B_n = B'$, $B_0 = B$, and $B_m = \mathrm{Im}\, Q_m \to Q_{m-1}$ for all $m \in [1, n-1]$. For each $m \in [0, n-1]$ we have an exact sequence $0 \to B_{m+1} \to Q_m \to B_m \to 0$ of left R-moduls. Applying $\mathrm{Ext}_R(-, C)$ gives an exact triangle which consists of exact sequences

(1) $\qquad \mathrm{Hom}_R(Q_m, C) \to \mathrm{Hom}_R(B_{m+1}, C) \to \mathrm{Ext}^1_R(B_m, C) \to 0,$

and

$$0 \to \mathrm{Ext}_R^i(B_{m+1}, C) \to \mathrm{Ext}_R^{i+1}(B_m, C) \to 0, \ i \geqslant 1.$$

Taking $(m, i) = (n-2, 1), (n-3, 2), \ldots, (0, n-1)$ we have isomorphisms

$$\mathrm{Ext}_R^1(B_{n-1}, C) \approx \mathrm{Ext}_R^2(B_{n-2}, C) \approx \cdots \approx \mathrm{Ext}_R^n(B_0, C) = \mathrm{Ext}_R^n(B, C).$$

Substituting into (1), taking $m = n-1$, gives the desired exact sequence. ∎

The next result shows in particular that $\mathrm{pd}_R B \leqslant n$ if and only if $\mathrm{Ext}_R^{n+1}(B, -) = 0$.

2.7 Corollary. *If* $0 \to C \to Q_n \to Q_{n-1} \cdots \to Q_0 \to B \to 0$ *is an exact sequence of left R-modules with each* Q_i *projective and* $\mathrm{Ext}_R^{n+1}(B, C) = 0$ *then* $\mathrm{pd}_R B \leqslant n$.

Proof. Let us treat C as a submodule of Q_n. By Lemma 2.6, there is an exact sequence $\mathrm{Hom}_R(Q_n, C) \to \mathrm{Hom}_R(C, C) \to \mathrm{Ext}_R^{n+1}(B, C) = 0$. In particular, the identity map on C factors through Q_n, so C is an R-summand of Q_n. Thus Q_n/C is projective, and

$$0 \to Q_n/C \to Q_{n-1} \cdots \to Q_0 \to B \to 0$$

is a projective resolution of B of length n. ∎

2.8 Definitions. We say that B is an FP_n left R-module if there is an exact sequence $Q_n \to Q_{n-1} \to \cdots \to Q_0 \to B \to 0$ of left R-modules such that each Q_i is finitely generated projective.

For example, FP_0 means finitely generated, and FP_1 means finitely presented.

It follows from Schanuel's Lemma IV.3.3 that B is FP_n if and only if for every $k \in [0, n]$ and every exact sequence

$$0 \to B_k \to Q'_{k-1} \to Q'_{k-2} \to \cdots \to Q'_0 \to B \to 0$$

of left R-modules, if the Q'_i are finitely generated projective, then B_k is finitely generated.

If B is FP_n for all $n \geqslant 0$ then we say B is an FP_∞ left R-module. This is equivalent to the existence of a projective R-resolution $\cdots \to P_1 \to P_0 \to B \to 0$ in which all the P_i are finitely generated.

If B is FP_∞ and $\mathrm{pd}_R B < \infty$ then B is said to be an FP left R-module. Thus B is FP if and only if there exists a finite projective left R-resolution $0 \to P_n \to P_{n-1} \to \cdots \to P_0 \to B \to 0$, where the P_i are finitely generated. Here we have a resolution $P \to B$, with P finitely generated, and we call P a *finitely generated projective R-resolution* of B. ∎

2.9 Corollary. *If B is a nonzero* FP *left R-module and* $\mathrm{pd}_R B = n$ *then* $\mathrm{Ext}_R^n(B, R) \neq 0$.

Proof. This is clear for $n = 0$, where B is a nonzero finitely generated projective left R-module. Thus we may assume $n \geqslant 1$, and let

$$0 \to Q_n \to Q_{n-1} \to \cdots \to Q_0 \to B \to 0$$

be a finitely generated projective R-resolution of B.

By Corollary 2.7, $\mathrm{Ext}_R^n(B, Q_n) \neq 0$. Since Q_n is a summand of a direct sum of finitely many copies of R, it follows that $\mathrm{Ext}_R^n(B, R) \neq 0$. ∎

We leave the converse of the following result as an exercise.

2.10 Theorem. *If* $\mathrm{Ext}_R(B, -)$ *commutes with direct limits, then B is an* FP_∞ *left R-module.*

Proof. Suppose that we have constructed an exact sequence of left R-modules

$$0 \to C \to Q_{n-1} \to \cdots \to Q_0 \to B \to 0$$

with each Q_m finitely generated projective.

We shall show that C is finitely generated, and the result then follows by induction. Let $(C_i | i \in I)$ be the directed system consisting of all finitely generated submodules of C, so $(C/C_i | i \in I)$ is a directed system of left R-modules with $\operatorname*{dir.lim.}_{i \in I}(C/C_i) = 0$. By the hypothesis, $\operatorname*{dir.lim.}_{i \in I} \mathrm{Ext}_R^n(B, C/C_i) = 0$.

By Lemma 2.6 we have a family of exact sequences

(2) $\qquad \mathrm{Hom}_R(Q_{n-1}, C/C_i) \to \mathrm{Hom}_R(C, C/C_i) \to \mathrm{Ext}_R^n(B, C/C_i) \quad (i \in I).$

For each $i \in I$, denote by α_i the image under $\mathrm{Hom}_R(C, C/C_i) \to \mathrm{Ext}_R^n(B, C/C_i)$ of the natural surjection $C \to C/C_i$. As i ranges over I, the sequences (2) form a directed system, so the $(\alpha_i | i \in I)$ determine an element of $\operatorname*{dir.lim.}_{i \in I}(\mathrm{Ext}_R^n(B, C_i)) = 0$, which means that $\alpha_i = 0$ for some $i \in I$. From the exactness of (2), we see that the surjective map $C \to C/C_i$ factors through the finitely generated module Q_{n-1}. Therefore C/C_i is finitely generated, and hence so is C. ∎

2.11 Definition. Let $P \to A$, $Q \to B$ be projective resolutions.

The homology of the complex $P \otimes_R B$ is denoted $\mathrm{Tor}^R(A, B) = \bigoplus_{i \geqslant 0} \mathrm{Tor}_i^R(A, B)$, a graded abelian group. Tor, like Ext, is independent of the choice of resolution, and $\mathrm{Tor}^R(-, -)$ is a bifunctor.

Let $P \otimes_R Q$ be made into a complex by specifying that for all homogeneous $p \in P$, $q \in Q$, $\deg(p \otimes q) = \deg p + \deg q$, and $\partial(p \otimes q) = \partial_P p \otimes q + (-1)^{\deg q} p \otimes \partial_Q q$. It can be shown that the maps $P \otimes_R Q \to P \otimes_R B$, $P \otimes_R Q \to A \otimes_R Q$ induce isomorphisms on homology. ∎

As for Ext we have exact triangles.

2.12 Theorem. *If* $0 \to A' \to A \to A'' \to 0$, $0 \to B' \to B \to B'' \to 0$, *are exact sequences of right and left R-modules, respectively, then there exist exact triangles*

$$\begin{array}{ccc} \mathrm{Tor}^R(A',B) & \longleftarrow & \mathrm{Tor}^R(A'',B) \\ & \searrow \quad \nearrow & \\ & \mathrm{Tor}^R(A,B) & \end{array} \qquad \begin{array}{ccc} \mathrm{Tor}^R(A,B') & \longleftarrow & \mathrm{Tor}^R(A,B'') \\ & \searrow \quad \nearrow & \\ & \mathrm{Tor}^R(A,B) & \end{array}$$

of graded abelian groups. ∎

2.13 Definition. An important homological tool used in the study of Poincaré duality is the *cap product* which is a natural multiplication map

(3) $\qquad \cap : \mathrm{Tor}^R(A, B) \otimes_Z \mathrm{Ext}_R(B, C) \to \mathrm{Tor}^R(A, C)$

defined as follows.

Let $\alpha: P \to A$ be a projective right R-resolution and $\beta: Q \to B$ a projective left R-resolution.

Any $\phi \in \mathrm{Hom}_R(Q, C)$ induces an additive map $P \otimes_R \phi: P \otimes_R Q \to P \otimes_R C$. There is thus determined a natural map

$$\cap : (P \otimes_R Q) \otimes_Z (\mathrm{Hom}_R(Q, C)) \to P \otimes_R C, \; (p \otimes q) \otimes \phi \mapsto p \otimes \phi q,$$

which we denote $x \otimes \phi \mapsto x \cap \phi$. This is called the *cap-product at the complex level.* If $x = p \otimes q$ is homogeneous then

$$\begin{aligned} &(\partial_{P \otimes Q} x) \cap \phi \; - \; \partial_{P \otimes C}(x \cap \phi) \\ &\quad = (\partial_P p \otimes q + (-1)^{\deg q} p \otimes \partial_Q q] \cap \phi - \partial_{P \otimes C}(p \otimes \phi q) \\ &\quad = \partial_P p \otimes \phi q + (-1)^{\deg q} p \otimes \phi \partial_Q q - \partial_P p \otimes \phi q = (-1)^{\deg q} p \otimes \phi \partial_Q q \\ &\quad = (-1)^{\deg q} p \otimes (\partial_{\mathrm{Hom}(Q,C)} \phi)(q) = (-1)^{\deg q} (x \cap \partial_{\mathrm{Hom}(Q,C)} \phi). \end{aligned}$$

If further ϕ is homogeneous and $\phi \partial_Q q \neq 0$ then $\deg \phi = \deg \partial_Q q = -1 + \deg q$; thus if $x \cap \partial_{\mathrm{Hom}(Q,C)} \phi \neq 0$ then $\deg \phi = -1 + \deg q$ and we can write

(4) $\qquad \partial_{P \otimes C}(x \cap \phi) = (\partial_{P \otimes Q} x \cap \phi) + (-1)^{\deg \phi} (x \cap \partial_{\mathrm{Hom}(Q,C)} \phi)$

for all x and all homogeneous ϕ. It follows readily that there is induced a map on homology

$$\cap : H(P \otimes_R Q) \otimes_Z H(\mathrm{Hom}_R(Q, C)) \to H(P \otimes_R C), \; \xi \otimes \eta \mapsto \xi \cap \eta.$$

Thus we have a map (3); we want to make sure that it does not vary with the choice of resolutions.

Suppose that $Q' \to B$ is some other projective left R-resolution of B and denote the resulting cap-product at the complex level by

$$\cap' : (P \otimes_R Q') \otimes_{\mathbb{Z}} (\mathrm{Hom}_R(Q', C)) \to P \otimes_R C.$$

Let $\Phi : Q \to Q'$, $\Phi' : Q' \to Q$ be lifts of the identity map on B. Then each of the composites $\Phi\Phi'$, $\Phi'\Phi$ is R-linearly homotopic to the appropriate identity map. Consider the diagram

$$\begin{array}{ccc} P \otimes_R Q \otimes_{\mathbb{Z}} \mathrm{Hom}_R(Q, C) & \xrightarrow{\cap} & P \otimes_R C \\ \downarrow{\scriptstyle (P\otimes\Phi)\otimes \mathrm{Hom}(\Phi', C)} & & \| \\ P \otimes_R Q' \otimes_{\mathbb{Z}} \mathrm{Hom}_R(Q', C) & \xrightarrow{\cap'} & P \otimes_R C. \end{array}$$

Along the upper route an element of the form $x \otimes \phi = (p \otimes q) \otimes \phi$ is sent to $p \otimes \phi q = x \cap \phi$; along the bottom route it is sent to $(p \otimes \Phi q) \cap' \phi\Phi' = p \otimes \phi\Phi'\Phi q = ((P \otimes \Phi'\Phi)(x)) \cap \phi$. Since $\Phi'\Phi$ is R-homotopic to the identity map on P, it follows that $P \otimes \Phi'\Phi$ is homotopic to the identity map on $P \otimes Q$ and so induces the identity map on homology. Thus if x is any cycle in $P \otimes_R Q$ then x and $(P \otimes \Phi'\Phi)x$ define the same element of $H(P \otimes_R Q)$. This proves that the cap-product is independent of the choice of resolution of B.

A similar argument shows that the cap-product is also independent of the choice of resolution for A.

The cap-product consists of a family of maps

$$\cap : \mathrm{Tor}_m^R(A, B) \otimes_{\mathbb{Z}} \mathrm{Ext}_R^i(B, C) \to \mathrm{Tor}_{m-i}^R(A, C)$$

where $m, i \in \mathbb{Z}$; for if $x \in P_{m-i} \otimes_R Q_i$ and $\phi \in \mathrm{Hom}_R(Q_i, C)$, then $x \cap \phi \in P_{m-i} \otimes_R C$. ∎

2.14 Proposition. *If $A_1 \to A_2$ is a right R-linear map and $C_1 \to C_2$ a left R-linear map then*

$$\begin{array}{ccc} \mathrm{Tor}^R(A_1, B) \otimes_{\mathbb{Z}} \mathrm{Ext}_R(B, C_1) & \to & \mathrm{Tor}^R(A_2, B) \otimes_{\mathbb{Z}} \mathrm{Ext}_R(B, C_2) \\ \downarrow{\scriptstyle \cap} & & \downarrow{\scriptstyle \cap} \\ \mathrm{Tor}^R(A_1, C_1) & \to & \mathrm{Tor}^R(A_2, C_2) \end{array}$$

commutes.

Proof. Denote the maps by $\alpha : A_1 \to A_2$, $\gamma : C_1 \to C_2$.

Let $\xi \in \mathrm{Tor}^R(A_1, B)$, $\eta \in \mathrm{Ext}_R(B, C_1)$.

Let $P_1 \to A_1$, $P_2 \to A_2$, $Q \to B$ be projective R-resolutions. Lift $\alpha : A_1 \to A_2$

to $\Phi:P_1\to P_2$, lift ξ back to a cycle $\sum p_i\otimes q_i \in P_1\otimes_R Q$, and η back to a cocycle $\beta:Q\to C_1$ in $\mathrm{Hom}_R(Q, C_1)$.

At the cycle level, $(\sum p_i\otimes q_i)\otimes\beta$ is carried along the upper route to $(\sum\Phi p_i\otimes q_i)\otimes\gamma\beta$ and then to $\sum\Phi p_i\otimes\gamma\beta q_i$. Along the lower route it is carried to $\sum p_i\otimes\beta q_i$ and thence to $\sum\Phi p_i\otimes\gamma\beta q_i$. Thus we have commutativity even at the cycle level. ∎

2.15 Proposition. *If $B_1\to B_2$ is a left R-linear map then*

$$\begin{array}{ccc} \mathrm{Tor}^R(A,B_1)\otimes_{\mathbb{Z}}\mathrm{Ext}_R(B_2,C) & \longrightarrow & \mathrm{Tor}^R(A,B_1)\otimes_{\mathbb{Z}}\mathrm{Ext}_R(B_1,C) \\ \downarrow & & \downarrow{\scriptstyle\frown} \\ \mathrm{Tor}^R(A,B_2)\otimes_{\mathbb{Z}}\mathrm{Ext}_R(B_2,C_2) & \xrightarrow{\frown} & \mathrm{Tor}^R(A,C) \end{array}$$

commutes.

Proof. Similar to 2.14. ∎

2.16 Proposition. *If $0\to A'\to A\to A''\to 0$ is an exact sequence of right R-modules then*

$$\begin{array}{ccc} \mathrm{Tor}^R(A'',B)\otimes_{\mathbb{Z}}\mathrm{Ext}_R(B,C) & \to & \mathrm{Tor}^R(A',B)\otimes_{\mathbb{Z}}\mathrm{Ext}_R(B,C) \\ \downarrow{\scriptstyle\frown} & & \downarrow{\scriptstyle\frown} \\ \mathrm{Tor}^R(A'',C) & \to & \mathrm{Tor}^R(A',C) \end{array}$$

commutes, where the horizontal arrows arise from the connecting maps.

Proof. Similar to 2.17. ∎

2.17 Proposition. *If $0\to B'\to B\to B''\to 0$ is an exact sequence of left R-modules then*

$$\begin{array}{ccc} \mathrm{Tor}^R(A,B'')\otimes_{\mathbb{Z}}\mathrm{Ext}^n_R(B',C) & \to & \mathrm{Tor}^R(A,B'')\otimes_{\mathbb{Z}}\mathrm{Ext}^{n+1}_R(B'',C) \\ \downarrow & & \downarrow{\scriptstyle\frown} \\ \mathrm{Tor}^R(A,B')\otimes_{\mathbb{Z}}\mathrm{Ext}^n_R(B',C) & \xrightarrow{\frown} & \mathrm{Tor}^R(A,C) \end{array}$$

commutes with sign $(-1)^{n+1}$, where the first arrows arise from connecting maps.

Proof. Let $\eta\in\mathrm{Tor}^R(A,B'')$, $\xi\in\mathrm{Ext}^n_R(B',C)$.

By Lemma 2.4 we can lift the exact sequence $0\to B'\to B\to B''\to 0$ to an exact sequence $0\to Q'\xrightarrow{\beta'}Q\xrightarrow{\beta''}Q''\to 0$ of projective left R-resolutions. This is split so induces an exact sequence $0\to\mathrm{Hom}_R(Q'',C)\to\mathrm{Hom}_R(Q,C)\to\mathrm{Hom}_R(Q',C)\to 0$ of differential graded abelian groups.

Lift η back to a cycle $\phi' \in \mathrm{Hom}_R(Q'_n, C)$. Choose $\phi \in \mathrm{Hom}_R(Q_n, C)$, $\phi'' \in \mathrm{Hom}_R(Q''_{n+1}, C)$ such that $\phi\beta' = \phi'$, $\phi''\beta'' = \phi\partial_Q$.

Let $P \to A$ be a projective right R-resolution, so there is an exact sequence $0 \to P \otimes_R Q' \to P \otimes_R Q \to P \otimes_R Q'' \to 0$ of differential graded abelian groups. Lift ξ back to a cycle $x'' \in P \otimes_R Q''$. Choose $x' \in P \otimes_R Q'$, $x \in P \otimes_R Q$ such that $\beta'' x = x''$, $\beta' x' = \partial_{P \otimes Q} x$.

At the cycle level, $x'' \otimes \phi'$ is sent along the upper route to $x'' \otimes \phi''$ and then to $x'' \cap \phi'' = \beta'' x \cap \phi'' = x \cap \phi''\beta'' = x \cap \phi\partial_Q = x \cap \partial_{\mathrm{Hom}(Q,C)}\phi$.

Along the bottom route it is sent to $x'' \cap \phi'$ and then to $x' \cap \phi' = x' \cap \phi\beta' = \beta' x' \cap \phi = \partial_{P \otimes Q} x \cap \phi = \partial_{P \otimes C}(x \cap \phi) - (-1)^{\deg\phi}(x \cap \partial_{\mathrm{Hom}(Q,C)}\phi)$. Clearly $\partial_{P \otimes C}(x \cap \phi)$ is a boundary, so the diagram commutes with sign $-(-1)^{\deg\phi} = (-1)^{n+1}$. ∎

2.18 Proposition. *If $0 \to C' \to C \to C'' \to 0$ is an exact sequence of left R-modules then*

$$\begin{array}{ccc} \mathrm{Tor}^R(A,B) \otimes_{\mathbb{Z}} \mathrm{Ext}^n_R(B,C'') & \to & \mathrm{Tor}^R(A,B) \otimes_{\mathbb{Z}} \mathrm{Ext}^{n-1}_R(B,C') \\ \downarrow \cap & & \downarrow \cap \\ \mathrm{Tor}^R(A,C'') & \to & \mathrm{Tor}^R(A,C') \end{array}$$

commutes with sign $(-1)^n$, where the horizontal arrows arise from the connecting maps.

Proof. Similar to 2.17. ∎

2.19 Definitions. By the *R-dual* $(-)^* = \mathrm{Hom}_R(-, R)$ we mean the contravariant functor $\mathrm{Hom}_R(-, R)$ from the category of left R-modules to the category of right R-modules, and the same contravariant functor in the opposite direction.

Thus, for any left R-module B, the R-dual of B, $B^* = \mathrm{Hom}_R(B, R)$, is a right R-module with $(\phi r)(b) = (\phi b)r$ for all $\phi \in B^*$, $r \in R$, $b \in B$. And, for any left R-linear map $\beta: B \to C$, the *R-dual* of β, $\beta^* = \mathrm{Hom}_R(\beta, R): C^* \to B^*$, has $\beta^*(\phi) = \phi\beta \in B^*$ for all $\phi \in C^*$. If $\beta: B \to C$ and $\gamma: C \to D$ are left R-linear then, for all $\phi \in B^*$, $(\gamma\beta)^*(\phi) = \phi(\gamma\beta) = \phi\gamma\beta = \beta^*(\phi\gamma) = \beta^*(\gamma^*(\phi))$ so $(\gamma\beta)^* = \beta^*\gamma^*$.

And, similarly, starting with right R-modules.

Recall that R^n denotes the left R-module of all $1 \times n$ matrices over R. We write ${}^nR$ for the right R-module of all $n \times 1$ matrices over R. For typographical reasons we write the elements of the latter in the form $(r_1, \ldots, r_n)^t$, where t denotes the transpose.

There are natural identifications $(R^n)^* = {}^nR$, $({}^nR)^* = R^n$ and $(M \oplus N)^* = M^* \oplus N^*$. From this it follows that if P is a finitely generated projective left R-module then P^* is a finitely generated projective right R-module, and $P^{**} = P$. Also, a left R-linear map $\beta: R^n \to R^m$ corresponds to right multiplication by an $n \times m$ matrix, and the R-dual $\beta^*: {}^mR \to {}^nR$ corresponds to left multiplication by the same matrix. ∎

The following well-known result is easy to prove.

2.20 Lemma. *If P is a finitely generated projective left R-module and P^* the R-dual, then the natural maps*

$$A \otimes_R P \to \mathrm{Hom}_R(P^*, A), \quad a \otimes p \mapsto (\phi \mapsto a\phi(p))$$
$$P^* \otimes_R C \to \mathrm{Hom}_R(P, C), \quad \phi \otimes c \mapsto (p \mapsto \phi(p)c)$$

are bijective. ∎

We can now study an abstract version of Poincaré duality which will apply to groups and also to sets with group actions.

2.21 Theorem. *For any nonzero right R-module A, nonzero left R-module B, and integer $n \geqslant 0$, the following are equivalent:*

(*a*) *There exists an exact sequence*

$$(5) \qquad 0 \to Q_n \to \cdots \to Q_0 \xrightarrow{\beta} B \to 0$$

of left R-modules with each Q_i finitely generated projective, such that applying $(-)^* = \mathrm{Hom}_R(-, R)$ gives an exact sequence

$$(6) \qquad 0 \leftarrow A \xleftarrow{\alpha} Q_n^* \leftarrow \cdots \leftarrow Q_0^* \leftarrow 0$$

of right R-modules.

(*b*) *There is an element $\xi \in \mathrm{Tor}_n^R(A, B)$ such that*

$$\xi \cap - : \mathrm{Ext}_R(B, C) \to \mathrm{Tor}^R(A, C)$$

is an isomorphism for every left R-module C.

(*c*) *For each integer k, $\mathrm{Ext}_R^k(B, -) \approx \mathrm{Tor}_{n-k}^R(A, -)$, as functors of left R-modules.*

(*d*) *B is an FP left R-module and, as graded right R-module, $\mathrm{Ext}_R(B, R)$ is A concentrated in degree n.*

Proof. (*a*)⇒(*b*). Suppose we are given the exact sequences (5) and (6) and write $Q = (\oplus Q_i, \partial)$. Grade $Q^* = \oplus (Q_i)^*$ so that $(Q_i)^* = (Q^*)_{n-i}$. We then have projective resolutions $\alpha: Q^* \to A$, $\beta: Q \to B$.

The map $\alpha \otimes Q: Q^* \otimes_R Q \to A \otimes_R Q$ induces an isomorphism on homology. By Lemma 2.20, $A \otimes_R Q \to \mathrm{Hom}_R(Q^*, A)$, $a \otimes q \mapsto (\phi \mapsto a\phi(q))$, is an isomorphism, and it commutes with cycles, so induces an isomorphism on homology, although degree k is mapped to degree $n-k$. Now $H(\mathrm{Hom}_R(Q^*, A)) = \mathrm{Ext}_R(A, A)$. In degree zero this is $\mathrm{Hom}_R(A, A)$, so we can lift the identity map on A back to a homogeneous cycle $y = \sum \phi_j \otimes q_j \in Q^* \otimes_R Q$ of degree n. Let $\xi \in \mathrm{Tor}_n^R(A, B)$ denote the class of y at the homology level. We shall verify that ξ has the required property. The map $y \cap - : \mathrm{Hom}_R(Q, R) \to Q^* \otimes_R R$, $\phi \mapsto \sum \phi_j \otimes \phi(q_j)$, can be rewritten as $y \cap - : Q^* \to Q^*$, $\phi \mapsto \sum \phi_j \phi(q_j)$. This is a chain map of degree 0 since y is homogeneous.

The image of y in $A \otimes_R Q$ is $\sum \alpha(\phi_j) \otimes q_j$, and the image of the latter in $\mathrm{Hom}_R(Q^*, A)$ is $\phi \mapsto \sum \alpha(\phi_j)\phi(q_j) = \alpha(\sum \phi_j \phi(q_j)) = \alpha(y \cap \phi)$. Thus the homology class of $\alpha(y \cap -) : Q^* \to A$ is the identity map on A, which means that $y \cap - : Q^* \to Q^*$ is a lift of the identity on A, so is R-homotopic to the identity on Q^*.

Consider the composite

$$Q^* \otimes_R C \to \mathrm{Hom}_R(Q, C) \xrightarrow{y \cap -} Q^* \otimes_R C,$$

$$\phi \otimes c \mapsto \quad (q \mapsto \phi(q)c) \mapsto \sum \phi_j \otimes \phi(q_j)c = \sum \phi_j \phi(q_j) \otimes c = (y \cap \phi) \otimes c.$$

This is just $(y \cap -) \otimes C$, which is homotopic to the identity map on $Q^* \otimes_R C$ so induces the identity map on homology. By Lemma 2.20, the left hand map is an isomorphism; as it commutes with the differentials, it induces an isomorphism on homology. Thus the identity map on $H(Q^* \otimes_R C) = \mathrm{Tor}^R(A, C)$ factors through an isomorphism to $H(\mathrm{Hom}_R(Q, C)) = \mathrm{Ext}^R(B, C)$ followed by $\xi \cap -$. Hence $\xi \cap -$ is an isomorphism, and (*b*) is proved.

(*b*)$\Rightarrow$(*c*) by the naturality of the cap-product.

(*c*)$\Rightarrow$(*d*). Suppose that (*c*) holds. Each $\mathrm{Ext}_R^k(B, -) \approx \mathrm{Tor}_{n-k}^R(B, -)$ commutes with direct limits, so B is FP_∞ by Theorem 2.10. Also $\mathrm{Ext}_R^{n+1}(B, -) \approx \mathrm{Tor}_{-1}^R(A, -) = 0$ so $\mathrm{pd}_R B \leqslant n$ by Corollary 2.7, and thus B is FP. Finally, as right R-module, $\mathrm{Ext}_R(B, R) \approx \mathrm{Tor}^R(A, R)$, and $\mathrm{Tor}^R(A, R)$ is A concentrated in degree 0, so $\mathrm{Ext}_R(B, R)$ is A concentrated in degree n.

(*d*)$\Rightarrow$(*a*). Suppose that (*d*) holds. Since B is FP and $\mathrm{Ext}_R(B, R)$ is concentrated in degree n, $\mathrm{pd}_R B = n$ by the contrapositive of Corollary 2.9. Thus there is an exact sequence (5). The homology of the dual complex $0 \leftarrow Q_n^* \leftarrow \cdots \leftarrow Q_0^* \leftarrow 0$ is $\mathrm{Ext}_R(B, R)$ by definition, and this is A concentrated in degree n by hypothesis. Thus there is an exact sequence (6). ∎

2.22 Definitions. If the equivalent conditions of Theorem 2.21 are satisfied then B is said to be a *Poincaré duality R-module* with *Poincaré R-dual A* and *dimension n*. An element $\xi \in \mathrm{Tor}_n^R(A, B)$ as in (*b*) is called a *fundamental class* for B. It will be made explicit in the next result that n is unique, A is unique up to isomorphism and ξ is unique up to an automorphism of B, or of A.

For brevity, we shall sometimes say that B is a PD R-module or PD^n R-module. ∎

2.23 Proposition. *Suppose B is a Poincaré duality R-module with Poincaré R-dual A and dimension n. Then* $n = \mathrm{pd}_R B$, $A \approx \mathrm{Ext}_R^n(B, R)$.

Let T be the ring of R-linear endomorphisms of A, written on the left, so A is a (T, R)-bimodule. Similarly, let S be the ring of R-linear endomorphisms of B written on the right, so B is an (R, S)-bimodule. Then for any $\xi \in \mathrm{Tor}_n^R(A, B)$ *the following are equivalent:*

(*a*) ξ *is a fundamental class for B;*

(*b*) ξ *is a right S-basis of* $\mathrm{Tor}_n^R(A, B)$;

(*c*) ξ *is a left T-basis of* $\mathrm{Tor}_n^R(A, B)$;

(*d*) $\xi \cap - : \mathrm{Ext}_R^n(B, R) \to A$ *is an isomorphism.*

In this event there is a ring isomorphism $T \to S, t \mapsto t^{\xi}$, *such that if* $t \in T$ *then* $t\xi = \xi t^{\xi}$ *in* $\mathrm{Tor}_n^R(A, B)$.

Proof. The first part is clear from conditions (*a*) and (*d*) of Theorem 2.21.

Let $\xi \in \mathrm{Tor}_n^R(A, B)$.

(*a*)⇒(*b*). Suppose ξ is a fundamental class for B. In particular, $\xi \cap - : \mathrm{Ext}_R^0(B, B) \to \mathrm{Tor}_n^R(A, B)$ is an isomorphism, and it is S-linear by Proposition 2.14. Now $\mathrm{Ext}_R^0(B, B) = \mathrm{Hom}_R(B, B) = S$ is a free right S-module with basis 1, and it follows from the definition of the cap-product that $\xi \cap 1 = \xi$. Hence (*b*) holds.

(*b*)⇒(*a*) is clear from (*a*)⇒(*b*) and the fact that right multiplying by a unit of S, that is applying an R-linear automorphism to B, will carry a fundamental class to a fundamental class, by Proposition 2.15.

(*a*)⇒(*c*). There is a simple category theoretic argument using the isomorphism of the functors $\mathrm{Ext}_R(B, -)$ and $\mathrm{Tor}^R(A, -)$, or one can argue concretely as follows. Suppose ξ is a fundamental class for B. Since we know ξ is unique up to an automorphism of B we may assume ξ is as constructed in the proof of Theorem 2.21 (*a*)⇒(*b*). With the notation of that proof, any $t \in T$ lifts to a map $Q^* \to Q^*$, which dualizes to a map $Q \to Q$, which determines an element of S. In this way we get a map $\phi: T \to S$, which is easily seen to be an isomorphism of rings.

By Lemma 2.20, we have an isomorphism $Q^* \otimes_R B \to \mathrm{Hom}_R(Q, B)$,

$\psi \otimes b \mapsto (q \mapsto \psi(q)b)$. It induces an isomorphism on homology $\mathrm{Tor}^R(A,B) \to \mathrm{Ext}_R(B,B)$, and it is easy to see this is an isomorphism of right S-modules, and that for left actions it respects ϕ. In particular we have an isomorphism. $\mathrm{Tor}_n^R(A,B) \to \mathrm{Ext}_R^0(B,B) \approx S$ and by construction ξ maps to 1. Thus $t\xi$ and $\xi\phi(t)$ map to $\phi(t)1 = 1\phi(t)$, so $t\xi = \xi\phi(t)$. Hence (*c*) holds. Notice we have proved the final statement of the proposition also.

(*c*)$\Rightarrow$(*a*) is proved in the same way as (*b*)$\Rightarrow$(*a*).

(*a*)$\Rightarrow$(*d*) clearly.

(*d*)$\Rightarrow$(*c*). Suppose (*d*) holds. By (*a*)$\Rightarrow$(*c*) we can express ξ in the form $t\xi'$, where ξ' is a fundamental class, and $t \in T$. By Proposition 2.14, the isomorphism $\xi \cap -: \mathrm{Ext}^R(B,R) \to A$ is given by the isomorphism $\xi' \cap -: \mathrm{Ext}^R(B,R) \to A$ followed by $t: A \to A$. Thus t is a unit of T, so (*c*) holds. ∎

We shall have occasion to meet the following unusual-looking situation.

2.24 Theorem. *Let* $0 \to A' \xrightarrow{\alpha'} A \xrightarrow{\alpha''} A'' \to 0$, $0 \to B' \xrightarrow{\beta'} B \xrightarrow{\beta''} B'' \to 0$ *be exact sequences of right and left R-modules, respectively.*

Suppose there exist functions $\Phi: \mathrm{Tor}_n^R(A'',B') \to \mathrm{Tor}_n^R(A',B'')$ *and* $\Delta: \mathrm{Tor}_n^R(A'',B) \to \mathrm{Tor}_n^R(A,B)$ *such that the composite*

$$\mathrm{Tor}_n^R(A'',B) \to \mathrm{Tor}_n^R(A,B) \xrightarrow{\alpha''} \mathrm{Tor}_n^R(A'',B)$$

is the identity and the diagram

$$(7)\qquad \begin{array}{ccccc} \mathrm{Tor}_n^R(A'',B') & & \xrightarrow{\Phi} & & \mathrm{Tor}_n^R(A',B'') \\ \downarrow{\scriptstyle \beta'} & & & & \downarrow{\scriptstyle \alpha'} \\ \mathrm{Tor}_n^R(A'',B) & \xrightarrow{\Delta} & \mathrm{Tor}_n^R(A,B) & \xrightarrow{\beta''} & \mathrm{Tor}_n^R(A,B'') \end{array}$$

commutes.

Let $\xi \in \mathrm{Tor}_n^R(A'',B')$. *If* B', B'' *are* PD^n *R-modules with Poincaré dual R-modules* A'', A' *and fundamental classes* $\xi, \Phi\xi$, *respectively, then B is a* PD^n *R-module with Poincaré dual R-module A, and fundamental class* $\Delta\beta'\xi$. *Moreover there is a natural commuting diagram*

$$(8)\qquad \begin{array}{ccccccc} 0 \to \mathrm{Ext}_R^n(B'',R) & \to & \mathrm{Ext}_R^n(B,R) & \to & \mathrm{Ext}_R^n(B',R) & \to 0 \\ \downarrow{\scriptstyle \Phi\xi\cap -} & & \downarrow{\scriptstyle \Delta\beta'\xi\cap -} & & \downarrow{\scriptstyle \xi\cap -} & \\ 0 \to A' & \to & A & \to & A'' & \to 0 \end{array}$$

in which the rows are exact, and the columns are isomorphisms.

Proof. By hypothesis $\mathrm{Ext}^R(B'',R)$ and $\mathrm{Ext}^R(B',R)$ are A' and A'' concentrated in degree n. Applying $\mathrm{Ext}^R(-,R)$ to the exact sequence

$0 \to B' \to B \to B'' \to 0$ we see that $\mathrm{Ext}^R(B,R)$ is concentrated in degree n, and $0 \to \mathrm{Ext}^n_R(B'',R) \to \mathrm{Ext}^n_R(B,R) \to \mathrm{Ext}^n_R(B',R) \to 0$ is exact.

Viewing $\mathrm{Tor}^R_0(-,R)$ as the identity functor we have a diagram (8) in which the rows are exact.

We now show commutativity of (8) by showing that each of the squares is made up of two commuting triangles.

Consider the left square of (8). By Proposition 2.15, the diagonal given by $\beta''\Delta\beta'\xi \cap -$, makes the upper triangle commute. By Proposition 2.14 the diagonal given by $\alpha'\Phi\xi \cap -$ makes the lower triangle commute. But (7) commutes so $\beta''\Delta\beta'\xi = \alpha'\Phi\xi$. Thus the left square of (8) commutes.

In the right square the argument is similar and uses the fact that $\alpha''\Delta\beta'\xi = \beta'\xi$ since $\mathrm{Tor}^R_n(A'',B) \xrightarrow{\Delta} \mathrm{Tor}^R_n(A,B) \xrightarrow{\alpha''} \mathrm{Tor}^R_n(A'',B)$ is the identity.

Hence we have a commuting diagram (8). By hypothesis, the left and right columns are isomorphisms, so the middle column is an isomorphism by Theorem 2.2, which is this special case is known as the 5-lemma.

Finally, $\Delta\beta'\xi$ is a fundamental class for B, by Proposition 2.23 $(d) \Rightarrow (a)$. ∎

The next result is proved in an entirely similar fashion.

2.25 Theorem. *Let* $0 \to A' \xrightarrow{\alpha'} A \xrightarrow{\alpha''} A'' \to 0$, $0 \to B' \xrightarrow{\beta'} B \xrightarrow{\beta''} B'' \to 0$ *be exact sequences of right and left R-modules, respectively.*

Suppose there exist functions $\Phi: \mathrm{Tor}^R_{n+1}(A'',B'') \to \mathrm{Tor}^R_n(A',B)$ *and* $\Delta: \mathrm{Tor}^R_n(A'',B') \to \mathrm{Tor}^R_n(A,B')$ *such that the composite* $\mathrm{Tor}^R_n(A'',B') \xrightarrow{\Delta} \mathrm{Tor}^R_n(A,B') \xrightarrow{\alpha''} \mathrm{Tor}^R_n(A'',B')$ *is the identity and the diagram*

$$\begin{array}{ccc} \mathrm{Tor}^R_{n+1}(A'',B'') & \xrightarrow{\Phi} & \mathrm{Tor}^R_n(A',B) \\ \downarrow{\scriptstyle\beta} & & \downarrow{\scriptstyle\alpha'} \\ \mathrm{Tor}^R_n(A'',B') \xrightarrow{\Delta} & \mathrm{Tor}^R_n(A,B') & \xrightarrow{\beta'} \mathrm{Tor}^R_n(A,B) \end{array}$$

commutes, where β *is the connecting map.*

Let $\xi \in \mathrm{Tor}^R_{n+1}(A'',B'')$. *If* B'', B *are* PD *R-modules of dimension* $n+1$, n *with Poincaré dual R-modules* A'', A' *and fundamental classes* ξ, $\Phi\xi$, *respectively, then* B' *is a* PD^n *R-module with Poincaré dual R-module* A, *and fundamental class* $\Delta\beta\xi$. *Moreover there is a natural commuting diagram*

$$\begin{array}{ccccccccc} 0 \to & \mathrm{Ext}^n_R(B,R) & \to & \mathrm{Ext}^n_R(B',R) & \to & \mathrm{Ext}^{n+1}_R(B'',R) & \to 0 \\ & \downarrow{\scriptstyle \Phi\xi\cap-} & & \downarrow{\scriptstyle \Delta\beta\xi\cap-} & & \downarrow{\scriptstyle (-1)^{n+1}\xi\cap-} & \\ 0 \to & A' & \to & A & \to & A'' & \to 0 \end{array}$$

in which the rows are exact, and the columns are isomorphisms. ∎

2.26 Definitions. By an *involution* $^\circ$ of R, we mean an additive map $R \to R$, $x \mapsto x^\circ$, such that $1^\circ = 1$, and for all $x, y \in R$, $(xy)^\circ = y^\circ x^\circ$, $x^{\circ\circ} = x$.

For any left R-module M, we write M° for the right R-module having underlying abelian group M, and R-action such that $mr^\circ = rm$. This can be viewed as the action induced via $^\circ : R \to R$. We denote the identity map $M \to M^\circ$ by $m \mapsto m^\circ$, so $m^\circ r^\circ = (rm)^\circ$. In particular, there is a natural isomorphism $\mathrm{Tor}^R(C^\circ, B) \to \mathrm{Tor}^R(B^\circ, C)$ which will be denoted $\xi \mapsto \xi^\circ$.

The same notation will apply for right modules. ∎

It is straightforward to verify the following.

2.27 Proposition. *If* $^\circ$ *is an involution of* R, *and* B *is a* PD^n R*-module with Poincaré dual* C° *and fundamental class* ξ, *then* C *is a* PD^n R*-module with Poincaré dual* B° *and fundamental class* ξ°. ∎

3 PD-groups

In this section we introduce the objects of study for the chapter and note some elementary properties.

3.1 Definition. We say that G is FP *over* R if R is FP as left RG-module; if G is FP over $\mathbb{Z}$ we say simply that G is FP, and in this event G is FP over any nonzero ring. Similar terminology applies for FP_∞ and FP_n. ∎

3.2 Proposition. (i) *G is* FP_0 *over* R.

(ii) *G is* FP_1 *over R if and only if G is finitely generated.*

(iii) *If $G = F/N$ for some finitely generated free group F and normal subgroup N of F then G is* FP_2 *over R if and only if* $R \otimes_{\mathbb{Z}} N^{ab}$ *is finitely generated as left RG-module, where the G-action on* N^{ab} *arises from the F-action on N by left conjugation.*

Proof. (i) is clear.

(ii) Let S be a subset of G and X the Cayley graph with respect to S. By Lemma I.6.3, X is connected if and only if the augmented cellular chain complex $0 \to R[G \times S] \to RG \to R \to 0$ is exact at R. It follows that

S generates G if and only if $\omega RG = \sum_{s \in S} RG(s-1)$. Hence, G is finitely generated if and only if ωRG is finitely generated, which proves (ii).

(iii) Let $G = F/N$ with F of finite rank, and take S to be the image in G of a free generating set for F. Applying $R \otimes_{\mathbb{Z}} -$ to the $\mathbb{Z}$-split sequence of Theorem I.9.2 gives an exact sequence

$$0 \to R \otimes_{\mathbb{Z}} N^{ab} \to R[G \times S] \to RG \to R \to 0$$

of left RG-modules. Now (iii) is clear. ∎

3.3 Definition. Let $n \geqslant 0$ be an integer and K° a right KG-module which is free of rank one as right K-module. We say G is a *Poincaré duality group over* K of *dimension* n with *orientation module* K° if K is a PD^n KG-module with Poincaré dual KG-module K°. By Definition 2.22, this means that the following equivalent conditions are satisfied:

(*a*) There exists an exact sequence

$$0 \to \; P_n \to \cdots \to P_0 \to K \to 0$$

of left KG-modules, with each P_i finitely generated projective, such that applying $(-)^* = \mathrm{Hom}_{KG}(-, KG)$ gives rise to an exact sequence

$$0 \leftarrow K^\circ \leftarrow P_n^* \to \cdots \leftarrow P_0^* \leftarrow 0$$

of right KG-modules.

(*b*) There exists $\xi \in \mathrm{Tor}_n^{KG}(K^\circ, K)$ such that

$$\xi \cap - : \mathrm{Ext}_{KG}(K, C) \to \mathrm{Tor}^{KG}(K^\circ, C)$$

is an isomorphism for all left KG-modules C.

(*c*) For each integer k, $\mathrm{Ext}_{KG}^k(K, -) \approx \mathrm{Tor}_{n-k}^{KG}(K^\circ, -)$ as functors of left KG-modules.

(*d*) G is FP over K, and, as graded right KG-module, $\mathrm{Ext}_{KG}(K, KG)$ is K° concentrated in degree n.

For brevity we say G is a PD-group of dimension n over K, or a PD^n-group over K. Where the phrase 'over K' is omitted, we understand $K = \mathbb{Z}$.

The G-module structure of K° gives a group (anti-)homomorphism from G to the group of K-linear automorphisms of K, or, equivalently, to UK, the group of units of K. The resulting homomorphism $\varepsilon^\circ : G \to UK$ is called the *orientation map over* K. The K-linear map $KG \to KG$, $x \mapsto x^\circ$, with $g^\circ = \varepsilon^\circ(g) g^{-1}$ for all $g \in G$, is an involution of KG, called the *orientation involution*. Notice that the notation K° is compatible with that of Definition 2.26, and $\varepsilon(g^\circ) = \varepsilon^\circ(g)$.

If ε° is trivial we say that G is *orientable* over K.

The element ξ as in (*b*) is called a *fundamental class* for G. ∎

3.4 Remarks. Let the notation be as in Definition 3.3.

(i) By Proposition 2.23, K and G determine the dimension and the orientation map uniquely, and determine the fundamental class up to multiplication by a unit of K.

(ii) Since $\mathrm{cd}_K G = n$, every finite subgroup of G has order invertible in K by Proposition IV.3.8.

(iii) Since G is FP over K, and hence FP_1 over K, G is finitely generated by Proposition 3.2(ii). ∎

3.5 Notation. Let A be a right KG-module and C a left KG-module.

Choose a projective $\mathbb{Z}G$-resolution $Q \to \mathbb{Z}$ of $\mathbb{Z}$.

Since $KG \otimes_{\mathbb{Z}G} - = K \otimes_{\mathbb{Z}} -$ and $Q \to \mathbb{Z}$ is $\mathbb{Z}$-split we see that the homology of $KG \otimes_{\mathbb{Z}G} Q$ is K, so $KG \otimes_{\mathbb{Z}G} Q \to K$ is a projective KG-resolution of K.

Recall that the homology of the complex $A \otimes_{\mathbb{Z}G} Q$ is $H_*(G, A) = \bigoplus_{i \in \mathbb{Z}} H_i(G, A)$, and the homology of the complex $A \otimes_{KG} (KG \otimes_{\mathbb{Z}G} Q)$ is $\mathrm{Tor}^{KG}(A, K)$. But $A \otimes_{KG} (KG \otimes_{\mathbb{Z}G} Q) = A \otimes_{\mathbb{Z}G} Q$, so $H_*(G, A) = \mathrm{Tor}^{KG}(A, K)$ and $H_i(G, A) = \mathrm{Tor}_i^{KG}(A, K)$.

Similar $\mathrm{Ext}_{KG}(K, C) = H^*(G, C)$ and $H^i(G, C) = \mathrm{Ext}^i_{KG}(K, C)$.

In the preceding chapter, for any G-module M, we defined $H^1(G, M)$ to be the abelian group of derivations modulo inner derivations. To see the connection with the above definition, observe that $\mathrm{Ext}_{\mathbb{Z}G}(-, M)$ applied to the exact sequence $0 \to \omega\mathbb{Z}G \to \mathbb{Z}G \to \mathbb{Z} \to 0$ gives an exact sequence

$$\mathrm{Hom}_{\mathbb{Z}G}(\mathbb{Z}G, M) \to \mathrm{Hom}_{\mathbb{Z}G}(\omega\mathbb{Z}G, M) \to \mathrm{Ext}^1_{\mathbb{Z}G}(\mathbb{Z}G, M) \to 0.$$

The derivation $G \to \omega\mathbb{Z}G$, $g \mapsto g - 1$, composed with any $\mathbb{Z}G$-linear map $\omega\mathbb{Z}G \to M$, gives a derivation $G \to M$. Conversely, any derivation $d: G \to M$ determines a $\mathbb{Z}G$-linear map $\omega\mathbb{Z}G \to M$, $\sum_{g \in G} n_g g \mapsto \sum_{g \in G} n_g d(g)$. This gives a bijective correspondence between $\mathbb{Z}G$-linear homomorphisms $\omega\mathbb{Z}G \to M$ and derivations $G \to M$.

Further, the image of $\mathrm{Hom}_{\mathbb{Z}G}(\mathbb{Z}G, M) \to \mathrm{Hom}_{\mathbb{Z}G}(\omega\mathbb{Z}G, M)$ corresponds to the inner derivations.

Thus $\mathrm{Ext}^1_{\mathbb{Z}G}(\mathbb{Z}G, M)$ agrees with our previous understanding of $H^1(G, M)$. ∎

3.6 Remark. In Definition 3.3(*b*), we have $\xi \in \mathrm{Tor}_n^{KG}(K^{\circ}, K) = H_n(G, K^{\circ})$ such that $\xi \cap -: \mathrm{Ext}_{KG}(K, C) \to \mathrm{Tor}^{KG}(K^{\circ}, C)$ is an isomorphism for all left KG-modules C. There are natural identifications $H^*(G, C) = \mathrm{Ext}_{KG}(K, C)$, $H_*(G, K^\circ \otimes_K C) = \mathrm{Tor}^{KG}(K^\circ, C)$ so we can rewrite (*b*) as:

(*b'*) There exists $\xi \in H_n(G, K^\circ)$ such that $\xi \cap - : H^*(G, C) \to H_*(G, K^\circ \otimes_K C)$ is an isomorphism for all left KG-modules C.

For the case $K = \mathbb{Z}$, the orientation module is a G-module which is infinite cyclic as abelian group, usually denoted Ω, and here we have a $\xi \in H_n(G, \Omega)$ such that $\xi \cap - : H^*(G, C) \to H_*(G, \Omega \otimes_{\mathbb{Z}} C)$ is an isomorphism for all G-modules C. This is the notation used in Theorem 1.1. ∎

The next result shows that a PD-group over $\mathbb{Z}$ is a PD-group over every commutative ring.

3.7 Proposition. *If $K \to F$ is a homomorphism of commutative rings and G is a* PD^n*-group over K then G is a* PD^n*-group over F, with induced orientation.*

Proof. Let K° be the orientation module and $F^\circ = K^\circ \otimes_K F$ with diagonal G-action. There exist exact sequences

$$0 \to P_n \to \cdots \to P_0 \to K \to 0$$
$$0 \leftarrow K^\circ \leftarrow P_n^* \leftarrow \cdots \leftarrow P_0^* \leftarrow 0$$

as in Definition 3.3(*a*). These sequences are K-split, so remain exact under $F \otimes_K -$ and we have exact sequences

$$0 \to F \otimes_K P_n \to \cdots \to F \otimes_K P_0 \to F \to 0$$
$$0 \leftarrow F^\circ \leftarrow P_n^* \otimes_K F \leftarrow \cdots \leftarrow P_0^* \otimes_K F \leftarrow 0$$

of left, and right, respectively, FG-modules.

The $F \otimes_K P_i = FG \otimes_{KG} P_i$ are finitely generated projective left FG-modules.

For any left KG-module P, there is a natural map of right FG-modules $\mathrm{Hom}_{KG}(P, KG) \otimes_K F = \mathrm{Hom}_{KG}(P, KG) \otimes_{KG} FG = \mathrm{Hom}_{KG}(P, KG) \otimes_{KG} \mathrm{Hom}_{KG}(KG, FG) \to \mathrm{Hom}_{KG}(P, FG) = \mathrm{Hom}_{FG}(FG \otimes_{KG} P, FG)$. This construction respects finite direct KG-sums, and yields an isomorphism if $P = KG$, so yields an isomorphism if P is a finitely generted projective left KG-module. The result now follows. ∎

3.8 Definition. If G acts properly, and with compact quotient, on a contractible n-manifold X without boundary such that the G-stabilizer of each compact subset of X has order invertible in K, we say that G is an *n-dimensional geometric duality group over K*, or simply a GD^n-group over K. Here X has two possible orientations and there is a *geometric orientation* *map* $\varepsilon^\circ: G \to \{1, -1\}$ where, for each $g \in G$, $\varepsilon^\circ(g)$ is 1 or -1 depending as g preserves or reverses the orientation, respectively.

The GD-groups over $\mathbb{Z}$ are precisely the GD-groups of Section 1 so these are PD-groups by Theorem 1.1. Even in this case, there are examples of Davis (1983) with $n = 4$ where the contractible n-manifold X cannot be taken to be $\mathbb{R}^n$.

In general we do know if being GD^n over K implies, or is implied by, being PD^n over K. In the next section we state what we know of the cases $n = 0, 1, 2$, where the manifold X can be taken to be $\mathbb{R}^n$. ∎

4 Low-dimensional examples

We begin with a laboured statement of the zero-dimensional case.

4.1 Theorem. *The following are equivalent:*

(*a*) *G is a* GD^0*-group over K.*

(*b*) *G is a* PD^0*-group over K.*

(*c*) *G acts on a point with finite stabilizer having order invertible in K.*

(*d*) *G is finite with order invertible in K.*

(*e*) *All finite subgroups of G have order invertible in K, and G has a trivial subgroup of finite index.*

Here the geometric and algebraic orientation maps are trivial.

Proof. (*a*)$\Leftrightarrow$(*c*) is clear since the only contractible zero-manifold is $\mathbb{R}^0$.

(*c*)$\Leftrightarrow$(*d*)$\Leftrightarrow$(*e*) obviously.

(*b*)$\Rightarrow$(*d*). If (*b*) holds then $\mathrm{cd}_K G = 0$ so (*d*) holds by Proposition IV.3.5.

(*d*)$\Rightarrow$(*b*). Suppose (*d*) holds, and let K° be the right KG-module with trivial G-action. Now $K \approx KGe$, where $e = |G|^{-1} \sum_{g \in G} g$, so G is FP over K of cohomological dimension zero. The structure of $H^*(G, KG)$ as graded right KG-module is $\mathrm{Hom}_{KG}(K, KG) \approx \mathrm{Hom}_{KG}(KGe, KG) \approx eKG \approx K^\circ$ concentrated in degree zero. Thus G is a PD^0-group over K, with orientation module K°. ∎

Before considering the one-dimensional case we make some preliminary observations.

4.2 Remark. If W is G-finite G-set with finite stabilizers and $RW^* = \mathrm{Hom}_{RG}(RW, RG)$ then there is a right RG-module homomorphism $RW^* \to RW, \phi \mapsto \sum_{w \in W} \mathrm{tr}(\phi(w))w$, where $\mathrm{tr}: RG \to R$ sends $\sum_{g \in G} r_g g$ to r_1.

This can be shown to be an isomorphism with inverse $RW \to RW^*$ given by

$$\sum_{w \in W} r_w w \mapsto \left(\sum_{w \in W} s_w w \mapsto \sum_{g \in G} \sum_{w \in W} s_{gw} r_w g \right).$$

The corresponding nondegenerate bilinear form $RW \otimes_{RG} RW \to RG$ sends $w \otimes w'$ to the sum of all those $g \in G$ such that $gw = w'$. ∎

4.3 Remarks. Suppose that G is finitely generated and $\mathrm{cd}_R G = 1$, or, equivalently, that there exists an incompressible infinite G-finite G-tree T with stabilizers being finite and having order invertible in R; see Theorem IV.3.13. and Proposition IV.7.4.

Here the cellular chain complex for T gives a projective left RG-resolution $0 \to RE \xrightarrow{\partial} RV \to R \to 0$, so there is an exact sequence $RV^* \xrightarrow{\partial^*} RE^* \to H^1(G, RG) \to 0$ of right RG-modules where * is RG-dual.

By Remark 4.2 we can write this as $RV \xrightarrow{\partial^*} RE \to H^1(G, RG) \to 0$, and it is not difficult to show that $\partial^*(v) = \sum_{e \in \tau^{-1}(v)} e - \sum_{e \in \iota^{-1}(v)} e$.

At each vertex of T choose a cyclic ordering of the incident edges. Let W be the set of doubly infinite paths such that at each vertex the outgoing edge is the immediate successor of the incoming edge, in the cyclic ordering.

Of course, W is closed under the G-action if and only if the ordering is respected by the G-action, and this is possible if and only if each vertex stabilizer either cyclically permutes the incident edges or leaves them stabilized.

Since T is incompressible, it has no vertices of valency one. Hence, for each edge e, W contains a unique doubly infinite path in which e occurs oriented correctly, and we denote it $\tau^* e$. Similarly, we write $\iota^* e$ for the unique doubly infinite path in W having e oriented incorrectly. This determines a map $\delta: RE \to RW$, $e \mapsto \tau^* e - \iota^* e$, and it is not difficult to show that

(1) $$0 \to RV \xrightarrow{\partial^*} RE \xrightarrow{\delta} RW \xrightarrow{\varepsilon} R \to 0$$

is an exact sequence of R-modules.

Hence $H^1(G, RG) \approx \omega RW$ as right R-modules. If W is a G-set, this is an isomorphism of right RG-modules.

In particular, $H^1(G, RG)$ is a free R-module, of rank one or ∞ depending as G has two or infinitely many ends. (In fact this holds for any finitely generated group with more than one end; see Swan, 1969.)

Let us remark that (1) occurs as the cellular chain complex of a contractible two-dimensional CW-complex Y which can be constructed as follows.

Embed T in the plane so that at each vertex the edges are arranged clockwise in increasing order, and so that the ends of T go to the line at infinity. In this way the plane is divided into a set of regions, each one having a unique element of W as its boundary.

For the purposes of illustration one can think of a finite tree embedded in a disc with vertices of valency one on the boundary, see Fig. V.3. We then take Y to be the dual structure with one vertex for each region, and one edge crossing each edge of T joining the vertices in the regions, as illustrated in Fig. V.4.

Another way to construct Y is to thicken T and so construct a contractible two-manifold X with the elements of W being the components of ∂X. Contracting each of these boundary components to a point gives Y. ∎

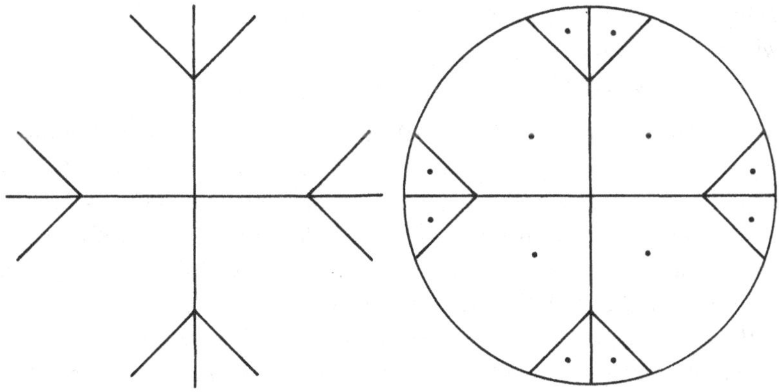

Fig. V.3

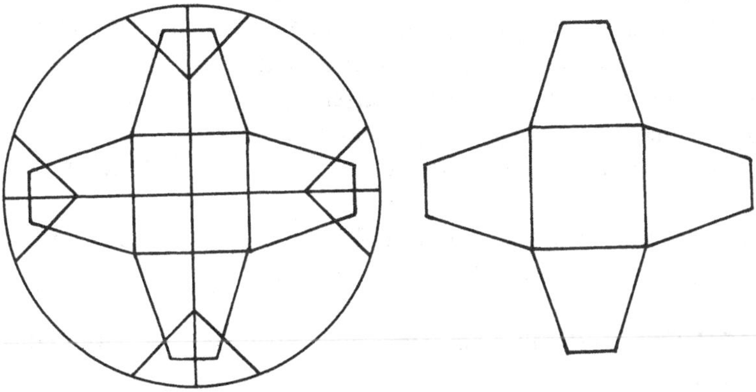

Fig. V.4

4.4 Theorem. *The following are equivalent:*

(*a*) *G is a* GD^1*-group over K.*

(*b*) *G is a* PD^1*-group over K.*

(*c*) *G acts on a tree T homeomorphic to* $\mathbb{R}$ *with one edge orbit, and stabilizers finite with order invertible in K.*

(*d*) *Either* $G = N \rtimes C_\infty$ *and N is finite with order invertible in K or* $G = A \underset{N}{*} B$ *where* $(A:N) = (B:N) = 2$ *and A, B are finite with order invertible in K.*

(*e*) *The order of every finite subgroup of G is invertible in K, and G has an infinite cyclic subgroup of finite index.*

Here the algebraic orientation map is given by the geometric orientation map.

Proof. (*a*)$\Leftrightarrow$(*c*) follows from the fact that the only contractible 1-manifold-without-boundary is $\mathbb{R}^1$.

(*c*)$\Rightarrow$(*b*). Suppose (*c*) holds. We have a finitely generated projective KG-resolution $0 \to KE \xrightarrow{\partial} KV \to K \to 0$, so G is FP over K. Let W consist of the two doubly infinite paths in T, so W is a G-set. Let $\varepsilon^\circ: G \to UK$, with $\varepsilon^\circ(g) = 1$ or -1, depending as g stabilizes or interchanges the two elements of W, and let K° be the resulting right KG-module, so $\omega KW \approx K^\circ$. It follows from Remark 4.3 that $H^1(G, KG) \approx \omega KW \approx K^\circ$ as right KG-modules, so G is a PD^1-group over K, with orientation map ε°.

(*b*)$\Rightarrow$(*c*). Suppose (*b*) holds. Since $\mathrm{cd}_K G = 1$, and $H^1(G, KG)$ is a free K-module of rank one, we see from Remark 4.3 that G acts on a tree T with two ends, with finite stabilizers having order invertible in K. It follows that (*c*) holds.

(*e*)$\Leftrightarrow$(*d*)$\Leftrightarrow$(*c*) follows from Theorem IV.6.12.

Now suppose the above equivalent conditions hold.

Here G has a finite normal subgroup N such that G/N is either C_∞ or $C_2 * C_2$. In the former case G is orientable, and in the latter case the geometric orientation map ε° is given by the map from $G/N = C_2 * C_2$ which maps each copy of C_2 isomorphically to $\{1, -1\}$. The algebraic orientation map is then given by the natural map to UK. ∎

There is some evidence that a similar result holds in dimension two.

We shall require a rather technical concept.

4.5 Definition. Recall from Zieschang, Vogt and Coldewey (1980), for example, that G is an *infinite discontinuous planar group* if there are integers

$$P \geqslant 0, \quad S \geqslant 0, \quad T \geqslant 0,$$
$$R_p \geqslant 1, \quad (p \in [1, P]),$$
$$m_s \geqslant 2, \quad (s \in [1, S]),$$
$$m_{p,r} \geqslant 2, \quad (p \in [1, P], r \in [1, R_p]),$$

such that the *Euler characteristic*

$$2 - P - T - S + \sum_{s \in [1,S]} 1/m_s - \sum_{p \in [1,P]} \sum_{r \in [1,R_p]} (m_{p,r} - 1)/2m_{p,r}$$

is not positive, and G has a presentation with

generators:

$\pi_p, \rho_{p,r}, \sigma_s, \tau_t (p \in [1, P], r \in [1, R_p], s \in [1, S], t \in [1, T])$,

relators:

$(\rho_{p,r})^2, (\rho_{p,r-1}\rho_{p,r})^{m_{p,r}}, (p \in [1, P], r \in [1, R_p])$, where $\rho_{p,0}$ denotes $\pi_p^{-1} \rho_{p,R_p} \pi_p$,

$\sigma_s^{m_s}, \quad (s \in [1, S])$,

and either (A) $\pi_1 \cdots \pi_p \sigma_1 \cdots \sigma_s [\tau_1, \tau_2] \cdots [\tau_{T-1}, \tau_T]$ and T is even, or (B) $\pi_1 \cdots \pi_p \sigma_1 \cdots \sigma_s \tau_1^2 \cdots \tau_T^2$ and $T \geqslant 1$. ∎

4.6 Conjecture. The following are equivalent:

(*a*) G is a GD^2-group over K.

(*b*) G is a PD^2-group over K.

(*c*) G acts on an oriented CW-complex homeomorphic to $\mathbb{R}^2$, with finite quotient, and cell stabilizers finite with order invertible in K.

(*d*) G has a nice presentation.

(*e*) The order of every finite subgroup of G is invertible in K, and G has an infinite surface group as a subgroup of finite index.

Discussion. (*a*)⇔(*c*) is true; see Zieschang, Vogt and Coldewey (1980).

(*c*)⇔(*d*) is true in the sense that G is a GD^2-group if and only if there is a finite normal subgroup N such that G/N is as in Definition 4.5, and the integers $|N|$, m_s and $2m_{p,r}$ are all invertible in K. Moreover, the orientation map ε° on G/N is described by $\varepsilon^\circ(\pi_p) = 1$, $\varepsilon^\circ(\rho_{p,r}) = -1$, $\varepsilon^\circ(\sigma_s) = 1$ and $\varepsilon^\circ(\tau_t)$ is 1 in case (A) and -1 in case (B).

The presentation for G/N arises from the structure theorem for groups acting on simply connected CW-complexes mentioned in Remark I.9.5, and it is theoretically possible to give a presentation for G for the same reason.

$(a) \Rightarrow (b)$ is true, and here condition 3.3(a) arises naturally. Since $(a) \Rightarrow (c)$ we can give $\mathbb{R}^2$ the structure of an oriented two-dimensional CW-complex X on which G acts.

Let X° be the dual CW-complex constructed as follows. In each face f of X choose a vertex f°. Each edge e of X is common to two faces f, f', and we introduce an edge e° joining f°, f'° crossing only the edge e. For each vertex v of X, let v° be the region whose boundary is given by the edges e° such that e is incident to v; see Fig. V.4, page 161.

Orient X° as follows. For each edge e of X, orient e° so that it crosses e from the right to the left, relative to the given orientation of e. For each vertex v of X, orient v° clockwise.

Since $\mathbb{R}^2$ is contractible, the augmented cellular chain complexes of X, X°

(2) $$0 \to KF \xrightarrow{\partial} KE \xrightarrow{\partial} KV \to K \to 0$$

(3) $$0 \to K[V^\circ] \xrightarrow{\partial^\circ} K[E^\circ] \xrightarrow{\partial^\circ} K[F^\circ] \to K \to 0$$

are exact.

For each face f of X, $\partial f = \sum_{e \in E} \langle f, e \rangle e$ where $\langle f, e \rangle$ is $1, -1$, or 0 depending as e, e^{-1}, or neither, respectively, occurs in the path around the boundary of f in the direction given by the orientation of f. Similarly, for each edge e of X, $\partial e = \sum_{v \in V} \langle e, v \rangle v$, where $\langle e, v \rangle$ is $1, -1$, or 0 depending as $\tau e, \iota e$, or neither, respectively, is v.

By our choice of orientation of X° we see that, for each face v° of X°, $\partial^\circ(v^\circ) = \sum_{e^\circ \in E^\circ} \langle e, v \rangle e^\circ$, and, for each edge e° of X°, $\partial^\circ(e^\circ) = \sum_{f^\circ \in F^\circ} \langle f, e \rangle \varepsilon(f) f^\circ$, where $\varepsilon(f) = 1$ or -1 depending as f is oriented anti-clockwise or clockwise, respectively. This is illustrated in Fig. V.5.

We now look at the implications for KG-modules.

Let $\varepsilon^\circ : G \to \{\pm 1\}$ with $\varepsilon^\circ(g)$ being 1 or -1 if g respects or reverses the two orientations of $\mathbb{R}^2$, respectively. Let K° be the corresponding KG-module.

In a natural way G acts on X°, and as G-sets $V^\circ = V$, $E^\circ = E$, $F^\circ = F$. However, G need not respect the orientation of X°; specifically, if $\varepsilon^\circ(g) = -1$ then g reverses the orientations of edges and faces in X°. Thus (3) can be viewed as an exact sequence of KG-modules

$$0 \to K^\circ \otimes_K K[V^\circ] \xrightarrow{\partial^\circ} K^\circ \otimes_K K[E^\circ] \xrightarrow{\partial^\circ} K[F^\circ] \to K \to 0.$$

Applying $K^\circ \otimes_K -$ we get another exact sequence of KG-modules

$$0 \to K[V^\circ] \xrightarrow{\partial^\circ} K[E^\circ] \xrightarrow{\partial^\circ} K^\circ \otimes_K K[F^\circ] \to K^\circ \to 0.$$

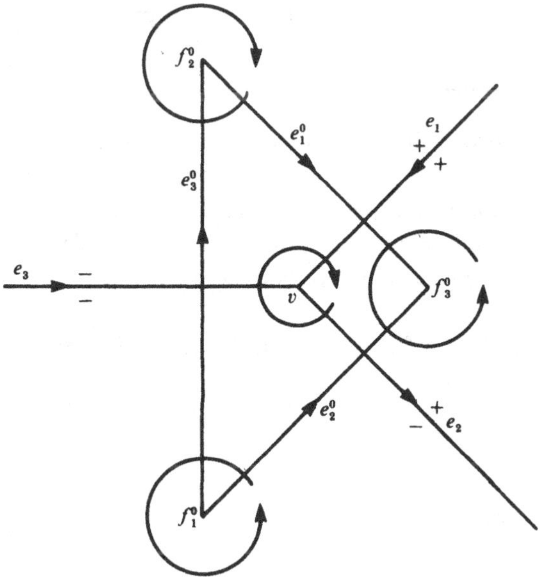

Fig. V.5

Let us identify the KG-modules $K[V^\circ] = KV$, $K[E^\circ] = KE$ in the obvious way and $K^\circ \otimes_K K[F^\circ] \mathrel{\bar{=}} KF$ via $1 \otimes \varepsilon^\circ(f) f^\circ = f$. Then we get an exact sequence of KG-modules

(4) $\qquad 0 \to KV \xrightarrow{\partial^*} KE \xrightarrow{\partial^*} KF \to K^\circ \to 0,$

where $\partial^*(v) = \sum_{e \in E} \langle e, v \rangle e$, $\partial^*(e) = \sum_{f \in F} \langle f, e \rangle f$.

Since G respects the orientation of X, (2) is a sequence of KG-modules, and it is a projective KG-resolution since the cell-stabilizers are finite with order invertible in K. By Remark 4.2, the KG-dual of the maps in (2) are as in (4), and hence $H(G, KG)$ is K° concentrated in degree 0.

$(a) \Rightarrow (e)$ is true. For a proof that every discontinuous planar group has a surface subgroup of finite index see, for example, Zieschang, Vogt and Coldewey (1980, section 4.10). It is then an exercise in combinatorial group theory to verify that every finite-by-discontinuous planar group has a surface subgroup of finite index.

$(b) \Rightarrow (a)$ is known in the case where G acts on a tree without stabilizing a vertex, see Eckmann and Müller (1982); this includes the case $K = \mathbb{Z}$ which will be proved in Theorem 12.4.

$(b) \Rightarrow (e)$ is known only in the cases where $(b) \Rightarrow (a)$ is known.

$(e) \Rightarrow (a)$ is true in the case where the centralizer of the surface subgroup in the ambient group is just the centre of the surface group, by the theorem of Kerckhoff (1983).

(*e*)$\Rightarrow$(*b*) is true by Theorem 5.5. ∎

Our objective is to prove the torsion-free case of Conjecture 4.6, where the groups are precisely the infinite surface groups. It is of some interest to have a purely algebraic analysis of the infinite surface groups, and the remainder of this introduction will be devoted to this. We first give an algebraic construction of the cellular chain complex (2).

4.7 Definitions. Let F be a free group on $x_1, \ldots, x_m$. For each $i \in [1, m]$, there exists a unique derivation

$$\frac{\partial}{\partial x_i}: F \to \mathbb{Z}F$$

such that for each $j \in [1, m]$,

$$\frac{\partial x_j}{\partial x_i} = \delta_{ij},$$

the Kronecker delta. Using the uniqueness of these derivations, it is easy to show that

$$r - 1 = \sum_{i=1}^{m} \frac{\partial r}{\partial x_i}(x_i - 1),$$

for all $r \in F$. To make this more suggestive we define $\partial r = r - 1$, and the preceding formula becomes

$$\partial r = \sum_{i=1}^{m} \frac{\partial r}{\partial x_i} \partial x_i.$$

Consider any group G generated by elements $x_1, \ldots, x_m$, and any word r in $x_1, \ldots, x_m$ and their inverses. Then there is a corresponding element of F, again denoted r, and hence there are elements $\partial r/\partial x_i$ of $\mathbb{Z}F$. The images of these elements under the natural map $\mathbb{Z}F \to \mathbb{Z}G$ are also denoted $\partial r/\partial x_i$. Since the context is generally clear, there is usually no ambiguity. ∎

4.8 Lemma. *If* $G = \langle x_1, \ldots, x_m | r \rangle$, *where* $m \geqslant 2$ *and* $r = x_1^{-1} x_2^{-1} x_1 x_2^{\pm 1} w$ *for some word* w *in* $x_3, \ldots, x_m$ *then there is an exact sequence of left* $\mathbb{Z}G$-modules

$$(5) \qquad 0 \to \mathbb{Z}G \xrightarrow{(p_1, \ldots, p_m)} \mathbb{Z}G^m \xrightarrow{(q_1, \ldots, q_m)^t} \mathbb{Z}G \to \mathbb{Z} \to 0$$

with $p_i = \partial r/\partial x_i$ *and* $q_i = x_i - 1$ *for each* $i \in [1, m]$.

Proof. Notice that $\sum_i p_iq_i = 0$ in $\mathbb{Z}G$, so (5) is a complex.

Let A be the free group on $x_2, x_3, \ldots, x_m$ and C the free subgroup on x_2. There is an embedding $x_1: C \to A$, $x_2 \mapsto x_2^{\pm 1}w$, and we can identify G with the resulting HNN-extension $G = A \underset{C}{*} x_1$. In particular, A is a subgroup of G.

Consider the commuting diagram

$$
\begin{array}{ccccc}
 & 0 & & 0 & \\
 & \downarrow & & \downarrow & \\
 & \mathbb{Z}G & \xrightarrow{(p_2,\ldots,p_m)} & \mathbb{Z}G^{m-1} & \\
 & \downarrow{\scriptstyle -p_1} & & \downarrow{\scriptstyle (q_2,\ldots,q_m)^t} & \\
 & \mathbb{Z}G & \xrightarrow{q_1} & \mathbb{Z}G & \\
 & \downarrow & & \downarrow & \\
0\to & \mathbb{Z}[G/C] & \xrightarrow{q_1} & \mathbb{Z}[G/A] & \to \mathbb{Z} \to 0 \\
 & \downarrow & & \downarrow & \\
 & 0 & & 0 &
\end{array}
$$

where the lower arrow labelled q_1 sends gC to $gq_1A = gx_1A - gA$. This map is well-defined because $gx_1A = x_1A$ for all $g \in C$. The diagram commutes because $\sum_i p_iq_i = 0$. Since C, A are free groups, their augmentation ideals are free; since $-p_1 = x_1^{-1} - x_1^{-1}x_2^{-1} = x_1^{-1}x_2^{-1}(x_2 - 1)$ we deduce that the vertical sequences are exact. The horizontal sequence is the exact sequence associated with the tree for the HNN-extension. The exactness of (5) now follows by diagram chasing. ∎

4.9 Theorem. (i) *If* $G = \langle x_1, \ldots, x_m | [x_1, x_2] \cdots [x_{m-1}, x_m] \rangle$ *with* $m \geqslant 2$ *and even, then* G *is an orientable* PD^2*-group and there is an exact sequence of left* $\mathbb{Z}G$*-modules* $0 \to \mathbb{Z}G \to \mathbb{Z}G^m \to \mathbb{Z}G \to \mathbb{Z} \to 0$.

(ii) *If* $G = \langle x_1, \ldots, x_m | x_1^2 \cdots x_m^2 \rangle$ *with* $m \geqslant 2$ *then* G *is a* PD^2*-group with* $\varepsilon^\circ(x_i) = -1$ *for each* $i \in [1, m]$, *and there is an exact sequence of left* $\mathbb{Z}G$*-modules* $0 \to \mathbb{Z}G \to \mathbb{Z}G^m \to \mathbb{Z}G \to \mathbb{Z} \to 0$.

Proof. (ii) Let $z_1 = x_1^{-1}$, $z_2 = x_2^{-1}x_1^{-1}$, so $x_1^2 \cdots x_m^2 = z_1^{-1}z_2^{-1}z_1z_2^{-1}x_3^2 \cdots x_m^2$. Here Lemma 4.8 applies to give an exact sequence of $\mathbb{Z}G$-modules, and using elementary matrix manipulations we can bring it to the form

$$(6) \qquad 0 \to \mathbb{Z}G \xrightarrow{(p_1,\ldots,p_m)} \mathbb{Z}G^m \xrightarrow{(q_1,\ldots,q_m)^t} \mathbb{Z}G \to \mathbb{Z} \to 0$$

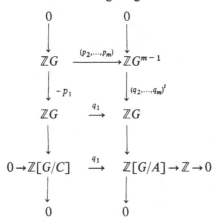

with $p_i = x_1^2 \cdots x_{i-1}^2(x_i + 1)$ and $q_i = x_i - 1$ for each $i \in [1, m]$.

For each $i \in [1, m]$, let $y_i = x_1^2 \cdots x_{i-1}^2 x_i x_{i-1}^{-2} \cdots x_1^{-2}$. We shall show by induction that for each $i \in [1, m]$

(7) $$x_1^2 \cdots x_i^2 = y_i \cdots y_1 x_1 \cdots x_i = y_i^2 \cdots y_1^2.$$

This is clear for $i = 1$ since $y_1 = x_1$. Now suppose that (7) holds for some $i \in [1, m]$ and let w_i denote the common value of the expressions in (7). Then $y_{i+1} = w_i x_{i+1} w_i^{-1}$ so $w_i x_{i+1} = y_{i+1} w_i$ and we see $w_i x_{i+1}^2 = y_{i+1} w_i x_{i+1} = y_{i+1}^2 w_i$. This readily gives the analogue of (7) with $i+1$ replacing i. By induction, (7) holds for all $i \in [1, m]$.

It follows from the case $i = m$ of (7) that we can define an anti-endomorphism $^\circ$ of $\mathbb{Z}G$ with $x_i^\circ = -y_i$ for each $i \in [1, m]$. Notice that each w_i is fixed by $^\circ$, so $x_{i+1}^{\circ\circ} = -(w_i^\circ)^{-1}(x_{i+1}^\circ)(w_i^\circ) = -w_i^{-1}(-y_{i+1})w_i = x_{i+1}$, and therefore $^\circ$ is an involution of $\mathbb{Z}G$.

From the left equality in (7) we see that $y_i \cdots y_1 = x_1^2 \cdots x_i^2 x_i^{-1} \cdots x_1^{-1} = x_1^2 \cdots x_{i-1}^2 x_i x_{i-1}^{-1} \cdots x_1^{-1}$, and thus

$$\begin{aligned}(x_1 \cdots x_{i-1} q_i)^\circ &= (x_1 \cdots x_i)^\circ - (x_1 \cdots x_{i-1})^\circ \\ &= (-1)^i x_1^2 \cdots x_{i-1}^2 x_i x_{i-1}^{-1} \cdots x_1^{-1} \\ &\quad - (-1)^{i-1} x_1^2 \cdots x_{i-1}^2 x_{i-1}^{-1} \cdots x_1^{-1} \\ &= (-1)^i x_1^2 \cdots x_{i-1}^2 (x_i + 1)(x_1 \cdots x_{i-1})^{-1} \\ &= (-1)^i p_i (x_1 \cdots x_{i-1})^{-1}.\end{aligned}$$

The sequence (6) remains exact if we replace each q_i with the element $a_i = x_1 \cdots x_{i-1} q_i$ and each p_i with the corresponding element $p_i(x_1 \cdots x_{i-1})^{-1} = (-1)^i a_i^\circ$. Hence, we have an exact sequence of left $\mathbb{Z}G$-modules

(8) $$0 \to \mathbb{Z}G \xrightarrow{(-a_1^\circ, \ldots, (-1)^m a_m^\circ)} \mathbb{Z}G^m \xrightarrow{(a_1, \ldots, a_m)^t} \mathbb{Z}G \to \mathbb{Z} \to 0.$$

Applying the involution $^\circ$ gives an exact sequence of right $\mathbb{Z}G$-modules

$$0 \to \mathbb{Z}G \xrightarrow{(-a_1, \ldots, (-1)^m a_m)^t} {}^m\mathbb{Z}G \xrightarrow{(a_1^\circ, \ldots, a_m^\circ)} \mathbb{Z}G \to \mathbb{Z}^\circ \to 0.$$

After adjusting signs we get

$$0 \to \mathbb{Z}G \xrightarrow{(a_1, \ldots, a_m)^t} {}^m\mathbb{Z}G \xrightarrow{(-a_1^\circ, \ldots, (-1)^m a_m^\circ)} \mathbb{Z}G \to \mathbb{Z}^\circ \to 0,$$

which is the $\mathbb{Z}G$-dual of the resolution in (8). By Definition 3.3(*a*), G is a PD^2-group with orientation module $\mathbb{Z}^\circ$. In particular $\varepsilon^\circ(x_i) = \varepsilon(x_i^\circ) = -1$ for each $i \in [1, m]$.

The proof of (i) is very similar to that for (ii) and proceeds as follows. It can be shown that there is an involution $^\circ$ of $\mathbb{Z}G$ such that for each even $i \in [0, m-2]$

$$x_{i+1}^{\circ} = w_{i+1} x_{i+2} w_{i+1}^{-1} \quad \text{with} \quad w_{i+1} = [x_1, x_2] \cdots [x_{i-1}, x_i] x_{i+1}^{-1},$$
$$x_{i+2}^{\circ} = w_{i+2} x_{i+1} w_{i+2}^{-1} \quad \text{with} \quad w_{i+2} = [x_1, x_2] \cdots [x_{i-1}, x_i] x_{i+1}^{-1} x_{i+2}^{-1}.$$

By Lemma 4.8, there is an exact sequence of left $\mathbb{Z}G$-modules (5), and this can be brought to the form

$$0 \to \mathbb{Z}G \xrightarrow{(-a_1^{\circ}, \ldots, (-1)^m a_m^{\circ})} \mathbb{Z}G^m \xrightarrow{(a_1, \ldots, a_m)^t} \mathbb{Z}G \to \mathbb{Z} \to 0,$$

where $a_i = x_1^{-1} x_2 x_3^{-1} x_4 \cdots x_i^{(-1)^i} - x_1^{-1} x_2 x_3^{-1} x_4 \cdots x_{i-1}^{(-1)^{i-1}}$ for each $i \in [1, m]$. It then follows that G is an orientable PD^2-group. ∎

For later reference we note that we have proved the following.

4.10 Corollary. *If G is an infinite surface group then G is a PD^2-group over K; moreover, for some $m \geqslant 2$ there is an exact sequence $0 \to KG \to KG^m \to KG \to K \to 0$ of left KG-modules.* ∎

Let us observe another application of the above arguments.

4.11 Theorem. *Let $m \geqslant 1$, and let G be the free group on $x_1, \ldots, x_m$. If $z_1, \ldots, z_m$ are elements of G such that either $z_1^2 \cdots z_m^2 = x_1^2 \cdots x_m^2$ or m is even and $[z_1, z_2] \cdots [z_{m-1}, z_m] = [x_1, x_2] \cdots [x_{m-1}, x_m]$ then $z_1, \ldots, z_m$ freely generate G.*

Proof. We consider only the case $z_1^2 \cdots z_m^2 = x_1^2 \cdots x_m^2$, since the other case is similar.

It suffices to show that the subgroup H of G generated by $z_1, \ldots, z_m$ is all of G.

Let $i \in [1, m]$, and consider the derivation

$$D = \frac{\partial}{\partial x_i} : G \to \mathbb{Z}_2 G$$

of Definition 4.7. Since $D(x_1^2 \cdots x_m^2) = D(z_1^2 \cdots z_m^2)$ it follows that $x_1^2 \cdots x_{i-1}^2 (x_i + 1) = \sum_{j=1}^{m} z_1^2 \cdots z_{j-1}^2 (z_j + 1) D(z_j)$. Under the natural map $\mathbb{Z}_2 G \to \mathbb{Z}_2 [H \backslash G]$ this becomes

$$\begin{aligned} H x_1^2 \cdots x_{i-1}^2 (x_i + 1) &= \sum_{j=1}^{m} H z_1^2 \cdots z_{j-1}^2 (z_j + 1) D(z_j) \\ &= \sum_{j=1}^{m} H(1+1) D(z_j) = 0. \end{aligned}$$

Hence $Hx_1^2\cdots x_{i-1}^2x_i = Hx_1^2\cdots x_{i-1}^2$ so $x_1^2\cdots x_{i-1}^2x_i(x_1^2\cdots x_{i-1}^2)^{-1}\in H$. As this holds for all $i\in[1,m]$ we see $H=G$. ▮

5 Subgroups of finite index

We want to show that Poincaré duality behaves well with respect to finite index.

5.1 Proposition. *If H is a subgroup of G of finite index then H is* FP_∞ *over R if and only if G is* FP_∞ *over R.*

Proof. Suppose that G is FP_∞ over R, so there exists an exact sequence of left RG-modules

(1) $\cdots\to P_1\to P_0\to R\to 0$

with each P_i a finitely generated projective RG-module. Then (1) is an exact sequence of left RH-modules, with each P_i finitely generated projective as left RH-module. Thus H is FP_∞ over R.

Conversely, suppose that H is FP_∞ over R, and that, for some $k\geqslant 0$, we have an exact sequence of left RG-modules

$$0\to B_k\to Q_{k-1}\to\cdots\to Q_0\to R\to 0,$$

with each Q_i a finitely generated projective RG-module. Then the Q_i are finitely generated projective as RH-modules, so B_k is finitely generated as RH-module, so is finitely generated as RG-module. Thus H is FP_∞ over R. ▮

The main result of this section is a corresponding statement for cohomological dimension, and here we must restrict to a commutative coefficient ring. Recall the following.

5.2 Serre's Construction. Let H be a subgroup of G of finite index, say $n=(G:H)$. We now review a method for constructing a certain exact sequence of left KG-modules out of an exact sequence of left KH-modules

$$\cdots\xrightarrow{\partial}K[W_1]\xrightarrow{\partial}K[W_0]\to K\to 0,$$

where the W_d are H-sets.

Write $W=\bigvee_{d\geqslant 0}W_d$ and $V=W^n$, the Cartesian product of $n=(G:H)$ copies of W.

Throughout the discussion we let v represent an arbitrary element

$(w_1, \dots, w_n)$ of V, with $w_i \in W_{d_i}$, $i = 1, \dots, n$. There is a natural grading on V given by setting $\deg c = d_1 + \cdots + d_n$ We identify the resulting graded K-module KV with the nth tensor power $KW^{\otimes n}$ by identifying $v = w_1 \otimes \cdots \otimes w_n$. Since K is commutative, ∂ is a K-bimodule map, so there is a differential on KV given by

$$\partial v = \sum_{i=1}^{n} (-1)^{d_1 + \cdots + d_{i-1}} w_1 \otimes \cdots \otimes \partial w_i \otimes \cdots \otimes w_n,$$

where ∂ vanishes on W_0. It follows by induction, using the assertions in Definition 2.11, that the homology of the complex KV is simply K concentrated in degree zero.

The next step is to extend the K-module structure of KV to a KG-module structure. Choose a transversal $x_1, \dots, x_n$ for the right H-action on G, with $x_1 = 1$. Consider any $g \in G$; for each $i \in \{1, \dots, n\}$, there is a unique expression $gx_i = x_{gi}h_i$, where $gi \in \{1, \dots, n\}$ and $h_i \in H$. Thus G acts on $\{1, \dots, n\}$. We define

$$g(w_1 \otimes \cdots \otimes w_n) = (-1)^{\sum d_i d_j} h_{g^{-1}1} w_{g^{-1}1} \otimes \cdots \otimes h_{g^{-1}n} w_{g^{-1}n},$$

where the summation is over $\{(i, j) : 1 \leqslant i < j \leqslant n, gi > gj\}$. It is a straightforward tedious task to verify that this defines a KG-action on KV; notice that the unsigned G-action can be viewed as coming from the embedding $V \to \mathbb{Z}[G \otimes_H W]$, $w_1 \otimes \cdots \otimes w_n \mapsto x_1 \otimes w_1 + \cdots + x_n \otimes w_n$. It is clear that the action respects the grading, and it is another task to show that it respects the differential. The G-action on $H_*(KV)$ is easily seen to be trivial, so we have an exact sequence of left KG-modules $\cdots \xrightarrow{\partial} K[V_1] \xrightarrow{\partial} K[V_0] \to K \to 0$.

To conclude, we comment on the KG-module structure of KV. By the definition of the G-action, $V \cup -V$ is a G-set. Thus KV is a direct sum of KG-modules of the form $KGv \approx KG \otimes_{KG_{\pm v}} Kv$, where $G_{\pm v} = \{g \in G \mid gv \in \{v, -v\}\}$. Let $\gamma_v : G_{\pm v} \to \{\pm 1\}$ so that $\gamma_v(g)v = gv$.

Notice that we have an H-map $V \cup -V \to W$, $\pm w_1 \otimes \cdots \otimes w_n \mapsto w_1$, because H centralizes $x_1 = 1$. In particular, if W has finite H-stabilizers then $V \cup -V$ also has finite H-stabilizers, and therefore has finite G-stabilizers.

If the order of $G_{\pm v}$ happens to be finite and invertible in K then the $KG_{\pm v}$-linear map $KG_{\pm v} \to Kv$, $r \mapsto rv$, is split by the $KG_{\pm v}$-linear map $Kv \to KG_{\pm v}$, $rv \mapsto |G_{\pm v}|^{-1} \sum_{g \in G_{\pm v}} r\gamma_v(g)g$, so Kv is a projective $KG_{\pm v}$-module and KGv is a projective KG-module.

We remark that, even if W is H-finite, V will not be G-finite unless G is finite. ∎

5.3 Serre's Extension Theorem. *If the order of each finite subgroup of G is invertible in K, and H is a subgroup of G of finite index then* $\mathrm{cd}_K H = \mathrm{cd}_K G$.

Proof. By Lemma IV.3.18(i), it suffices to show that if $\mathrm{cd}_K H$ is finite then so is $\mathrm{cd}_K G$. By adding identity maps of large free modules to a finite length projective KH-resolution of K we may assume that we have a free KH-resolution of K of finite length, m say. We are now in the situation of Serre's Construction 5.2, where the W_d are free H-sets, and are empty for $d > m$. Thus we obtain a projective KG-resolution of K of length $m(G:H)$, and so $\mathrm{cd}_K G$ is finite, as desired. ∎

We can combine Proposition 5.1 and Theorem 5.3.

5.4 Theorem. *If H is a subgroup of G of finite index then G is* FP *over K if and only if H is* FP *over K and the order of each finite subgroup of G is invertible in K.*

5.5 Theorem. *Let H be a subgroup of G of finite index.*

Then G is a PD^n*-group over K if and only if K contains the inverse of the order of each finite subgroup of G, and H is a* PD^n*-group over K.*

In this event, H has the orientation induced from G.

Proof. By Remark 4.2, the (KG, KH)-bimodule map

$$\mathrm{Hom}_{KH}({}_{KH}KG_{KG}, KH) \to KG, \phi \mapsto \sum_{gH \in G/H} g\phi(g^{-1})$$

is an isomorphism. Thus, for any left KG-module M, there are natural isomorphisms of right KH-modules

$$\begin{aligned}\mathrm{Hom}_{KG}(M, KG) &\approx \mathrm{Hom}_{KG}(M, \mathrm{Hom}_{KH}(KG, KH))\\ &\approx \mathrm{Hom}_{KH}(KG \otimes_{KG} M, KH) \approx \mathrm{Hom}_{KH}(M, KH).\end{aligned}$$

Let $P \to K \to 0$ be a projective KG-resolution of K. By the naturality of the above isomorphisms, we have an isomorphism of differential graded right KH-modules $\mathrm{Hom}_{KG}(P, KG) \approx \mathrm{Hom}_{KH}(P, KH)$. On taking homology we see $\mathrm{Ext}_{KG}(K, KG) \approx \mathrm{Ext}_{KH}(K, KH)$, because $P \to K \to 0$ is also a projective KH-resolution of K.

Now suppose that G is a PD^n-group over K with orientation module K°. Since G is FP over K, H is clearly FP over K. From the isomorphism of graded right KH-modules, $\mathrm{Ext}_{KH}(K, KH) \approx \mathrm{Ext}_{KG}(K, KG) \approx K^\circ$, we see that H is a PD^n-group over K with induced orientation.

Conversely, suppose that H is a PD^n-group over K with orientation module K°. By Theorem 5.4, G is FP over K, and we have seen that there is an isomorphism of graded KH-modules $\mathrm{Ext}_{KG}(K, KG) \approx \mathrm{Ext}_{KH}(K, KH)$, and the latter is K° concentrated in degree n. Hence, G is a PD^n-group over K. ∎

6 Subgroups of infinite index

In this section it will be shown that the subgroups of infinite index in a PD^n-group over K have cohomological dimension over K at most $n-1$.

We begin with a result on G-trees.

6.1 Theorem. *If T is a G-free locally finite G-finite G-tree such that $G\backslash VT$ is infinite then the R-linear map $\partial^*: RVT \to RET$, $v \mapsto \sum_{e \in \tau^{-1}(v)} e - \sum_{e \in \iota^{-1}(v)} e$, is an RG-split embedding.*

Proof. Let us write V, E for VT, ET, respectively.

It is clear that ∂^* is RG-linear.

Let Y be a fundamental G-transversal in T, with subtree Y_0. By hypothesis the set $V_0 = VY_0 = VY$ is infinite, so Y_0 is an infinite locally finite tree, and $Y - Y_0$ consists of edges which have only one vertex in Y.

Choose a vertex v_0 of Y_0 in such a way that all components of $Y_0 - (\{v_0\} \cup \mathrm{star}(v_0))$ are infinite; such a choice is possible, for if the property fails for some vertex of Y_0 then Y_0 must have a vertex of valency one, which we can take as our v_0.

For each $v \in V_0$ we choose a reduced infinite path P_v in Y_0 with initial vertex v, and by the choice of v_0 we can arrange that if $v \neq v_0$ then P_v does not pass through v_0.

Because G acts freely on V, we can define $P_{gv} = gP_v$ for all $(g, v) \in G \times V_0$.

In summary, for each vertex v of the original tree T, we have a reduced infinite path P_v in T, with initial vertex v. Further, $gP_v = P_{gv}$ for all $g \in G$, $v \in V$, and if $v \in VY_0 - \{v_0\}$ then P_v does not pass through v_0.

We claim that each edge e of T belongs to P_v for only finitely many $v \in V$. Suppose $v \in V$ such that $e \in P_v$. There is a unique $g \in G$ such that $gv \in Y_0$, and then $ge \in gP_v = P_{gv}$. Hence $ge \in EY_0$, and gv lies in the geodesic joining ge to v_0. Therefore g is the unique element of G such that $ge \in EY_0$, and v lies in the geodesic joining e to $g^{-1}v_0$. Thus there are only finitely many possibilities for v, as desired.

Construct a function $\varepsilon: E \times V \to R$ by setting $\varepsilon(e, v) = -1, 1$, or 0 if e^1, e^{-1}, or neither, occurs in P_v, respectively.

Consider any edge e of T. Since e occurs in P_v for only finitely many vertices v, there are only finitely many $v \in V$ such that $\varepsilon(e, v) \neq 0$. Hence there exists an R-linear map $\phi: RE \to RV$ defined by $\phi(e) = \sum_{v \in V} \varepsilon(e, v)v$. For all $g \in G$, $v \in V$, $e \in E$, the fact that $P_{gv} = gP_v$ implies that $\varepsilon(g^{-1}e, v) = \varepsilon(e, gv)$, and hence $\phi: RE \to RV$ is left RG-linear.

From the definition, it is straightforward to check that if $w, v \in V$ then $\sum_{e \in \tau^{-1}(w)} \varepsilon(e, v) - \sum_{e \in \iota^{-1}(w)} \varepsilon(e, v) = \delta_{v,w}$, so $\phi(\partial^*(w)) = w$. Thus $\phi: RE \to RV$ is a left inverse of ∂. ∎

This has a consequence for groups.

6.2 Theorem. *Let G be generated by a finite set $X = \{x_1, \ldots, x_n\}$, and let $\partial^*: RG \to RG^n$ be right multiplication by the row $(x_1 - 1, \ldots, x_n - 1)$. If H is a subgroup of G of infinite index then ∂^* is a right RH-split embedding.*

Proof. Consider first the case where G is freely generated by X.

Let T be the Cayley graph of G with respect to X^{-1}, so T has vertex set $V = G$ and edge set $E = G \times X$, where $\iota(g, x) = g$, $\tau(g, x) = gx^{-1}$. Notice that T is a locally finite G-finite G-free G-tree.

There are identifications $RV = RG$, $RE = RG^n$, with $(g, x_i) \in E$ corresponding to the row in RG^n with g in the ith place and zeros elsewhere. With this identification, for any $g \in V$, $\partial^*(g) = (gx_1 - g, \ldots, gx_n - g) = \sum_{e \in \tau^{-1}(g)} e - \sum_{e \in \iota^{-1}(g)} e$.

If H is any subgroup of G of infinite index then H acts freely on T and $H \backslash V$ is infinite. By Theorem 6.1, there exists an RH-linear map $\phi: RG^n \to RG$ which is a left inverse of ∂^*.

Now consider any normal subgroup N of G contained in H and let $\bar{G} = G/N$, $\bar{H} = H/N$. There is a natural identification $R\bar{H} \otimes_{RH} RG = R \otimes_{RN} RH \otimes_{RH} RG = R \otimes_{RN} RG = R\bar{G}$ so $R\bar{H} \otimes_{RH} \phi: R\bar{G} \to R\bar{G}^n$ may be viewed as the map given by right multiplication by $(x_1 - 1, \ldots, x_n - 1)$. But $R\bar{H} \otimes_{RH} \partial^*$ has an $R\bar{H}$-linear left inverse $R\bar{H} \otimes_{RH} \phi$. This encompasses the general case. ∎

6.3 Theorem. *If G is a PD^n-group over K and H a subgroup of G of infinite index then $\mathrm{cd}_K H \leqslant n - 1$.*

Proof. If $n = 0$ then G is finite by Theorem 4.1, and this case is vacuous. Thus we may assume that $n \geqslant 1$.

Let $X = \{x_1, \ldots, x_m\}$ be a finite generating set of G and let $^\circ$ be the orientation involution.

Let $\partial: KG^m \to KG$ be right multiplication by the column vector $(x_1 - 1, .., x_m - 1)^t$. Then the cokernel of ∂ is K, and we can extend ∂ to a resolution of K by finitely generated projective left KG-modules

$$0 \to P_n \to \cdots \to P_2 \to KG^m \xrightarrow{\partial} KG \to K \to 0.$$

By Poincaré duality, there is a resolution of K° by finitely generated projective right KG-modules

$$0 \to {}^mKG \xrightarrow{\partial^*} KG \to P_2^* \to \cdots \to P_n^* \to K^\circ \to 0,$$

and ∂^* is left multiplication by $(x_1 - 1, \ldots, x_m - 1)^t$. By the left–right dual of Theorem 6.2, ∂^* is a KH-split embedding, so $\mathrm{pd}_{KH} K^\circ \leqslant n - 1$. On applying the involution we see $\mathrm{pd}_{KH} K \leqslant n - 1$, that is $\mathrm{cd}_K H \leqslant n - 1$. ∎

Suppose that G is a PD^2-group over K, and that H is subgroup of G. If H has finite index then H is a PD^2-group over K by Theorem 5.5. If H has infinite index then $\mathrm{cd}_K H \leqslant 1$ by the above, and here we know the structure of H by Theorem IV.3.13. We record the torsion-free case, which includes the case $K = \mathbb{Z}$.

6.4 Corollary. *If G is a torsion-free* PD^2*-group over K and H a subgroup of G then either H has finite index and is a* PD^2*-group over K, or H has infinite index and is free.* ∎

Using Corollary 4.10, we get a nonhomological consequence.

6.5 Corollary. *In a surface group, every subgroup of infinite index is free.* ∎

7 PD-pairs

Throughout this section let W be a G-set, $n \geqslant 0$ an integer, K° a right KG-module which is free of rank one as right K-module, $\varepsilon^\circ: G \to UK$ the corresponding map and $^\circ$ the corresponding involution of KG.

7.1 Definition. We say (G, W) is a *Poincaré duality pair over K of dimension n with orientation module K°* if either W is empty and K is a PD^n KG-module with Poincaré dual KG-module K°, or W is nonempty, $n \geqslant 1$, and ωKW is a PD^{n-1} KG-module with Poincaré dual KG-module K°.

The former case gives the PD-groups studied in Section 3; the latter case means that the following equivalent conditions are satisfied:

(a) There exists an exact sequence

$$0 \to P_n \to \cdots \to P_1 \to KW \to K \to 0$$

of left KG-modules, with each P_i finitely generated projective, such that applying $(-)^* = \mathrm{Hom}_{KG}(-, KG)$ gives rise to an exact sequence

$$0 \leftarrow K^\circ \leftarrow P_n^* \leftarrow \cdots \leftarrow P_1^* \leftarrow 0$$

of right KG-modules.

(b) There exists $\xi \in \mathrm{Tor}_{n-1}^{KG}(K^\circ, \omega KW)$ such that

$$\xi \cap - : \mathrm{Ext}_{KG}(\omega KW, C) \to \mathrm{Tor}^{KG}(K^\circ, C)$$

is an isomorphism for all left KG-modules C.

(c) For each integer k, $\mathrm{Ext}_{KG}^{k+1}(\omega KW, -) \approx \mathrm{Tor}_{n-k}^{KG}(K^\circ, -)$ as functors of left KG-modules.

(d) G is FP over K, and, as graded right KG-module, $\mathrm{Ext}_{KG}(\omega KW, KG)$ is K° concentrated in degree $n-1$.

For brevity we say (G, W) is a PD-pair over K, or a PD^n-pair over K. Where the phrase 'over K' is omitted, we understand $K = \mathbb{Z}$.

We call $\varepsilon^\circ : G \to UK$ the *orientation map* and $^\circ$ the *orientation involution* of KG. If ε° is trivial we say that (G, W) is *orientable* over K.

The element ξ as in (b) is called a *fundamental class* for (G, W). ∎

7.2 Remarks. Suppose (G, W) is a PD^n-pair over K.

(i) By Proposition 2.23, K and (G, W) determine the dimension and the orientation map uniquely, and determine the fundamental class up to multiplication by a unit of K.

(ii) If W is empty then $\mathrm{cd}_K G = n$. If W is nonempty then $\mathrm{cd}_K G = \mathrm{pd}_{KG} K = \mathrm{pd}_{KG} K^\circ = n-1$. Thus every finite subgroup of G has order invertible in K by Proposition IV.3.8.

(iii) G is FP over K. Hence it is FP$_1$ over K, so G is finitely generated by Proposition 3.2(ii).

(iv) ωKW is an FP left KG-module. In particular, it is finitely generated, and so is KW, and thus W is G-finite. ∎

7.3 Notation. Let A be a right KG-module and C a left KG-module.

For the empty set we define $H^k((G, \varnothing), C) = H^k(G, C) = \mathrm{Ext}_{\mathbb{Z}G}^k(\mathbb{Z}, C) = \mathrm{Ext}_{KG}^k(K, C)$ and $H_k((G, \varnothing), A) = H_k(G, A) = \mathrm{Tor}_k^{\mathbb{Z}G}(A, \mathbb{Z}) = \mathrm{Tor}_k^{KG}(A, K)$.

If W is nonempty we define $K^k((G, W), C) = \mathrm{Ext}_{\mathbb{Z}G}^{k-1}(\omega \mathbb{Z} W, C) = \mathrm{Ext}_{KG}^{k-1}(\omega KW, C)$ and $H_k((G, W), A) = \mathrm{Tor}_{k-1}^{\mathbb{Z}G}(A, \omega \mathbb{Z} W) = \mathrm{Tor}_{k-1}^{KG}(A, \omega KW)$.

Thus in Definition 7.1(b), we have $\xi \in \mathrm{Tor}_{n-1}^{KG}(K^\circ, \omega KW)$ such that $\xi \cap - : \mathrm{Ext}_{KG}^{k-1}(\omega WK, C) \to \mathrm{Tor}_{n-k}^{KG}(K^\circ, C)$ is an isomorphism for all integers k. Now $\mathrm{Tor}_{n-1}^{KG}(K^\circ, \omega KW) = H_n((G,W),K^\circ)$, $\mathrm{Ext}_{KG}^{k-1}(\omega KW, C) = H^k((G,W),C)$, and there is a natural identification $\mathrm{Tor}_{n-k}^{KG}(K^\circ, C) = \mathrm{Tor}_{n-k}^{KG}(K^\circ \otimes_K C, K) = H_k(G, K^\circ \otimes_K C)$. Thus $\xi \cap - : H^k((G,W),C) \to H_{n-k}(G, K^\circ \otimes_K C)$ is an isomorphism for all integers k. Moreover, the involution determines an isomorphism $\mathrm{Tor}_{n-1}^{KG}(K^\circ, \omega KW) \approx \mathrm{Tor}_{n-1}^{KG}(\omega KW^\circ, K)$, $\xi \mapsto \xi^\circ$, and by Proposition 2.27, K is a PD^{n-1} KG-module with Poincaré dual KG-module ωKW° and fundamental class ξ°. Sometimes we treat the map $\mathrm{Tor}_{n-1}^{KG}(K^\circ, \omega KW) \approx \mathrm{Tor}_{n-1}^{KG}(\omega KW^\circ, K)$, $\xi \mapsto \xi^\circ$, as an identification and write $\xi^\circ = \xi$, and then $\xi \cap - : H^k(G, C) \to H_{n-k}((G,W), K^\circ \otimes_K C)$ is an isomorphism for all integers k.

This allows a unified description of PD-pairs.

For any $\xi \in H_n((G,W),K^\circ)$ the following are equivalent:

(*a*) (G,W) is a PD^n-pair over K with orientation module K° and fundamental class ξ.

(*b*) $\xi \cap - : H^k((G,W),C) \to H_{n-k}(G, K^\circ \otimes_K C)$ is an isomorphism for all left KG-modules C, and all integers k.

(*c*) $\xi \cap - : H^k(G,C) \to H_{n-k}((G,W), K^\circ \otimes_K C)$ is an isomorphism for all left KG-modules C, and all integers k.

For $K = \mathbb{Z}$ we write the orientation module as Ω, giving us the notation of Theorem 1.3. ∎

Using the same proof as for Proposition 3.7, we see that a PD-pair over $\mathbb{Z}$ is a PD-pair over each commutative ring.

7.4 Proposition. *If $K \to F$ is a homomorphism of commutative rings and (G,W) is a* PD^n*-pair over K then (G,W) is a* PD^n*-pair over F with induced orientation.* ∎

7.5 Definition. If G acts properly, and with compact quotient, on a contractible n-manifold X whose boundary components are contractible, such that the G-stabilizer of each compact subset of X has order invertible in K, and W is the G-set of components of ∂X then we say that (G,W) is an *n-dimensional geometric duality pair over* K, or simply a GD^n-pair *over* K. Here X has two possible orientations and there is a *geometric orientation map* $\xi^\circ : G \to \{1, -1\}$, where, for each $g \in G$, $\varepsilon^\circ(g)$ is 1 or -1 depending as g preserves or reverses the orientation, respectively. ∎

7.6 Remarks. The GD^n-pairs over $\mathbb{Z}$ are precisely the GD^n-pairs of Section 1 so these are PD^n-pairs by Theorem 1.3. We shall see that the converse is true for $n \leqslant 2$.

For arbitrary K, we do not know if (G,W) being GD^n over K implies, or is implied by being PD^n over K, although much is known for $n \leqslant 2$.

Being GD^0 over K is equivalent to being PD^0 over K. Here X is a point and W is empty. See Theorem 4.1.

Being GD^1 over K is equivalent to being PD^1 over K. Here X is a circle or a line segment, and W is empty or consists of two endpoints, respectively. See Theorems 4.4 and 7.7.

If W is nonempty, then being GD^2 over K is equivalent to being PD^2 over K, by the arguments of Eckmann and Müller (1982); the case where W is empty was discussed in Conjecture 4.6.

It follows from Eckmann and Müller (1982) that a GD^2-pair over K can be described algebraically as follows: G has a finite normal subgroup N such that $\bar{G} = G/N$ has a presentation which looks like that of Definition 4.5 but with the m_s and $m_{p,r}$ being allowed to take the value ∞; $|N|$ and all the m_s, $2m_{p,r}$ are invertible in K (with the convention that ∞ is invertible); G is infinite (that is, either some m_s or $m_{p,r}$ is ∞, or $\bar{G}$ is as in Definition 4.5);

$$W = \bigvee_{s:m_s=\infty} \bar{G}/\langle \sigma_s \rangle \vee \bigvee_{(p,r):m_{p,r}=\infty} \bar{G}/\langle \rho_{p,r-1}, \rho_{p,r} \rangle.$$

As in Zieschang, Vogt and Coldewey (1980), section 4.7, one can construct a two-manifold X which is a CW-complex and satisfies the conditions of Definitions 7.5. In fact, by treating half-planes as if they were two-cells for the 'infinite relations', one can attach half-planes along each boundary component of X to yield a plane. Thus, depending as W is empty or not, X is a plane or a thickened tree lying in the plane; cf. Example 1.4(iii) and Remark 4.3. ∎

7.7 Theorem. *If W is nonempty then (G, W) is a* PD^1*-pair over K if and only if W consists of two elements, and G is finite with order invertible in K.*

Proof. Suppose that (G, W) is a PD^1-pair over K. Here $\mathrm{cd}_K G = 0$, so G is finite with order invertible in K. Also, there are K-module isomorphisms $\omega KW \approx \mathrm{Ext}_{KG}(K, KG) = \mathrm{Hom}_{KG}(K, KG) \approx K$, so ωKW is a free K-module of rank one. Since K is commutative and nonzero, W must have exactly two elements.

Conversely suppose that G is finite with order invertible in K and W consists of two elements, v, w. Then ωKW is a free K-module on $v - w$. Define $\varepsilon^\circ: G \to UK$ by setting $\varepsilon^\circ(g)$ to be 1 if g stabilizes v and w, and -1 if g interchanges v and w. Let $e = |G|^{-1} \sum_{g \in G} \varepsilon^\circ(g) g \in KG$. Then $\omega KW \approx KGe$ so $\mathrm{Ext}_{KG}(\omega KW, KG) \approx \mathrm{Ext}_{KG}(KGe, KG) = \mathrm{Hom}_{KG}(KGe, KG) \approx$

$eKG \approx K^\circ$ concentrated in degree zero. Hence (G, W) is a PD^1-pair over K with orientation map ε°. ∎

To study PD^2-pairs we need to develop some theory.

7.8 Lemma. *For any left KG-modules B, C such that B is free as K-module there is a natural isomorphism* $\mathrm{Tor}^{KG}(B^\circ, C) \approx \mathrm{Tor}^{KG}(K^\circ, B \otimes_K C)$, *where $B \otimes_K C$ has diagonal G-action.*

Proof. Consider the isomorphism of K-modules $B \otimes_K KG \to KG \otimes_K B$, $b \otimes g \mapsto g \otimes g^{-1}b$. The diagonal G-action on $B \otimes_K KG$ is carried to the action on $KG \otimes_K B$ in which G acts trivially on B, that is the KG-module structure induced from the K-module structure of B. But B is free as K-module, so $KG \otimes_K B$ is free as KG-module. Thus $B \otimes_K KG$ is free as KG-module.

Let $Q \to C$ be a free KG-resolution. The map $K^\circ \otimes_{KG}(B \otimes_K Q) \to B^\circ \otimes_{KG} Q$, $\lambda^\circ \otimes (b \otimes q) \to (\lambda b)^\circ \otimes q$ is easily seen to be an isomorphism of complexes. Since B is free as right K-module, $B \otimes_K Q \to B \otimes_K C$ is exact, and, by the foregoing, $B \otimes_K Q$ is a free KG-resolution of $B \otimes_K C$. Hence, $\mathrm{Tor}^{KG}(K^\circ, B \otimes_K C) \approx H(K^\circ \otimes_{KG}(B \otimes_K Q)) \approx H(B^\circ \otimes_{KG} Q) \approx \mathrm{Tor}^{KG}(B^\circ, C)$. ∎

7.9 Definition. A left KG-module B is said to be *W-graded over* K if there is specified a K-module decomposition $B = \bigoplus_W B_w$, where each B_w is a KG_w-submodule of B, and $gB_w = B_{gw}$ for all $g \in G$, $w \in W$.

The *diagonal* map $\Delta: B \to KW \otimes_K B$ is the unique additive map such that if $b \in B_w$ then $\Delta b = w \otimes b$.

For example, KW is W-graded over K, with $(KW)_w = Kw$, and $\Delta w = w \otimes w$.

Notice Δ is KG-linear, where G acts diagonally on $KW \otimes_K B$, and the composite $B \xrightarrow{\Delta} KW \otimes_K B \xrightarrow{\varepsilon \otimes 1} B$ is the identity.

Let $w \in W$.

There is a *projection* map $B \to KG \otimes_{KG_w} B_w$ which sends gb to $g \otimes b$ for all $g \in G$, $b \in B_w$, and vanishes on $B_{w'}$ for all $w' \in W - Gw$.

The inclusion map $B_w \to B$ and the projection $KG \to KG_w$ together induce a map $\mathrm{Ext}_{KG}(B, KG) \to \mathrm{Ext}_{KG_w}(B_w, KG_w)$ which we denote $\eta \mapsto \eta_w$.

Let $g \in G$. There is a ring isomorphism $KG_w \approx KG_{gw}$ by left conjugation, and a module isomorphism $B_w \approx gB_w = B_{gw}$ by left multiplication by g, with the module isomorphism respecting the ring isomorphism.

We denote the resulting canonical isomorphism of K-modules $\mathrm{Ext}_{KG_w}(B_w, KG_w) \to \mathrm{Ext}_{KG_{gw}}(B_{gw}, KG_{gw})$ by $\eta \mapsto \eta g^{-1}$. This defines a right KG-module structure on $\bigoplus_W \mathrm{Ext}_{KG_w}(B_w, KG_w)$, and it is W-graded over K since $\mathrm{Ext}_{KG_w}(B_w, KG_w)g^{-1} = \mathrm{Ext}_{KG_{gw}}(B_{gw}, KG_{gw})$. ∎

7.10 Lemma. *Let B be a left KG-module which is W-graded over K, and $\xi \in \mathrm{Tor}^{KG}(K^\circ, B)$.*

(i) *Applying $\mathrm{Tor}^{KG}(K^\circ, -)$ to the diagonal map $\Delta: B \to KW \otimes_K B$ induces a map $\mathrm{Tor}^{KG}(K^\circ, B) \to \mathrm{Tor}^{KG}(KW^\circ, B)$, $\xi \mapsto \Delta\xi$.*

(ii) *For each $w \in W$, applying $\mathrm{Tor}^{KG}(K^\circ, -)$ to the projection map $B \to KG \otimes_{KG_w} B_w$ induces a map $\mathrm{Tor}^{KG}(K^\circ, B) \to \mathrm{Tor}^{KG_w}(K^\circ, B_w)$, $\xi \mapsto \xi_w$,*

(iii) *Suppose B_w is nonzero for all $w \in W$. Then B is FP_n as left KG-module if and only if W is G-finite and B_w is FP_n as left KG_w-module for each $w \in W$. In this event there is an isomorphism of right KG-modules $\mathrm{Ext}^n_{KG}(B, KG) \approx \bigoplus_W \mathrm{Ext}^n_{KG_w}(B_w, KG_w)$, $\eta \mapsto \sum_W \eta_w$.*

(iv) *B is a PD^n KG-module with Poincaré dual KG-module KW° and fundamental class $\Delta\xi$ if and only if W is G-finite and, for each $w \in W$, B_w is a PD^n KG_w-module with Poincaré dual KG_w-module K° and fundamental class ξ_w. In this event there is a commuting diagram of isomorphisms of right KG-modules*

$$\begin{array}{ccc} \mathrm{Ext}^n_{KG}(B, KG) & \approx & \bigoplus_W \mathrm{Ext}^n_{KG_w}(B_w, KG_w) \\ \wr\downarrow \Delta\xi \cap - & & \wr\downarrow \bigoplus_W (\xi_w \cap -) \\ KW^\circ & \approx & \bigoplus_W K^\circ. \end{array}$$

Proof. (i), (ii) are clear.

(iii). Let W_0 be a G-transversal in W so $B \approx \bigoplus_{W_0} KG \otimes_{KG_w} B_w$.

We consider first the case $n = 0$.

If each B_w is finitely generated as KG_w-module and W_0 is finite, then clearly B is finitely generated as left KG-module. Conversely, if B is finitely generated as left KG-module then each B_w is finitely generated, and, since each B_w is nonzero, we see that W_0 is finite. In this event we have K-module isomorphisms

$$\mathrm{Hom}_{KG}(B, KG) \approx \mathrm{Hom}_{KG}\Big(\bigoplus_{W_0} KG \otimes_{KG_w} B_w, KG\Big)$$

$$\approx \bigoplus_{W_0} \mathrm{Hom}_{KG}(KG \otimes_{KG_w} B_w, KG), \text{ since } W_0 \text{ is finite}$$

$$\approx \bigoplus_{W_0} \mathrm{Hom}_{KG_w}(B_w, KG)$$

$$= \bigoplus_{w \in W_0} \mathrm{Hom}_{KG_w}\left(B_w, \bigoplus_{gw \in Gw} KG_w g^{-1} \right)$$

$$\approx \bigoplus_{w \in W_0} \bigoplus_{gw \in Gw} \mathrm{Hom}_{KG_w}(B_w, KG_w g^{-1}),$$

since B_w is finitely generated,

$$\approx \bigoplus_{w \in W_0} \bigoplus_{gw \in Gw} \mathrm{Hom}_{KG_{gw}}(B_{gw}, KG_{gw})$$

$$= \bigoplus_{w \in W} \mathrm{Hom}_{KG_w}(B_w, KG_w).$$

Thus $\mathrm{Hom}_{KG}(B, KG) \approx \bigoplus_{w \in W} \mathrm{Hom}_{KG_w}(B_w, KG_w)$. It is straightforward to verify that this isomorphism has the form described, and respects the right KG-module structure. This completes the proof for $n = 0$.

If for each $w \in W_0$, we have a projective KG_w-resolution $P_w \to B_w$ then $\bigoplus_{W_0} KG \otimes_{KG_w} P_w \to B$ is a projective KG-resolution; by using Lemma 2.6 we get the result for general n.

(iv) We shall work with Theorem 2.21(d).

Let $P \to K^\circ$ be a projective right KG-resolution, and form a projective right KG-resolution $\bigoplus_W P \to KW^\circ$. As in (iii), we can form a projective left KG-resolution $\bigoplus_W Q_w \to B$ such that, for each $w \in W$, $Q_w \to B_w$ is projective left KG_w-resolution, and $gQ_w = Q_{gw}$. Lift ξ back to a cycle in $P \otimes_{KG}\left(\bigoplus_W Q_w \right)$.

The diagonal map $P \otimes_{KG}\left(\bigoplus_W Q_w \right) \to \left(\bigoplus_W P \right) \otimes_{KG}\left(\bigoplus_W Q_w \right)$, $p \otimes q_w \mapsto p_w \otimes q_w$, gives a lift of $\Delta\xi$.

Let $w \in W$. The map $P \otimes_{KG}\left(\bigoplus_W Q_w \right) \to P \otimes_{KG}(KG \otimes_{KG_w} Q_w) \approx P \otimes_{KG_w} Q_w$ gives a lift of ξ_w.

It is now straightforward to verify that the diagram

$$\begin{array}{ccc} \mathrm{Ext}_{KG}(B, KG) & \to & \mathrm{Ext}_{KG_w}(B_w, KG_w) \\ \downarrow{\scriptstyle \Delta\xi \cap -} & & \downarrow{\scriptstyle \xi_w \cap -} \\ KW^\circ & \to & w^\circ K = K^\circ \end{array}$$

commutes.

If B is an FP-module we then have a commuting diagram

$$\begin{array}{ccc} \mathrm{Ext}_{KG}(B, KG) & \approx & \bigoplus_W \mathrm{Ext}_{KG_w}(B_w, KG_w) \\ \downarrow{\scriptstyle \Delta\xi \cap -} & & \downarrow{\scriptstyle \bigoplus_W (\xi_w \cap -)} \\ KW^\circ & \approx & \bigoplus_W K^\circ. \end{array}$$

The left column is an isomorphism if and only if the right column is an isomorphism and here each B_w is nonzero since K° is nonzero. The result now follows easily. ∎

7.11 Theorem. *If W is a nonempty G-set such that (G, W) is a PD^n-pair over K then for each $w \in W$, G_w is a PD^{n-1}-group over K with induced orientation. Moreover, the natural K-linear map*

$$H^{n-1}(G, KG) \to \bigoplus_W H^{n-1}(G_w, KG_w) \approx \bigoplus_W K \approx KW$$

is injective and has image ωKW.

Proof. Let $^\circ$ be the orientation involution for (G, W) over K, and $\xi \in \mathrm{Tor}_{n-1}^{KG}(K^\circ, \omega KW)$ a fundamental class for (G, W).

We shall apply Theorem 2.24 to the exact sequences

$$0 \to \omega KW^\circ \xrightarrow{\alpha'} KW^\circ \xrightarrow{\alpha''} K^\circ \to 0 \quad \text{and}$$

$$0 \to \omega KW \xrightarrow{\beta'} KW \xrightarrow{\beta''} K \to 0.$$

There is a natural commuting diagram

$$\begin{array}{ccccc} K \otimes_K \omega KW & & \to & & \omega KW \otimes_K K \\ \downarrow & & & & \downarrow \\ K \otimes_K KW & \xrightarrow{\Delta} & KW \otimes_K KW & \to & KW \otimes_K K. \end{array}$$

of left KG-modules with diagonal G-action, where $\Delta w = w \otimes w$. Applying $\mathrm{Tor}^{KG}(K^\circ, -)$ and, using Lemma 7.8, we have a commuting diagram

$$\begin{array}{ccccc} \mathrm{Tor}^{KG}(K^\circ, \omega KW) & & \xrightarrow{\circ} & & \mathrm{Tor}^{KG}(\omega WK^\circ, K) \\ \downarrow{\scriptstyle \beta'} & & & & \downarrow{\scriptstyle \alpha'} \\ \mathrm{Tor}^{KG}(K^\circ, KW) & \xrightarrow{\Delta} & \mathrm{Tor}^{KG}(KW^\circ, KW) & \xrightarrow{\beta''} & \mathrm{Tor}^{KG}(KW^\circ, K). \end{array}$$

The composite $\mathrm{Tor}^{KG}(K^\circ, KW) \xrightarrow{\Delta} \mathrm{Tor}^{KG}(KW^\circ, KW) \xrightarrow{\alpha''} \mathrm{Tor}^{KG}(K^\circ, KW)$ is the identity.

The top arrow arises from the involution, and sends ξ to ξ°. By Proposition 2.27, we know that K is a PD^{n-1} KG-module with Poincaré dual KG-module ωKW° and fundamental class ξ°.

Thus Theorem 2.24 applies, and KW is a PD^{n-1} KG-module with Poincaré KG-dual KW° and fundamental class $\Delta\beta'\xi$.

By Lemma 7.10(iv), for each $w \in W$, K is a PD^{n-1} KG_w-module with

Poincaré KG_w-dual K° and fundamental class $(\beta'\xi)_w$. Thus G_w is a PD^{n-1}-group over K with induced orientation, and fundamental class $(\beta'\xi)_w$.

Also by Theorem 2.24 there is a commuting diagram

$$\begin{array}{ccccccccc} 0 \to & \mathrm{Ext}^{n-1}_{KG}(K, KG) & \to & \mathrm{Ext}^{n-1}_{KG}(KW, KG) & \to & \mathrm{Ext}^{n-1}_{KG}(\omega KW, KG) & \to 0 \\ & \wr\downarrow \xi^\circ\cap - & & \wr\downarrow \Delta\beta'\xi\cap - & & \wr\downarrow \xi\cap - & \\ 0 \to & \omega KW^\circ & \to & KW^\circ & \to & K^\circ & \to 0, \end{array}$$

which by Lemma 7.10(iv) can be rewritten

$$\begin{array}{ccccccccc} 0 \to & \mathrm{Ext}^{n-1}_{KG}(K, KG) & \to & \bigoplus\limits_W \mathrm{Ext}^{n-1}_{KG_w}(K, KG_w) & \to & \mathrm{Ext}^{n-1}_{KG}(\omega KW, KG) & \to 0 \\ & \wr\downarrow \xi^\circ\cap - & & \wr\downarrow \bigoplus((\beta'\xi)_w\cap -) & & \wr\downarrow \xi\cap - & \\ 0 \to & \omega KW^\circ & \to & \bigoplus\limits_W K^\circ & \to & K^\circ & \to 0, \end{array}$$

The commutativity of the left square gives the final part of the theorem. ∎

8 Trees and Poincaré duality

We now want to see how PD-groups acting on trees give rise to PD-pairs.

Throughout this section let n be an integer and T a G-tree, and write $E = ET$, $V = VT$.

8.1 Lemma. *There is a V-grading over K, $\omega KE = \bigoplus\limits_{v\in V} \omega KE_v$, where $E_v = \mathrm{star}(v)$ for each $v \in V$, and hence there is a diagonal map $\Delta_V: \omega KE \to KV \otimes_K \omega KE$.*

If E is nonempty, then for any right KG-module A there is a commuting diagram

(1)
$$\begin{array}{ccccc} \mathrm{Tor}^{KG}_n(A, K) & \xrightarrow{\delta_V} & \mathrm{Tor}^{KG}_{n-1}(A, KE) & \xrightarrow{-\Delta_E} & \mathrm{Tor}^{KG}_{n-1}(A, KE\otimes_K KE) \\ & \delta_E & & & \downarrow \partial \\ \mathrm{Tor}^{KG}_{n-1}(A, \omega KE) & \xrightarrow{\Delta_V} & \mathrm{Tor}^{KG}_{n-1}(A, KV\otimes_K \omega KE) & \to & \mathrm{Tor}^{KG}_{n-1}(A, KV\otimes_K KE) \end{array}$$

where δ_E, δ_V are the respective connecting maps arising from the exact sequences

(2) $0 \to \omega KE \to KE \to K \to 0,$

(3) $0 \to KE \to KV \to K \to 0.$

Proof. If $v_0, e_1^{\varepsilon_1}, v_1, \ldots, v_{m-1}, e_m^{\varepsilon_m}, v_m$ is any path in T then $e_m - e_1 = \sum\limits_{i=1}^{m-1} (e_{i+1} - e_i)$ with $e_{i+1} - e_i \in \omega KE_{v_i}$. One can then use this to construct an inverse to the natural map $\bigoplus\limits_{v\in V} \omega KE_v \to \omega KE$.

Now assume E is nonempty, so (2) is exact, and δ_E is defined.

Let $e \in E$. Write $v = \iota e$ and consider the diagram

$$\begin{array}{ccc} \omega KG \xrightarrow{\cdot v} \omega KV \approx KE & \xrightarrow{\Delta_E} & KE \otimes_K KE \\ \downarrow{\cdot e} & & \downarrow{\partial \otimes 1} \\ \omega KE \xrightarrow{\Delta_V} KV \otimes_K \omega KE & \rightarrow & KV \otimes_K KE \end{array} \tag{4}$$

where $\cdot m$ indicates the map $r \mapsto rm$. We shall show that the sum of the upper and lower routes is $\cdot(v \otimes e)$.

Let $g \in G$, and construct a path of the form $v_0 = v$, $e_1 = e$, $v_1, e_2^{\varepsilon_2}, \ldots$, $v_{m-1} = gv$, $e_m = ge$, v_m in T.

Along the upper route

$$g - 1 \mapsto gv - v \mapsto \sum_{i=1}^{m-1} \varepsilon_i e_i \mapsto \sum_{i=1}^{m-1} \varepsilon_i e_i \otimes e_i$$

$$\mapsto \sum_{i=1}^{m-1} \varepsilon_i(\tau e_i - \iota e_i) \otimes e_i = \sum_{i=1}^{m-1} (v_i - v_{i-1}) \otimes e_i.$$

Along the lower route

$$g - 1 \mapsto ge - e \mapsto \sum_{i=1}^{m-1} v_i \otimes (e_{i+1} - e_i).$$

The sum is

$$\sum_{i=1}^{m-1} (v_i - v_{i-1}) \otimes e_i + \sum_{i=1}^{m-1} v_i \otimes (e_{i+1} - e_i)$$
$$= -v_0 \otimes e_1 + v_{m-1} \otimes e_m$$
$$= -v \otimes e + gv \otimes ge = (g-1) \cdot (v \otimes e).$$

Applying $\mathrm{Tor}^{KG}(A, -)$ to the exact sequence

$$0 \rightarrow \omega KG \rightarrow KG \rightarrow K \rightarrow 0 \tag{5}$$

induces another connecting map $\delta: \mathrm{Tor}_n^{KG}(A, K) \rightarrow \mathrm{Tor}_{n-1}^{KG}(A, \omega KG)$. Now $\cdot e$ induces a map from (5) to (2) giving a commuting diagram

$$\begin{array}{ccc} \mathrm{Tor}_n^{KG}(A, K) & \xrightarrow{\delta} & \mathrm{Tor}_{n-1}^{KG}(A, \omega KG) \\ \| & & \downarrow{\cdot e} \\ \mathrm{Tor}_n^{KG}(A, K) & \xrightarrow{\delta_E} & \mathrm{Tor}_{n-1}^{KG}(A, \omega KE). \end{array}$$

Thus $(\cdot e)\delta = \delta_E$. Similarly $(\cdot v)\delta = \delta_V$.

Applying $\mathrm{Tor}_{n-1}^{KG}(A, -)$ to (4), we get a diagram

$$\begin{array}{ccccc} \mathrm{Tor}_{n-1}^{KG}(A, \omega KG) & \xrightarrow{\cdot v} & \mathrm{Tor}_{n-1}^{KG}(A, KE) & \xrightarrow{\Delta_E} & \mathrm{Tor}_{n-1}^{KG}(A, KE \otimes_K KE) \\ \downarrow{\cdot e} & & & & \downarrow \\ \mathrm{Tor}_{n-1}^{KG}(A, \omega KE) & \xrightarrow{\Delta_V} & \mathrm{Tor}_{n-1}^{KG}(A, KV \otimes_K \omega KE) & \rightarrow & \mathrm{Tor}_{n-1}^{KG}(A, KV \otimes KE) \end{array} \tag{6}$$

and the sum of the two routes factors through

$$\mathrm{Tor}_{n-1}^{KG}(A,\omega KG)\to \mathrm{Tor}_{n-1}^{KG}(A,KG)\xrightarrow{\cdot(v\otimes e)}\mathrm{Tor}_{n-1}^{KG}(A,KV\otimes_K KE).$$

Since the composite

$$\mathrm{Tor}_n^{KG}(A,K)\xrightarrow{\delta}\mathrm{Tor}_{n-1}^{KG}(A,\omega KG)\to \mathrm{Tor}_{n-1}^{KG}(A,KG)$$

is zero, the sum of the two routes vanishes on the image of δ. Composing with δ and changing one sign we get the commuting diagram (1). ∎

8.2 Theorem. *Let T be an incompressible G-tree. If G is a* PD^n*-group over K, and G_e is a* PD^{n-1}*-group over K for each edge e of T, then $(G_v, \mathrm{star}(v))$ is a* PD^n*-pair over K for each vertex v of VT.*

Moreover all the orientations are induced from G.

Proof. Since G is a PD^n-group over K, we have an orientation involution $^\circ$ and a fundamental class $\xi\in\mathrm{Tor}_n^{KG}(K^\circ,K)$.

Let the notation be as in Lemma 8.1.

If E is empty the result is vacuous, so we may assume E is nonempty.

In outline the proof consists of applying Theorem 2.25 to the exact sequences

$$0\to KE^\circ\xrightarrow{\alpha'}KV^\circ\xrightarrow{\alpha''}K^\circ\to 0 \quad\text{and}\quad 0\to\omega KE\xrightarrow{\beta'}KE\xrightarrow{\beta''}K\to 0,$$

by constructing a certain commuting diagram

$$\begin{array}{ccccccccc}
0\to & \mathrm{Ext}_{KG}^{n-1}(KE,KG) & \to & \mathrm{Ext}_{KG}^{n-1}(K\omega KE,KG) & \to & \mathrm{Ext}_{KG}^{n}(K,KG) & \to 0\\
& \wr\downarrow\ \Delta_E\delta_V\xi\cap - & & \wr\downarrow\ \Delta_V\beta\xi\cap - & & \wr\downarrow\ (-1)^n\xi\cap - & \\
0\to & KE^\circ & \to & KV^\circ & \to & K^\circ & \to 0
\end{array}$$

of right KG-modules, which can then be thought of as

$$\begin{array}{ccccccc}
0\to & \bigoplus_E \mathrm{Ext}_{KG_e}^{n-1}(K,KG_e) & \to & \bigoplus_V \mathrm{Ext}_{KG_v}^{n-1}(\omega KE_v,KG_v) & \to & \mathrm{Ext}_{KG}^{n}(K,KG) & \to 0\\
& \wr\downarrow\ \bigoplus_E(-(\delta_V\xi)_e\cap -) & & \wr\downarrow\ \bigoplus_V((\beta\xi)_V\cap -) & & \wr\downarrow\ (-1)^n\xi\cap - & \\
0\to & \bigoplus_E K^\circ & \to & \bigoplus_V K^\circ & \to & K^\circ & \to 0
\end{array}$$

To begin, let us apply Lemma 8.1 with $A=K^\circ$ and then Lemma 7.8 to get a commuting diagram

$$\begin{array}{ccccc}
\mathrm{Tor}_n^{KG}(K^\circ,K) & \xrightarrow{\delta_V} & \mathrm{Tor}_{n-1}^{KG}(K^\circ,KE) & \xrightarrow{-\Delta_E} & \mathrm{Tor}_{n-1}^{KG}(KE^\circ,KE)\\
\downarrow\beta & & & & \downarrow\alpha'\\
\mathrm{Tor}_{n-1}^{KG}(K^\circ,\omega KE) & \xrightarrow{\Delta_V} & \mathrm{Tor}_{n-1}^{KG}(KV^\circ,\omega KE) & \xrightarrow{\beta'} & \mathrm{Tor}_{n-1}^{KG}(KV^\circ,KE),
\end{array}\tag{7}$$

where β, δ_V denote the connecting maps arising from (2), (3).

The composite

$$\mathrm{Tor}_{n-1}^{KG}(K^\circ, \omega KE) \xrightarrow{\Delta_V} \mathrm{Tor}_{n-1}^{KG}(KV^\circ, \omega KE) \xrightarrow{\alpha''} \mathrm{Tor}_{n-1}^{KG}(K^\circ, \omega KE)$$

is the identity.

The remaining hypothesis to check in order to apply Theorem 2.25 is that KE is a $\mathrm{PD}^{n-1}KG$-module with Poincaré KG-dual KE° and fundamental class $-\Delta_E\delta_V\xi$ or equivalently $\Delta_E\delta_V\xi$. For this we shall need Lemma 7.10(iv).

Since G is finitely generated and T is incompressible, E is G-finite.

Now let $e \in E$, and consider the KG-linear projection $KE \to KGe$. This gives a map $\mathrm{Tor}_{n-1}^{KG}(K^\circ, KE) \to \mathrm{Tor}_{n-1}^{KG}(K^\circ, KGe) \approx \mathrm{Tor}_{n-1}^{KG_e}(K^\circ, K)$, and we denote the image of $\delta_V\xi$ by $(\delta_V\xi)_e$. We want to show that K is a PD^{n-1} KG_e-module with Poincaré KG-dual K°, and fundamental class $(\delta_V\xi)_e$.

By Propositions 2.18 and 2.14, we have a commuting diagram

$$\begin{array}{ccccc}
\mathrm{Hom}_{KG}(K,K) & \xrightarrow{\delta_V} & \mathrm{Ext}^1_{KG}(K,KE) & \to & \mathrm{Ext}^1_{KG}(K,KGe) \\
\wr\downarrow \xi\cap - & & \wr\downarrow \xi\cap - & & \wr\downarrow \xi\cap - \\
\mathrm{Tor}_n^{KG}(K^\circ,K) & \xrightarrow{\delta_V} & \mathrm{Tor}_{n-1}^{KG}(K^\circ,KE) & \to & \mathrm{Tor}_{n-1}^{KG}(K^\circ,KGe) \\
 & & & & \approx \mathrm{Tor}_{n-1}^{KG_e}(K^\circ,K)
\end{array}$$

where δ_V denotes the connecting maps arising from the exact sequence (3).

Consider the element $1 \in \mathrm{Hom}_{KG}(K,K)$ and denote its images along the top row as η and η_e. Thus the situation can be depicted

$$(8) \qquad \begin{array}{ccccc}
1 \mapsto & \eta & \mapsto \eta_e & & \\
\downarrow & \downarrow & \downarrow & & \\
\xi \mapsto & (\delta_V(\xi)) \mapsto & \xi\cap\eta_e \mapsto & (\delta_V\xi)_e. &
\end{array}$$

It is not difficult to show from the definition of the connecting map $\delta_V: \mathrm{Hom}_{KG}(K,K) \to \mathrm{Ext}^1_{KG}(K,KE)$ that $\eta = \delta_V(1) \in \mathrm{Ext}^1_{KG}(K,KE)$ is the class of a derivation $d: G \to KE$ determined by $g \mapsto gv - v \in \omega KV \approx KE$, where $v \in V$, and the class does not depend on the choice of v.

Recall that Ge is the edge set of a G-tree T_e whose vertices are the components of $T - Ge$. Again, $\eta_e \in \mathrm{Ext}^1_{KG}(K,KGe)$ is the class of the derivation $d_e: G \to KGe$ determined by $g \mapsto gv - v \in \omega KVT_e \approx KET_e = KGe$, where $v \in VT_e$.

Consider any $r \in K$.

If $\eta_e r = 0$ in $\mathrm{Ext}^1_{KG}(K,KGe)$ then $d_e r$ is inner, that is, there exists some

$p \in KGe = \omega KVT_e$ such that $d_e(g)r = gp - p$ for all $g \in G$. Thus G stabilizes $vr - p \in KVT_e$. Since T is incompressible, no vertex of T_e is stabilized by G, so, by Corollary I.4.8, the only element of KVT_e stabilized by G is zero. Hence $vr = p \in \omega KVT_e$. This can happen only if $r = 0$.

Similarly, if $\eta_e \in \mathrm{Ext}^1_{KG}(K, KGe)r$ then there exists a derivation $d': G \to KGe$ and an element $p \in KGe$ such that $d'(g)r + gp - p = d_e(g) = gv - v$ for all $g \in G$. Hence, $(g-1)(v_e - p) \in Kr$ for all $g \in G$. Passing to the quotient ring $\bar{K} = K/Kr$ we see that G stabilizes the image of $v_e - p$ in KVT_e. As before, this image must be zero, so $v_e - p \in KVT_e r$. Applying the augmentation map, we see $1 - 0 \in Kr$. Hence r is a unit of K.

Let $(-)^e: KG_e \to KG_e$ be the orientation involution for G_e. We have isomorphisms of K-modules

$$\begin{aligned} \mathrm{Ext}^1_{KG}(K, KGe) &\approx \mathrm{Tor}^{KG}_{n-1}(K^\circ, KGe) \text{ by Poincaré duality for } G,\\ &\approx \mathrm{Tor}^{KG_e}_{n-1}(K^\circ, K)\\ &\approx \mathrm{Tor}^{KG_e}_{n-1}(K^e, K^{\circ e}) \text{ applying the involution } {}^e,\\ &\approx \mathrm{Hom}_{KG_e}(K, K^{\circ e}) \text{ by Poincaré duality for } G_e. \end{aligned}$$

Now $\mathrm{Hom}_{KG_e}(K, K^{\circ e})$ can be identified with the set of elements of K stabilized by $g^{\circ e}$, or equivalently annihilated by $1 - g^{\circ e}g^{-1} \in K$, for all $g \in G_e$. But $\mathrm{Ext}^1_{KG}(K, KGe)$ contains the element η_e which has trivial K-annihilator. Thus ${}^{\circ e}$ is the identity automorphism of KG_e, and $K^e = K^\circ$ as KG_e-module. This proves that the Poincaré KG_e-dual of K is K°.

We next check the fundamental class. As K-module, $\mathrm{Ext}^1_{KG}(K, KGe) \approx \mathrm{Hom}_{KG_e}(K, K^{\circ e}) = \mathrm{Hom}_{KG_e}(K, K) \approx K$. Since η_e is not a proper K-multiple of any element of $\mathrm{Ext}^1_{KG}(K, KGe)$, η_e is a K-basis of $\mathrm{Ext}^1_{KG}(K, KGe)$. By (8), the isomorphisms $\mathrm{Ext}^1_{KG}(K, KGe) \approx \mathrm{Tor}^{KG}_{n-1}(K^\circ, KGe) \approx \mathrm{Tor}^{KG_e}_{n-1}(K^\circ, K)$ carry η_e to $(\delta_V\xi)_e$, so the latter is a K-basis of $\mathrm{Tor}^{KG_e}_{n-1}(K^\circ, K)$, and hence a fundamental class for the Poincaré KG_e-module K, as desired.

Now Lemma 7.10(iv) shows that KE is a PD^{n-1} KG-module with Poincaré KG-dual KE° and fundamental class $\Delta_E\delta_V\xi$.

By Theorem 2.25, ωKE is a PD^{n-1} KG-module with Poincaré dual KV^0, and fundamental class $\Delta_V\beta\xi$.

Let $v \in V$. By Lemma 7.10(iv), ωKE_v is a $\mathrm{PD}^{n-1} KG_v$-module with Poincaré dual K°. That is, (G_v, E_v) is a PD^n-pair over K with orientation module K°. ∎

The fact that gluing together two copies of a manifold-with-boundary by identifying the boundaries produces a manifold-without-boundary motivates the following construction.

8.3 Corollary. *Let Y be a connected graph consisting of two vertices and finitely many edges joining them. Let $(G(-), Y)$ be a graph of groups such that each vertex group is G, each edge group is a proper subgroup of G, which is a PD^{n-1}-group over K, and all maps are given by inclusion. If the fundamental group of $(G(-), Y)$ is a PD^n-group over K, then $(G, \bigvee_{e\in EY} G/G(e))$ is a PD^n-pair over K with induced orientation.* ∎

8.4 Lemma. *If $a, b, c \in G$ then $abab^{-1} = (ab)^2(b^{-1})^2$ and $a^2[b,c] = (a^2b^{-1}c^{-1}a^{-1})^2(acba^{-1}c^{-1}a^{-1})^2(ac)^2$.* ∎

8.5 Example. Suppose (G, W) is a GD^2-pair with orientation map ε°. As stated in Example 1.4(iii), G is freely generated by a nonempty set $t_1, \ldots, t_p, z_1, \ldots, z_m$, and $W = \bigvee_{i=0}^{p} G/\langle t_i \rangle$, where t_0 is either

(A) $\quad t_1 \cdots t_p[z_1, z_2] \cdots [z_{m-1}, z_m]$ (with m even)

or

(B) $\quad t_1 \cdots t_p z_1^2 z_2^2 \cdots z_m^2$ (with $m \geqslant 1$).

We now verify algebraically that (G, W) is a PD^2-pair.

Let Y be the graph with $VY = \{\iota, \tau\}$, $EY = \{w_0, \ldots, w_m\}$ and the naturally suggested incidence maps. Take the maximal subtree Y_0 of Y to consist of the two vertices and the edge w_0.

Let $(G(-), Y)$ be the graph of groups with $G(\iota) = G(\tau) = G$, $G(w_i) = \langle t_i \rangle$ for all $i \in [0, p]$, and all maps being inclusion maps.

Then the fundamental group $P = \pi(G(-), Y, Y_0)$ is generated by $t_1, \ldots, t_p, z_1, \ldots, z_m, t_1', \ldots, t_p', z_1', \ldots, z_m', u_1, \ldots, u_p$, with relations $t_i' = t_i^{u_i}$, and either $t_1 \cdots t_p[z_1, z_2] \cdots [z_{m-1}, z_m] = t_1^{u_1} \cdots y_p^{u_p}[z_1', z_2'] \cdots [z_{m-1}', z_m']$ or $t_1 \cdots t_p z_1^2 z_2^2 \cdots z_m^2 = t_1^{u_1} \cdots t_p^{u_p} z_1'^2 z_2'^2 \cdots z_m'^2$. Thus the generators t_i' may be omitted and P is a 1-relator group with $2m + 2p \geqslant 2$ generators. Using Lemma 8.4 one can show easily that P is an infinite surface group. By Corollary 8.3, (G, W) is a PD^2-pair with induced orientation map. Here each $\varepsilon^\circ(t_i) = 1$ while each $\varepsilon^\circ(z_i) = +1$ or -1 depending as t_0 is (A) or (B), respectively. ∎

For PD-pairs there is a result analogous to Theorem 8.2; for the convenience of the reader we plough through the details.

8.6 Lemma. *Let W be a G-set and $\phi: W \to VT$ a G-map and write $\tilde{E} = E \vee W$.*

There is a V-grading over $K, \omega K\tilde{E} = \bigoplus_{v \in V} \omega K\tilde{E}_v$, *where* $\tilde{E}_v =$ star$(v) \cup \phi^{-1}(v)$ *for each* $v \in V$, *and hence a diagonal map* $\Delta_V : \omega K\tilde{E} \to KV \otimes_K \omega KE$.

The diagram

$$\begin{array}{ccc} \omega KW \xrightarrow{\phi} \omega KV \approx KE & \xrightarrow{-\Delta_E} & KE \otimes_K KE \\ \downarrow & & \downarrow \partial \\ \omega K\tilde{E} \xrightarrow{\Delta_V} KV \otimes_K \omega K\tilde{E} & \xrightarrow{\pi_E} & KV \otimes_K KE \end{array} \tag{9}$$

commutes, where $\pi_E : K\tilde{E} \to KE$ *is the projection onto KE which vanishes on W.*

Proof. Let $\tilde{T}$ be the G-tree containing T with vertex set $\tilde{V} = V \cup W$, and edge set $\tilde{E} = E \cup W$, such that for each $w \in W \subseteq \tilde{E}$, the initial vertex of w is $w \in W \subseteq \tilde{V}$, and the terminal vertex is $\phi(v) \in V \subseteq \tilde{V}$. By Lemma 8.1, $\omega K\tilde{E} = \bigoplus_{\tilde{V}} \omega K\tilde{E}_v$ which is $\bigoplus_{V} \omega K\tilde{E}_v$, for if $v \in \tilde{V} - V = W$ then v has valency one so $\omega K\tilde{E}_v = \tilde{0}$.

It remains to consider (9).

Suppose $w, w' \in W$ and consider a path of the form $\phi(w) = v_0, e_1^{\varepsilon_1}, v_1, \ldots, v_{m-1}, e_m^{\varepsilon_m}, v_m = \phi(w')$ in T. Then

$$\Delta_V(w' - w) = v_0 \otimes (e_1 - w) + \sum_{i=1}^{m-1} v_i \otimes (e_{i+1} - e_i) + v_m \otimes (w' - e_m),$$

so the image of $w' - w$ along the bottom route of (9) is

$$\begin{aligned} & v_0 \otimes e_1 + \sum_{i=1}^{m-1} v_i \otimes (e_{i+1} - e_i) - v_m \otimes e_m \\ &= \sum_{i=1}^{m} (v_{i-1} - v_i) \otimes e_i = \sum_{i=1}^{m} -\varepsilon_i(\tau e_i - \iota e_i) \otimes e_i. \end{aligned}$$

Since this is the image of $w' - w$ along the upper route of (9), we see that (9) commutes. ▮

8.7 Theorem. *Let* $\phi : W \to V$ *be a G-map such that* ϕW *generates T. If* (G, W) *is a* PD^n*-pair over K, and* G_e *is a* PD^{n-1}*-group over K for each edge e of T, then* $(G_v,$ star$(v) \cup \phi^{-1}(v))$ *is a* PD^n*-pair over K for each vertex v of T.*

Moreover all the orientations are induced from G.

Proof. Since (G, W) is a PD^n-pair over K, we have an orientation involution $^\circ$ and a fundamental class $\xi \in \mathrm{Tor}_{n-1}^{KG}(\omega KW^\circ, K)$.

Let the notation be as in Lemma 8.6.

If W is empty the result is trivial, so we may assume that W is nonempty.

We wish to apply Theorem 2.24 to the exact sequences

$$0 \to KE^\circ \xrightarrow{\alpha'} KV^\circ \xrightarrow{\alpha''} K^\circ \to 0 \quad \text{and}$$

$$0 \to \omega KW \xrightarrow{\beta'} \omega K\tilde{E} \xrightarrow{\beta''} KE \to 0,$$

where β'' comes from the projection which vanishes on W, and is surjective since W is nonempty.

Essentially we shall be constructing a certain commuting diagram

$$\begin{array}{ccccccccc} 0 \to & \mathrm{Ext}^{KG}_{n-1}(KE, KG) & \to & \mathrm{Ext}^{KG}_{n-1}(\omega K\tilde{E}, KG) & \to & \mathrm{Ext}^{KG}_{n-1}(\omega KW, KG) & \to 0 \\ & \wr\downarrow -\Delta_E\phi\xi\cap - & & \wr\downarrow \Delta_V\beta'\xi\cap - & & \wr\downarrow \xi\cap - & \\ 0 \to & KE^\circ & \to & KV^\circ & \to & K^\circ & \to 0 \end{array}$$

of right KG-modules, which can then be thought of as

$$\begin{array}{ccccccc} 0 \to \bigoplus_E \mathrm{Ext}^{n-1}_{KG_e}(K, KG_e) & \to & \bigoplus_V \mathrm{Ext}^{n-1}_{KG_v}(\omega K\tilde{E}_v; KG_v) & \to & \mathrm{Ext}^{n-1}_{KG}(\omega KW, KG) & \to 0 \\ \wr\downarrow \bigoplus_E(-(\phi\xi)e\cap -) & & \wr\downarrow \bigoplus_V((\beta'\xi)v\cap -) & & \wr\downarrow \xi\cap - & \\ 0 \to \bigoplus_E K^\circ & \to & \bigoplus_V K^\circ & \to & K^\circ & \to 0. \end{array}$$

To begin, let us apply $\mathrm{Tor}^{KG}_{n-1}(K^\circ, -)$ to (9) and use Lemma 7.8 to get a commuting diagram

$$\begin{array}{ccccc} \mathrm{Tor}^{KG}_{n-1}(K^\circ, \omega KW) \xrightarrow{\phi} \mathrm{Tor}^{KG}_{n-1}(K^\circ, KE) & & \xrightarrow{-\Delta_E} & & \mathrm{Tor}^{KG}_{n-1}(KE^\circ, KE) \\ \downarrow \beta' & & & & \downarrow \alpha' \\ \mathrm{Tor}^{KG}_{n-1}(K^\circ, \omega K\tilde{E}) & \xrightarrow{\Delta_V} & \mathrm{Tor}^{KG}_{n-1}(KV^\circ, \omega K\tilde{E}) & \xrightarrow{\Delta_V} & \mathrm{Tor}^{KG}_{n-1}(KV^\circ, KE). \end{array} \tag{10}$$

The composite

$$\mathrm{Tor}^{KG}_{n-1}(K^\circ, \omega K\tilde{E}) \xrightarrow{\Delta_V} \mathrm{Tor}^{KG}_{n-1}(KV^\circ, \omega K\tilde{E}) \xrightarrow{\alpha''} \mathrm{Tor}^{KG}_{n-1}(K^\circ, \omega K\tilde{E})$$

is the identity.

The remaining hypothesis to check in order to apply Theorem 2.24 is that KE is a PD^{n-1} KG-module with Poincaré KG-dual KE° and fundamental class $-\Delta_E\phi\xi$ or equivalently $\Delta_E\phi\xi$. For this we shall need Lemma 7.10(iv).

Since W is G-finite by Remark 7.2(iv), and ϕW generates T, E is G-finite.

Now let $e \in E$, and consider the KG-linear projection $KE \to KGe$. This gives a map $\mathrm{Tor}_{n-1}^{KG}(K^\circ, KE) \to \mathrm{Tor}_{n-1}^{KG}(K^\circ, KGe) \approx \mathrm{Tor}_{n-1}^{KG}(K^\circ, K)$, and we denote the image of $\phi\xi$ by $(\phi\xi)_e$. We want to show that K is a PD^{n-1} KG_e-module with Poincaré KG-dual K°, and fundamental class $(\phi\xi)_e$.

By Proposition 2.14, we have a commuting diagram

$$\begin{array}{ccccc}
\mathrm{Hom}_{KG}(\omega KW, \omega KW) & \xrightarrow{\phi} & \mathrm{Hom}_{KG}(\omega KW, KE) & \to & \mathrm{Hom}_{KG}(\omega KW, KGe) \\
\downarrow{\scriptstyle \cap -} & & \downarrow{\scriptstyle \xi\cap -} & & \downarrow{\scriptstyle \xi\cap -} \\
\mathrm{Tor}_{n-1}^{KG}(K^\circ, \omega KW) & \xrightarrow{\phi} & \mathrm{Tor}_{n-1}^{KG}(K^\circ, KE) & \to & \mathrm{Tor}_{n-1}^{KG}(K^\circ, KGe) \approx \mathrm{Tor}_{n-1}^{KG_e}(K^\circ, K).
\end{array}$$

Let ϕ_e denote the composite $\omega KW \xrightarrow{\phi} KE \to KGe$, so we have the picture

$$\begin{array}{ccccccc}
1 & \mapsto & \phi & \mapsto & \phi_e & & \\
\downarrow & & \downarrow & & \downarrow & & \\
\xi & \mapsto & \phi\xi & \mapsto & \xi \cap \phi_e & \mapsto & (\phi\xi)_e.
\end{array} \tag{11}$$

Since ϕW generates T, there exist w, w' in W such that the coefficient of e in $\phi(w - w')$ is 1. Hence, for any $r \in K$, if $\phi_e r = 0$ in $\mathrm{Hom}_{KG}(\omega KW, KGe)$ then $r = 0$, and, if $\phi \in \mathrm{Hom}_{KG}(\omega KW, KGe)r$, then r is a unit.

Let $(-)^e : KG_e \to KG_e$ be the orientation involution for G_e. We have isomorphisms of K-modules

$\mathrm{Hom}_{KG}(\omega KW, KGe)$

$\approx \mathrm{Tor}_{n-1}^{KG}(K^\circ, KGe)$ by Poincaré duality for (G, W),

$\approx \mathrm{Tor}_{n-1}^{KG_e}(K^\circ, K)$

$\approx \mathrm{Tor}_{n-1}^{KG_e}(K^e, K^{\circ e})$ applying the involution ^{e},

$\approx \mathrm{Hom}_{KG_e}(K, K^{\circ e})$ by Poincaré duality for G_e.

Now $\mathrm{Hom}_{KG_e}(K, K^{\circ e})$ can be identified with the set of elements of K stabilized by $g^{\circ e}$, or equivalently annihilated by $1 - g^{\circ e} g^{-1} \in K$, for all $g \in G_e$. But $\mathrm{Hom}_{KG}(\omega KW, KGe)$ contains the element ϕ_e which has trivial K-annihilator. Thus ${}^{\circ e}$ is the identity automorphism of KG_e, and $K^e = K^\circ$ as KG_e-module. This proves that the Poincaré KG_e-dual of K is K°.

We next check the fundamental class. As K-module, $\mathrm{Hom}_{KG}(\omega KW, KGe) \approx \mathrm{Hom}_{KG_e}(K, K^{\circ e}) = \mathrm{Hom}_{KG_e}(K, K) \approx K$. Since ϕ_e is not a proper K-multiple of any element of $\mathrm{Hom}_{KG}(\omega KW, KGe)$, ϕ_e is a K-basis of $\mathrm{Hom}_{KG}(\omega KW, KGe)$. By (11), the isomorphisms $\mathrm{Hom}_{KG}(\omega KW, KGe) \approx \mathrm{Tor}_{n-1}^{KG}(K^\circ, KGe) \approx \mathrm{Tor}_{n-1}^{KG_e}(K^\circ, K)$ carry ϕ_e to $(\phi\xi)_e \in \mathrm{Tor}_{n-1}^{KG_e}(K^\circ, K)$. Hence, $(\phi\xi)_e$ is a K-basis of $\mathrm{Tor}_{n-1}^{KG_e}(K^\circ, K)$, so is a fundamental class of the Poincaré KG_e-module K, as desired.

Now Lemma 7.10(iv) shows that KE is a PD^{n-1} KG-module with Poincaré KG-dual KE° and fundamental class $\Delta_E\phi\xi$.

By Theorem 2.24, $\omega K\tilde{E}$ is a PD^{n-1} KG-module with Poincaré dual KV° and fundamental class $\Delta_V\beta'\xi$. Let $v\in V$. By Lemma 7.10(iv), $\omega K\tilde{E}_v$ is a PD^{n-1} KG_v-module with Poincaré dual K°. That is, $(G_v, \tilde{E}_v)$ is a PD^n-pair over K with orientation module K°. ∎

9 Relative ends

The next result is a relative version of Stallings' Ends Theorem III.2.1 which will be of fundamental importance for PD^2-groups and pairs.

9.1 Theorem. *Let G be finitely generated, V a G-finite G-set, and A a nonzero abelian group.*

There exists an outer derivation $d:G\to AG$ which is inner on G_v for all $v\in V$ if and only if there exists a G-tree T, and an edge e of T, such that $ET = Ge$, G_e is finite, G stabilizes no vertex of T, and there exists a G-map $V\to VT$.

If these equivalent conditions hold then for any finitely generated subgroup H of G the following cardinals are finite and equal:

(a) the smallest number n for which there exist $g_1,\ldots,g_n\in G$, and d as above, such that $d(H)\subseteq AHg_1+\cdots+AHg_n$.

(b) the smallest number n for which there exist $g_1,\ldots,g_n\in G$, and T as above, and an H-subtree T_0 of T such that $\{g\in G \mid ge\in ET_0\} = Hg_1\cup\cdots\cup Hg_n$.

Proof. We modify the proof of Theorem III.2.1.

Consider any outer derivation $d:G\to AG$ which is inner on G_v for all $v\in V$. Since H is finitely generated, it is easy to see there is some finite set of cosets $Hg_1\ldots,Hg_n$ such that $d(H)\subseteq \sum_{i=1}^{n} AHg_i$.

By adding a copy of G to V, we may assume that V contains an element v_0 with trivial stabilizer. Since G is finitely generated and V is G-finite, we can form a G-finite connected G-graph X with vertex set V and G-free edge set. Since H is finitely generated, by adding finitely many G-orbits to EX we may further assume there is a connected H-subgraph X' with vertex set Hv_0.

For any function s on V, recall that the *coboundary* of s is $\delta s =$

$\{e \in EX \mid s(\iota e) \neq s(\tau e)\}$; let us write $n(s)$ for the number of cosets gH such that δs meets gX'. Recall also that $\mathscr{B}X$ is the Boolean subring of $(V, \mathbb{Z}_2)$ consisting of all s such that δs is finite. The first step in the proof will be to show that $n \geqslant n(s)$ for some $s \in \mathscr{B}X$ such that s is not almost constant.

For each $g \in G$, write $d(g) = \sum_{x \in G} a_{g,x} x$ with $a_{g,x} \in A$, and define $\tilde{v} \in (G, A)$ by setting $\tilde{v}(g) = -a_{g,g}$. Viewing AG as the almost equality class of 0 in (G, A), we have $g\tilde{v} - \tilde{v} = d(g)$ for all $g \in G$, by Proposition IV.2.3. For any $g \in G$, $\tilde{v}(g) \neq h\tilde{v}(g)$ for some $h \in H$, if and only if Hg is one of the Hg_i. Hence there are exactly n cosets Hg such that $\tilde{v} \mid Hg$ is not constant.

Let $\tilde{V} = \tilde{v} + AG$, the almost equality class of $\tilde{v}$. Then $\tilde{V}$ is G-stable, and for any $p \in AG$, $\tilde{v} + p \in \tilde{V}$, and if $g \in G$ then $(g-1)(\tilde{v}+p) = (d + \operatorname{ad} p)(g)$. Hence for every subgroup K of G, d is inner on K if and only if K stabilizes an element of $\tilde{V}$. This has two consequences.

Since d is outer, no element of $\tilde{V}$ is stabilized by G, and $\tilde{v}$ is not almost constant.

For each $v \in V$, d is inner on G_v so G_v stabilizes an element of $\tilde{V}$.

Thus we can construct a G-map $V \to \tilde{V} \subseteq (G, A)$, $v \mapsto v \mid G$, such that $v_0 \mid G = \tilde{v}$. There is then a dual G-map $G \to (V, A)$, $g \mapsto g \mid V$. Let $\tilde{s}$ denote the image of 1. For any edge e of X, since $\iota e \mid G =_a \tau e \mid G$ in (G, A), there are only finitely many $g \in G$ such that $\iota e(g) \neq \tau e(g)$, that is, $g^{-1}\iota e(1) = g^{-1}\tau e(1)$, or, equivalently, $g^{-1}e$ lies in $\delta\tilde{s}$. Thus $Ge \cap \delta\tilde{s}$ is finite. As EX is G-finite we see that $\delta\tilde{s}$ is finite. Hence $X - \delta\tilde{s}$ has only finitely many components so $\tilde{s}$ takes only finitely many values. As $v_0 \mid G = \tilde{v}$ is not almost constant, so $\tilde{s} \mid Gv_0$ is not almost constant, and hence $\tilde{s}$ is not almost constant. Thus there exists a component of $X - \delta\tilde{s}$ with vertex set $s \in (V, \mathbb{Z}_2)$ which is not almost constant. Further, $\delta s \subseteq \delta\tilde{s}$ so δs is finite and $n(s) \leqslant n(\tilde{s})$. For any coset Hg, $v_0 \mid Hg = \tilde{v} \mid Hg$ is constant if and only if $\tilde{s} \mid g^{-1}Hv_0$ is constant, which is equivalent to $\delta\tilde{s}$ meeting $g^{-1}X'$. Thus $n(\tilde{s}) = n$, so $n(s) \leqslant n$.

In summary, we have found an element $s \in \mathscr{B}X$ which is not almost constant and $n(s) \leqslant n$. We now forget all the previous notation since we shall want another way of assigning to each vertex of X a function on G.

Let n' be the smallest value of $n(s)$ as s ranges over all $s \in \mathscr{B}X$ which are not almost constant, so $n' \leqslant n$. Let m be the smallest value of $|\delta s|$ as s ranges over all $s \in \mathscr{B}X$ which are not almost constant and with $n(s) = n'$.

Choose $s \in \mathscr{B}X$ with $n(s) = n'$, $|\delta s| = m$. Clearly the same conditions are satisfied by all elements of $Gs \cup Gs^*$.

Since s is not almost constant, one of the finitely many components s' of s with $s' \leqslant s$ is such that s' is not almost constant. Moreover, $n(s') \leqslant n(s)$ since each gX' is connected. By minimality of n, $n(s') = n'$. By minimality of $|\delta s|$, we see $|\delta s'| \geqslant m$. But $\delta s' \subseteq \delta s$, so $\delta s' = \delta s$, and thus $s' = s$. So s is connected. Similarly, s^* is connected.

Replacing s with s^* if necessary, we may assume that s contains v_0.

We claim that we can choose s minimal under inclusion among all $s \in \mathscr{B}X$ such that s is not almost constant, $n(s) = n'$, $|\delta s| = m$, and $v_0 \in s$.

Suppose not, so there is an infinite descending chain $s_1 > s_2 > \cdots$ of elements of $\mathscr{B}X$ containing v_0, with $n(s_i) = n'$, $|\delta s_i| = m$. Thus v_0 belongs to the set $s_\omega = \bigcap_{i \geqslant 1} s_i$. Since s_1 is connected, there exists an edge e_1 joining $s_1 s_\omega^*$ to s_ω, and hence joining $s_1 s_i^*$ to s_ω for some $i_1 > 1$. Thus $e_1 \in \bigcap_{i \geqslant i_1} \delta s_i$. Similarly s_{i_1} is connected so there exists an edge e_2 joining $s_{i_1} s_{i_2}^*$ to s_ω for some $i_2 > i_1$. Notice e_2 has both vertices in s_{i_1} so $e_2 \neq e_1$. Continuing in this way we arrive at an i_{m+1} and $m+1$ distinct edges $e_1, \ldots, e_{m+1}$ lying in $\delta s_{i_{m+1}}$, as illustrated in Fig. II.3, page 58. But $|\delta s_{i_{m+1}}| = m$, a contradiction. Thus the claim is proved.

Let $g \in G$. We claim that some $r \in s \square gs$ is almost 0. Suppose not, so each $r \in s \square gs$ is not almost constant. By minimality, $n' \leqslant n(r)$ for all $r \in s \square gs$, so

$$4n' \leqslant \sum_{r \in s \square gs} n(r) \leqslant 2n(s) + 2n(gs) = 4n'$$

and hence $n(r) = n'$ for all $r \in s \square gs$. By minimality, $m \leqslant |\delta r|$ for all $r \in s \square gs$, so as in Fig. II.2, page 56,

$$4m \leqslant \sum_{r \in s \square gs} |\delta r| \leqslant 2|\delta s| + 2|\delta gs| = 4m,$$

and hence $|\delta r| = m$ for all $r \in s \square gs$. Consider the unique $r \in s \square gs$ containing v_0. Then $r \leqslant s$, and, by minimality under inclusion, $r = s$, which forces some element of $s \square gs$ to be 0, and the claim is proved.

Let Σ be the set of all permutations of V which move only finitely many elements. Then Σ acts in the usual way on $(V, \mathbb{Z}_2)$, and we have a map $(V, \mathbb{Z}_2) \to \Sigma \backslash (V, \mathbb{Z}_2)$, $s \mapsto \Sigma s$.

Since Σ is a normal subgroup of the group of all permutations of V, the G-action on $(V, \mathbb{Z}_2)$ induces a G-action on $\Sigma \backslash (V, \mathbb{Z}_2)$. The involution $*$ on $(V, \mathbb{Z}_2)$ induces an involution, again denoted $*$, on $\Sigma \backslash (V, \mathbb{Z}_2)$.

The partial order $\leqslant$ on $(V, \mathbb{Z}_2)$ arising from inclusion induces a partial order, again denoted $\leqslant$, on $\Sigma \backslash (V, \mathbb{Z}_2)$. Explicitly, if $e, f \in (V, \mathbb{Z}_2)$ then $\Sigma e \leqslant \Sigma f$ if and only if $\sigma e \leqslant f$ for some $\sigma \in \Sigma$.

Let $e = \Sigma s \in \Sigma \backslash (V, \mathbb{Z}_2)$, and let $E = Ge \subseteq \Sigma \backslash (V, \mathbb{Z}_2)$. The facts that, for all

$g \in G$, one element of $s \square gs$ is almost 0, and s, gs are not almost constant imply that E is nested, and we can form a G-graph $T = T(E)$ as in Definition II.1.7.

The surjection $G \to E$, $g \mapsto ge$, allows us to view $(E, \mathbb{Z}_2)$ as a subset of $(G, \mathbb{Z}_2)$, and we have $VT \subseteq (E, \mathbb{Z}_2) \subseteq (G, \mathbb{Z}_2)$.

We have a map $G \to (V, \mathbb{Z}_2)$, $g \mapsto gs$, and its dual $V \to (G, \mathbb{Z}_2)$, $v \mapsto v|G$. Let $\tilde{V}$ be the almost equality class of $v_0|G$ in $(G, \mathbb{Z}_2)$. For each edge e of X, there are only finitely many $g \in G$ such that $g^{-1}e \in \delta s$. Thus $(gs)(\iota e) = (gs)(\tau e)$ for almost all $g \in G$, so $\iota e|G =_a \tau e|G$. Since X is connected, $V|G$ lies in $\tilde{V}$. Moreover, $\tilde{V}$ is G-stable.

Hence $VT \subseteq (E, \mathbb{Z}_2) \subseteq (G, \mathbb{Z}_2) \supseteq \tilde{V}$; we claim that $\tilde{V}$ contains VT.

Choose $v \in s$, $v' \in s^*$.

We shall show that, for almost all $g \in G$, $\tau e(g) \in \{v(g), v'(g)\}$; since $v|G =_a v'|G$ we see that, for almost all $g \in G$, $\tau e(g) = v(g) = v'(g)$, so $\tau e|\dot{G} \in \tilde{V}$.

Let Y be a finite connected subgraph of X containing δs. Since Y and δs are finite and EX is G-free, we see Y is disjoint from $g\delta s$ for almost all $g \in G$. Consider any such g.

Then Y is a connected subgraph of $X - g\delta s$, so lies in one of the two components. Let X'' denote the component of $X - g\delta s$, disjoint from Y, so $X'' \subseteq X - Y \subseteq X - \delta s$. Here VX'' is gs or gs^*. If $VX'' = gs$ then any $u, u' \in gs$ are joined by a path in $X'' \subseteq X - \delta s$, so $s(u) = s(u')$. Thus s is constant on gs, so s, gs are nested. Similarly, if $VX'' = gs^*$ then s, gs are nested. Since

$$\tau e(g) = \tau e(ge) = \begin{cases} 1 \text{ if } ge \geqslant e \text{ or } ge > e^* \\ 0 \text{ if } ge < e \text{ or } ge \leqslant e^*, \end{cases}$$

one of the following holds:

$$\begin{array}{llll} gs \supseteq s & \text{so} \quad ge \geqslant e & \text{and} & gs(v) = 1 = \tau e(ge), \\ gs \supset s^* & \text{so} \quad ge > e^* & \text{and} & gs(v') = 1 = \tau e(ge), \\ gs \subset s & \text{so} \quad ge < e & \text{and} & gs(v') = 0 = \tau e(ge), \\ gs \subseteq s^* & \text{so} \quad ge \leqslant e^* & \text{and} & gs(v') = 0 = \tau e(ge). \end{array}$$

Hence, for almost all $g \in G$, $\tau e(g) = \tau e(ge) \in \{gs(v), gs(v')\} = \{v(g), v'(g)\}$, as claimed. Thus $\tau e|G \in \tilde{V}$. Similarly, $\iota e|G \in \tilde{V}$, and since $\tilde{V}$ is a G-subset, $VT \subseteq \tilde{V}$ as desired.

By Theorem II.1.8, T is a G-tree with costructure map given by the embedding $VT \subseteq (E, \mathbb{Z}_2) \subseteq (G, \mathbb{Z}_2)$.

Here $ET = E = Ge$, and G_e is finite, since for any $g \in G_e$, $\iota e(g) = \iota e(ge) = \iota e(e) \neq \tau e(e) = \tau e(ge) = \tau e(g)$, and $\iota e|G =_a \tau e|G$, so there are only finitely many such g.

Since V is G-finite and s is not almost constant, there is some $v \in V$ such that $s|Gv$ is not almost constant, so $v|G$ is not almost constant. Hence no element of $\tilde{V} \supseteq VT$ is stabilized by G.

Consider any $v \in V$ such that G_v is infinite. We shall show that $v|G \in VT$ by verifying that it respects the nesting. Consider any $s', s'' \in Gs \cup Gs^*$ such that $s's''$ is finite, as subset of V. Then $s's''(v') = 0$ for infinitely many neighbours v' of v in X, since G_v is infinite. Since $\delta(s's'') \subseteq \delta s' \cup \delta s''$ is finite, $s's''(v) = 0$, that is, $s'(v)s''(v) = 0$. For example, if $\sigma \in \Sigma$ and $\sigma s \in Gs \cup Gs^*$ then $(\sigma s)s^*$ and $(\sigma s)^*s$ are finite, so $(\sigma s)(v)s(v)^* = 0$ and $(\sigma s)(v)^*s(v) = 0$, and thus $\sigma s(v) = s(v)$. This is the condition for $v|G$ to lie in $(E, \mathbb{Z}_2)$. Further, by what we have seen, for any $e', e'' \in E \cup E^*$ if $e' \geqslant e''$ then $v(e')^*v(e'') = 0$, so the nesting of E is respected. By Theorem II.1.8, $v|G$ lies in VT.

Since every finite subgroup of G stabilizes a vertex of T, there exists a G-map $V \to VT$.

Let v be any vertex of T, and write E' for the set of all $ge \in E$ such that the map $H \to \mathbb{Z}_2$, $h \mapsto v(hg) = v(hge) = (h^{-1}v)(ge)$ is not almost constant, that is, the partition of Hv in $T - \{ge\}$ divides H into two infinite parts.

If E' is nonempty then any edge between two elements of E' lies in E', so E' is the edge set of an H-subtree of T. If E' is empty then, from the trichotomy of Theorem I.4.12, it is clear that H stabilizes a vertex of T.

In any event, E' is the edge set of an H-subtree of T.

Now if $Hge \subseteq E'$ then $v|Hg$ is not almost constant so $v_0|Hg$ is not almost constant, so $s|g^{-1}Hv_0$ is not almost constant, and hence not constant. The number of cosets Hg satisfying the latter is n', and $n' \leqslant n$. Hence the value in (*b*) is at most the value in (*a*).

Conversely, suppose T is a G-tree with an edge e such that $ET = Ge$, G_e is finite, G stabilizes no vertex of T, and there exists a G-map $V \to VT$. Let v_0 be a vertex of T, and let T' be the subtree spanned by Hv_0. Since H is finitely generated there are finitely many cosets $Hg_1, \ldots, Hg_n$ such that $\{g \in G \mid ge \in ET'\} = Hg_1 \cup \cdots \cup Hg_n$.

Choose any $a \neq 0$ in A and consider the map $d: G \to AG$, $g \mapsto \sum_{x \in G} \varepsilon(g, x)ax$, where $\varepsilon(g, x)$ is 1, -1 or 0, if the T-geodesic from v_0 to gv_0 contains xe, xe^{-1} or neither, respectively. Thus d is the derivation which arises from composing the three maps $G \to \omega AVT$, $g \mapsto agv_0 - av_0$, $\omega AVT \approx AET$, $\tau e - \iota e \mapsto e$, and $AET \to AG$, $age \mapsto \sum_{x \in G_e} agx$.

In the expectation of a contradiction, suppose d is inner, so, for some $p \in AG$, $d(g) = (g - 1)p$ for all $g \in G$. Since G is infinite, we may choose g such that p and gp have disjoint support. Since $gp - p$ lies in the image of $AET \to AG$, so does p. Hence, $G \to \omega AVT$, $g \mapsto agv_0 - av_0$, is inner. It follows that G stabilizes an element of $av_0 + \omega AVT$, so stabilizes a nonzero element of AVT. By Corollary I.4.8, G stabilizes an element of VT, a contradiction. Hence, d is outer.

Also, $G \to \omega AVT$, $g \mapsto agv_0 - av_0$, is inner on G_v for all $v \in VT$, so d is inner on G_v for all $v \in V$.

Finally, it is clear that $d(H) \subseteq \sum_{i=1}^{n} AHg_i$, so the value in (b) is at most the value in (a). ∎

9.2 Definition. For any subgroup H of G, there is a canonical map $H^1(G, KG) \to H^1(H, KH)$ which sends the class of a derivation $d: G \to KG$ to the class of the composite $H \to G \to KG \to KH$.

Let W be a G-set, so, for each $w \in W$, we have a map $H^1(G, KG) \to H^1(G_w, KG_w), \eta \mapsto \eta_\omega$. For $\eta \in H^1(G, KG)$, the (G, W)-*support* of η is the set of $w \in W$ such that $\eta_w \neq 0$.

Suppose w is an element of W such that some $\eta \neq 0$ in $H^1(G, KG)$ has its (G, W)-support lying in Gw. Then the smallest cardinal which occurs as the number of elements in the (G, W)-support of such an η is called the (G, W)-*weight* of w. ∎

9.3 Theorem. *Let G be finitely generated, $n \geqslant 0$ an integer, W a G-finite G-set such that stabilizers are finitely generated, w_0 an element of W, and write $G_0 = G_{w_0}$.*

If w_0 has (G, W)-weight n then n is the smallest number such that there exists a G-tree T, a G_0-subtree T_0 of T, and an edge e of T such that G stabilizes no vertex of T, there exists a G-map $(W - Gw_0) \to VT$, $ET = Ge$, G_e is finite, and there are exactly n cosets $G_0 g$ with $G_0 ge \subseteq ET_0$.

Proof. By our hypotheses, KW is an FP_1 KG-module. Recall from the proof of Lemma 7.10(iii) that, for any G-transversal W_0 in W, we have K-linear isomorphisms

$$\begin{aligned}
\mathrm{Ext}^1_{KG}(KW, KG) &\approx \bigoplus_{W_0} \mathrm{Ext}^1_{KG_w}(K, KG) \\
&\approx \bigoplus_{W_0} \mathrm{Ext}^1_{KG_w}(K, \bigoplus_{gw \in Gw} KG_w g^{-1}) \\
&\approx \bigoplus_{w \in W_0} \bigoplus_{gw \in Gw} \mathrm{Ext}^1_{KG_w}(K, KG_w g^{-1}) \\
&\approx \bigoplus_{w \in W_0} \bigoplus_{gw \in Gw} \mathrm{Ext}^1_{KG_{gw}}(K, KG_{gw}) \\
&\approx \bigoplus_{w \in W} \mathrm{Ext}^1_{KG_w}(K, KG_w).
\end{aligned}$$

The augmentation map $KW \to K$ induces a map

$$\mathrm{Ext}^1_{KG}(K, G) \to \mathrm{Ext}^1_{KG}(KW, KG) \approx \bigoplus_{W} \mathrm{Ext}^1_{KG_w}(K, KG_w)$$

and it is given by $\eta \mapsto \sum_W \eta_w$.

By hypothesis there exist n elements $g_1, \ldots, g_n \in G$, and an outer derivation $d: G \to KG$ such that for all $w \in W$, the composite $d_w: G_w \to G \to KG \to KG_w$ is outer if and only if $w \in \{g_1 w_0, \ldots, g_n w_0\}$. Moreover, n is smallest possible.

Let $V = W - Gw$, and let $v \in V$. Since G_v is finitely generated, $d(G_v)$ is contained in a finite sum of $KG_v g^{-1}$. Since each d_{gv} is inner, the above isomorphism shows that the composite $G_v \to G \to KG \to KG_v g^{-1}$ is inner, so d itself is inner on G_v.

Similarly, $d(G_0)$ is contained in a finite sum of $KG_0 g^{-1}$ and adjusting d by an inner derivation we can arrange that the summation is over $g \in \{g_1, \ldots, g_n\}$. Clearly we cannot get a smaller sum so the result follows by Theorem 9.1. ∎

9.4 Corollary. *Let G be finitely generated, W a G-finite G-set with finitely generated stabilizers, and w an element of W which has (G, W)-weight 1 and torsion-free stabilizer G_0. Then one of the following holds:*

(a) $G = A * B$ and $G_0 = A_0 * B_0$ for some $A_0 \leqslant A$, $B_0 \leqslant B$ with $A_0, B_0 \neq 1$. For each $v \in W - Gw$, G_v lies in a G-conjugate of A or B.

(b) $G = A * b$ and $G_0 = A_0 * A_1^b$ for some A_0, $A_1 \leqslant A$ with A_0, $A_1 \neq 1$. For each $v \in W - Gw$, G_v lies in a G-conjugate of A.

(c) $G = \langle b \rangle = G_0$ and $W - Gw$ is G-free.

Proof. By Theorem 9.3, 1 is the smallest integer n such that there exists a G-tree T and a G_0-subtree T_0 and an edge e of T such that G_e is finite, G stabilizes no vertex of T, there exists a G-map $(W - Gw) \to VT$, and there is exactly $n = 1$ coset $G_0 g$ with $G_0 g e \subseteq ET_0$.

Without loss of generality we assume $e \in ET_0$ so $ET_0 = G_0 e$.

If $g \in G_e$ then $G_0 g e = G_0 e = ET_0$, so by hypothesis $g \in G_0$; this implies $g = 1$ since G_0 is torsion-free. Thus $G_e = 1$.

Write $A = G_{\iota e}$, $B = G_{\tau e}$, $A_0 = G_0 \cap A$, $B_0 = G_0 \cap B$.

If VT has two G-orbits then VT_0 has two G_0-orbits and (*a*) holds; notice $A_0, B_0 \neq 1$ since $n = 1$ is minimal.

Thus we may assume VT has one G-orbit, so $\iota e = c \tau e$ for some $c \in G$ and $G = A * c$, $B = c^{-1} A c$.

If VT_0 has two G_0-orbits then $G_0 = A_0 * B_0$, and $A_0 \leqslant A$ and $cB_0 c^{-1} \leqslant cBc^{-1} = A$. Hence (*b*) holds.

It remains to consider the case where VT_0 also has one G_0-orbit, so we may assume $c \in G_0$. Here $G_0 = A_0 * c$ and we can write $G = A * \langle c \rangle$,

$G_0 = A_0 * \langle c \rangle$. If $A \neq 1$ then (*a*) holds; again $A_0 \neq 1$ by the minimality of $n = 1$. If $A = 1$ then (*c*) holds. ∎

9.5 Corollary. *Let G be finitely generated torsion-free, W a G-finite G-set with finitely generated stabilizers, and w an element of W which has (G, W)-weight 2, and infinite cyclic stabilizer generated by g. Then one of the following holds:*

(*a*) *$G = A * B$, $g = ab$ for some $a \in A$, $b \in B$ with $a, b \neq 1$, and $W = Gw \vee G \otimes_A W_A \vee G \otimes_B W_B$ for some A-set W_A and B-set W_B. If $A = \langle a \rangle$ then W_A is nonempty, and if $B = \langle b \rangle$ then W_B is nonempty.*

(*b*) *$G = A * b$, $g = aba'b^{-1}$ for some $a, a' \in A$ with $a, a' \neq 1$, and $W = Gw \vee G \otimes_A W_A$ for some A-set W_A.*

(*c*) *$G = \langle b \rangle$ and $g = b^2$ and $W - Gw$ is G-free.*

Proof. By Theorem 9.3, 2 is the smallest integer n such that there exists a G-tree T and a $\langle g \rangle$-subtree T' and an edge e of T such that G_e is finite, G stabilizes no vertex of T, there exists a G-map $(W - Gw) \to VT$, and there are exactly $n = 2$ cosets $\langle g \rangle x$ with $\langle g \rangle xe \subseteq ET'$.

Here the descriptions of W will be immediate from Lemma III.3.4. Notice that in (*a*) if $A = \langle a \rangle$ and W_A is empty then $G = B * \langle g \rangle$ and we could take $n = 0$, which is a contradiction, so W_A is nonempty. Similarly, if $B = \langle b \rangle$ then W_B is nonempty.

Notice $G_e = 1$ since G is torsion-free.

Now T' must be homeomorphic to $\mathbb{R}$, with a fundamental $\langle g \rangle$-transversal consisting of two edges, say e, be, and we may assume we have one of the following configurations:

(1) $$\iota e \xrightarrow{e} \tau e = b\iota e \xrightarrow{be} b\tau e = b^2\iota e = g\iota e \xrightarrow{ge} g\tau e,$$

(2) $$\iota e \xrightarrow{e} \tau e = b\tau e \xleftarrow{be} b\iota e = g\iota e \xrightarrow{ge} g\tau e.$$

Suppose (1) holds. Write $A = G_{\iota e}$, $B = \langle b \rangle$, and $a = b^{-2}g$ so $G = A * b = A * B$ and $a \in A$. If $A \neq 1$ then (*a*) holds, since $B \neq 1$ and $g = b^2 a$ with $b^2 \in B$, $a \in A$, and we must have $a, b^2 \neq 1$ for otherwise we could take $n = 0$. If $A = 1$ then (*c*) holds.

Now suppose (2) holds. Write $A = G_{\iota e}$, $B = G_{\tau e}$ and $a = b^{-1}g$ so $a \in A$, $b \in B$ and $a, b \neq 1$.

If VT has two G-orbits then $G = A * B$ and (*a*) holds.

It VT has one G-orbit then $\iota e = c\tau e$ for some $c \in G$, and $G = B * c$. Here $c^{-1}Ac = B$, so $b' = c^{-1}ac \in B$ and $g = ba = bcb'c^{-1}$ with $b, b' \in B$ and $b, b' \neq 1$ so (*b*) holds. ∎

10 PD^2-pairs

We now find that for nonempty G-sets, the PD^2-pairs are geometric.

10.1 Theorem. *Let G be torsion-free, (G, W) a PD^2-pair over K, $w_0 \in W$, and t_0 a generator of G_{w_0}.*

Then there exists a nonempty free generating set $t_1, \ldots, t_m, x_1, \ldots, x_n$ for G, such that $W = \bigvee_{i=0}^{m} G/\langle t_i \rangle$ and t_0 is either $t_1 \cdots t_m[x_1, x_2] \cdots [x_{n-1}, x_n]$ or $t_1 \cdots t_m x_1^2 \cdots x_n^2$.

Proof. By Remarks 7.2(iii), (iv), G is finitely generated and $\mathrm{cd}_K G = 1$. As it is torsion-free, G is a nontrivial finitely generated free group by Corollary IV.3.14 or Remark III.2.3(ii). We proceed by induction on the rank of G.

By Remark 7.2(iv), W is G-finite. By Theorem 7.11, for each $w \in W$, G_w is a PD^1-group over K and the K-linear map

$$(1) \qquad H^1(G, KG) \to \bigoplus_W H^1(G_w, KG_w) \approx \bigoplus_W K \approx KW$$

is injective and has image ωKW. By Theorem 4.4, each G_w is infinite cyclic.

Suppose rank $G = 1$. By Theorem 4.4, $H^1(G, KG)$ is a free K-module of rank 1; but there is a K-module isomorphism $\omega KW \approx K$, and thus W has exactly two elements, w_0, w say.

If w_0, w are stabilized by G then t_0 is a free generator of G, and $W = G/\langle t_0 \rangle \vee G/\langle t_0 \rangle$. Let $t_1 = t_0$. Then t_1 is a free generator of G such that $W = G/\langle t_0 \rangle \vee G/\langle t_1 \rangle$ and $t_0 = t_1$. This is one of the desired forms.

If w_0, w are interchanged by some element x_1 of G then x_1 is a free generator of G such that $W = G/\langle t_0 \rangle$ and $t_0 = x_1^2$. This also is one of the desired forms.

Now suppose that rank $G \geqslant 2$, and the theorem holds for groups of smaller rank.

For any $w \in W$, since rank $G \geqslant 2$, we can choose $g \in G$ with $gw \neq w$. Hence $gw - w$ is a nonzero element of ωKW whose support lies in Gw and has two elements, and clearly 2 is the smallest possible number. It follows from (1) that w has (G, W)-weight 2, and this applies in particular with $w = w_0$.

Since rank $G \geqslant 2$, by Corollary 9.5 there are two possibilities.

(*a*) $G = A * B$, $t_0 = ab$ with $a \in A$, $b \in B$, a, $b \neq 1$ and $W = G/\langle t_0 \rangle \vee G \otimes_A W_A \vee G \otimes_B W_B$.

Express G as the fundamental group of the graph of groups

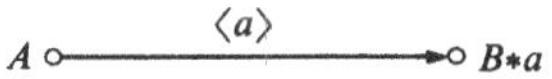

and let T denote the corresponding G-tree. Since $t_0 \in B*a$, there is a natural G-map $\phi: W \to VT$. Moreover, ϕW generates T, for if $A = \langle a \rangle$ then W_A is nonempty by Corollary 9.5. Now, by Theorem 8.7, $(A, A/\langle a \rangle \vee W_A)$ is a PD^2-pair over K. The induction hypothesis implies that there is a nonempty free generating set $t_1, \ldots, t_j, x_1, \ldots, x_k$ for A such that $W_A = \bigvee_{i=1}^{j} A/\langle t_i \rangle$ and $a = t_1 \cdots t_j q$, where $q = [x_1, x_2] \cdots [x_{k-1}, x_k]$ or $x_1^2 \cdots x_k^2$.

Similarly, there is a nonempty free generating set $t_{j+1}, \ldots, t_m, x_{k+1}, \ldots, x_n$ for B such that $W_B = \bigvee_{i=j+1}^{m} B/\langle t_i \rangle$ and $b = t_{j+1} \cdots t_m q'$, where q' is either $[x_{k+1}, x_{k+2}] \cdots [x_{m-1}, x_m]$ or $x_{k+1}^2 \cdots x_m^2$.

Combined, these give a nonempty free generating set $t_1, \ldots, t_m, x_1, \ldots, x_n$ for G such that $W = \bigvee_{i=0}^{m} G/\langle t_i \rangle$ and $t_0 = ab = t_1 \cdots t_j q t_{j+1} \cdots t_m q' = t_1 \cdots t_m q^{t_{j+1} \cdots t_m} q'$. By replacing x_i with $x_i^{t_{j+1} \cdots t_m}$ for $i \in [1, k]$ we arrive at the form $t_0 = t_1 \cdots t_m q q'$. If this is not already in the desired form, it can be adjusted by repeated use of Lemma 8.4.

(b) $G = A*b$, $t_0 = aba'b^{-1}$ with $a, a' \in A$, a, $a' \neq 1$, and $W = G/\langle t_0 \rangle \vee G \otimes_A W_A$.

In a natural way G is isomorphic to the fundamental group of the graph of groups

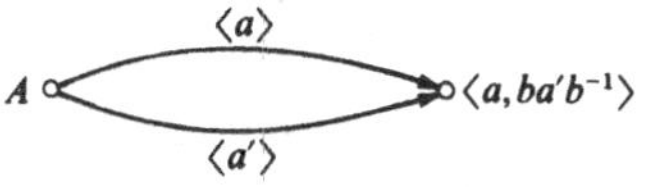

where a' maps to $ba'b^{-1}$. Let T be the corresponding G-tree. Since $t_0 \in \langle a, ba'b^{-1} \rangle$, there is a natural G-map $\phi: W \to VT$ and ϕW generates T. By Theorem 8.7, $(A, A/\langle a \rangle \vee A/\langle a' \rangle \vee W_A)$ is a PD^2-pair over K. By the induction hypothesis A has a nonempty free generating set $t_1, \ldots, t_{m+1}, x_1, \ldots, x_n$ such that $A/\langle a' \rangle \vee W_A = \bigvee_{i=1}^{m+1} A/\langle t_i \rangle$ and $a = t_1 \cdots t_{m+1} q$, where q is $[x_1, x_2] \cdots [x_{n-1}, x_n]$ or $x_1^2 \cdots x_n^2$. Thus $a' = t_i^{\pm c}$ for some i and some $c \in A$; we can arrange that $i = m+1$, by choosing new t_i's, since $t_i t_{i+1} = t_{i+1} t_i^{t_{i+1}}$. Thus A has a nonempty free generating set $t_1, \ldots, t_m, x_1, \ldots, x_n, a'^{\pm c}$ such that $W_A = \bigvee_{i=1}^{m} G/\langle t_i \rangle$ and $a = t_1 \cdots t_m a'^{\pm c} q$.

Hence there exists a nonempty free generating set $t_1, \ldots, t_m, x_1, \ldots, x_n$,

$x_{n+1} = a'^{cq}, x_{n+2} = bcq$ for G, such that $W = \bigvee_{i=0}^{m} G/\langle t_i \rangle$ and

$$t_0 = aba'b^{-1} = t_1 \cdots t_m a'^{\pm c} qba'b^{-1} = t_1 \cdots t_m q a'^{\pm cq}(bcq)a'^{cq}(bcq)^{-1}$$
$$= t_1 \cdots t_m q(x_{n+1})^{\pm 1} x_{n+2} x_{n+1} x_{n+2}^{-1},$$

where q is $[x_1, x_2] \cdots [x_{n-1}, x_n]$ or $x_1^2 \cdots x_n^2$. Using Lemma 8.4, we can bring this to the desired form. ∎

Let us combine this with Example 8.5

10.2 Theorem. *For any nonempty G-set W, (G, W) is a* GD^2*-pair if and only if G is torsion-free and (G, W) is a* PD^2*-pair over K. In this event the algebraic orientation map is given by the geometric orientation map.* ∎

We note the case $K = \mathbb{Z}$.

10.3 Theorem. *For any nonempty G-set W, (G, W) is a* GD^2*-pair if and only if it is a* PD^2*-pair, and then the orientation maps agree.* ∎

11 Projective modules over group rings

This section sets out the proofs of some classical results on projective modules over group rings which will be required in the next section.

Throughout N denotes a positive integer.

11.1 Definitions. Let $\mathcal{M}_N(KG)$ denote the ring of all $N \times N$ matrices over KG, and $e_{i,j}$ the matrix units. Let $a \in \mathcal{M}_N(KG)$.

There are elements $a_{g,i,j} \in K$ indexed by $(g, i, j) \in G \times [1, N] \times [1, N]$, and almost all zero, such that $a = \sum_{g \in G} \sum_{j \in [1,N]} g e_{i,j} a_{g,i,j}$. The map $\text{tr}: \mathcal{M}_N(KG) \to K, a \mapsto \sum_{i \in [1,N]} a_{1,i,i}$, is called the *trace*. Clearly tr is K-linear and $\text{tr}(ab) = \text{tr}(ba)$.

Now take $K = \mathbb{C}$.

For $z \in \mathbb{C}$ let $\bar{z}$ denote the complex conjugate of z, and $|z|$ the modulus $(z\bar{z})^{\frac{1}{2}} \in \mathbb{R}$.

Define $\bar{a} = \sum_{g \in G} \sum_{i,j \in [1,N]} g^{-1} e_{j,i} \bar{a}_{g,i,j}$. Then—is an involution of $\mathcal{M}_N(\mathbb{C}G)$ and $\text{tr}(\bar{a}) = \overline{\text{tr}(a)}$.

We write $\|a\| = (\text{tr}(a\bar{a}))^{\frac{1}{2}} = \left(\sum_{g,i,j} |a_{g,i,j}|^2\right)^{\frac{1}{2}}$ and $|a| = \sum_{g,i,j} |a_{g,i,j}|$, non-negative real numbers which are positive if $a \neq 0$. ∎

11.2 Lemma. *Let* $a, b \in \mathcal{M}_N(\mathbb{C}G)$.

(i) *If* $b \neq 0$ *and* $z = \operatorname{tr}(a\bar{b})\|b\|^{-2}$ *then* $\|a\|^2 - \|a - zb\|^2 = |\operatorname{tr}(a\bar{b})|^2\|b\|^{-2}$.

(ii) $|\operatorname{tr}(ab)| \leqslant \|a\|\,\|b\|$.

(iii) $\|a+b\| \leqslant \|a\| + \|b\|$.

(iv) $\|ab\| \leqslant \|a\|\,|b|$.

Proof. (i) $\|a - zb\|^2 = \operatorname{tr}((a - zb)(\bar{a} - \bar{z}\bar{b})) = \operatorname{tr}(a\bar{a}) - z\operatorname{tr}(b\bar{a}) - \bar{z}\operatorname{tr}(a\bar{b}) + z\bar{z}\operatorname{tr}(b\bar{b}) = \|a\|^2 - |\operatorname{tr}(a\bar{b})|^2\|b\|^{-2} - |\operatorname{tr}(a\bar{b})|^2\|b\|^{-2} + |\operatorname{tr}(a\bar{b})|^2\|b\|^{-2} = \|a\|^2 - |\operatorname{tr}(a\bar{b})|^2\|b\|^{-2}$.

(ii) We may assume $b \neq 0$. Since $\|a - zb\|^2 \geqslant 0$ it follows from (i) that $\|a\|^2 \geqslant |\operatorname{tr}(a\bar{b})|^2\|b\|^{-2}$ so $\|a\|\,\|b\| \geqslant |\operatorname{tr}(\bar{b})|$. Replacing b with $\bar{b}$ gives (ii).

(iii) Since $\operatorname{tr}(a\bar{b}) - \operatorname{tr}(\bar{a}b)$ is purely imaginary, its square is a real number which is at most zero. Hence $0 \leqslant (\operatorname{tr}(a\bar{b}) + \operatorname{tr}(b\bar{a}))^2 \leqslant 4\operatorname{tr}(a\bar{b})\operatorname{tr}(b\bar{a}) = 4|\operatorname{tr}(a\bar{b})|^2 \leqslant 4\|a\|^2\|b\|^2$, by (ii). Taking square roots we get $\operatorname{tr}(a\bar{b}) + \operatorname{tr}(b\bar{a}) \leqslant 2\|a\|\,\|b\|$; adding $\operatorname{tr}(a\bar{a}) + \operatorname{tr}(b\bar{b}) = \|a\|^2 + \|b\|^2$ to both sides and taking square roots again, we get the triangle inequality (iii).

(iv) For any $(g, i, j) \in G \times [1, N] \times [1, N]$, we have $0 \leqslant \|age_{i,j}\|^2 = \operatorname{tr}(age_{i,j}e_{j,i}g^{-1}\bar{a}) = \operatorname{tr}(ae_{ii}\bar{a})$, but $\sum_i \operatorname{tr}(ae_{ii}\bar{a}) = \operatorname{tr}(a\bar{a}) = \|a\|^2$ so $\|age_{i,j}\|^2 \leqslant \|a\|^2$. Now $\|ab\| = \left\| a \sum_{g,i,j} ge_{i,j}b_{g,i,j} \right\| \leqslant \sum_{g,i,j} \|age_{i,j}b_{g,i,j}\| = \sum_{g,i,j} \|age_{i,j}\|\,|b_{g,i,j}| \leqslant \sum_{g,i,j} \|a\|\,|b_{g,i,j}| = \|a\|\,|b|$, where the first inequality follows from (iii). ∎

11.3 Kaplansky's Theorem. *If* $e \in \mathcal{M}_N(\mathbb{C}G)$ *is a nonzero idempotent then* $\operatorname{tr}(e)$ *is a positive real number, and in fact* $\operatorname{tr}(e) \geqslant \|e\|^4\|e\bar{e}\|^{-2} \geqslant \|e\|^2|e|^{-2} > 0$.

Proof. Write $c = e\bar{e}$.

Set $a_0 = 1$. Suppose $n \geqslant 0$ and a_n has been defined. Notice

(1) $\quad \|a_n e\|^2 = \operatorname{tr}(a_n \overline{eea_n}) = \operatorname{tr}(a_n \overline{ca_n}) = \operatorname{tr}(a_n \overline{ca_n})$.

If $a_n c \neq 0$ define $s_n = \operatorname{tr}(a_n \overline{ca_n})\|a_n c\|^{-2} = \|a_n e\|^2\|a_n c\|^{-2} \in \mathbb{R}$ and $a_{n+1} = a_n - s_n a_n c$. Taking $r_n = |\operatorname{tr}(a_n \overline{ca_n})|\,\|a_n c\|^{-1} = \|a_n e\|^2\|a_n c\|^{-1} \in \mathbb{R}$ and applying Lemma 11.2(i) we get

(2) $\quad \|a_n\|^2 - \|a_{n+1}\|^2 = r_n^2 \geqslant 0 \quad$ and $\quad \|a_n c\| r_n = \|a_n e\|^2$.

If $a_n c = 0$, define $a_{n+1} = a_n$, $r_n = 0$; by (1), $\|a_n e\|^2 = 0$ so (2) holds.

In particular, $r_0 > 0$ since $a_0 e = e \neq 0$.

From (2) we see that $\|a_n\|$ is a decreasing sequence of non-negative real numbers, so converges to a limit. Hence $r_n^2 = \|a_n\|^2 - \|a_{n+1}\|^2 \to 0$, so $r_n \to 0$. By the second part of (2), and Lemma 11.2(iv), $0 \leqslant \|a_n e\|^2 = \|a_n c\| r_n \leqslant \|a_n\| |c| r_n \to 0$, and thus $\|a_n e\| \to 0$. By Lemma 11.2(ii), $0 \leqslant |\mathrm{tr}(a_n^2 e)| \leqslant \|a_n\| \|a_n e\| \to 0$ so $\mathrm{tr}(a_n^2 e) \to 0$.

It follows from the construction that a_n is a polynomial in c with real coefficients and constant term 1. Hence $\bar{a}_n = a_n$ since $\bar{c} = c$, so $\|a_n\|^2 = \mathrm{tr}(a_n \bar{a}_n) = \mathrm{tr}(a_n^2)$. Moreover, $e(1 - a_n^2) = (1 - a_n^2)$ since $ec = c$.

By (2), $\sum_{m=0}^{n-1} r_m^2 = \sum_{m=0}^{n-1} \|a_m\|^2 - \|a_{m+1}\|^2 = \|a_0\|^2 - \|a_n\|^2 = \mathrm{tr}(1) - \mathrm{tr}(a_n^2) = \mathrm{tr}(1 - a_n^2) = \mathrm{tr}(e(1 - a_n^2)) = \mathrm{tr}((1 - a_n^2)e) = \mathrm{tr}(e) - \mathrm{tr}(a_n^2 e) \to \mathrm{tr}(e)$.

Hence $\mathrm{tr}(e) = \sum_{n \geqslant 0} r_n^2 \geqslant r_0^2 > 0$, and thus $\mathrm{tr}(e)$ is a positive real number.

By Lemma 11.2(iv), $\|e\bar{e}\| \leqslant \|e\| |\bar{e}| = \|e\| |e|$ so

$$0 < \|e\| |e|^{-1} \leqslant \|e\|^2 \|e\bar{e}\|^{-1} = r_0 \leqslant \mathrm{tr}(e)^{1/2}. \quad \blacksquare$$

11.4 Definition. Let C be the set of conjugacy classes of G, and form the free K-module $K[C]$ on C. There is a map $\mathrm{Tr}: \mathscr{M}_N(KG) \to K[C]$, called the *Hattori–Stallings Trace*, defined by $\mathrm{Tr}(a) = \sum_{g,i} a_{g,i,i}[g]$, where $[g]$ denotes the conjugacy class of g.

Clearly Tr is K-linear, and since $[gh] = [hg]$ for all $h, g \in G$, it follows that $\mathrm{Tr}(ab) = \mathrm{Tr}(ba)$ for all $a, b \in \mathscr{M}_N(KG)$. $\blacksquare$

By identifying $[1] = 1$, we may treat the trace of Definition 11.1 as a map $\mathrm{tr}: \mathscr{M}_N(KG) \to K[C]$, and we shall be particularly interested in idempotents e with $\mathrm{Tr}(e) = \mathrm{tr}(e)$.

If M is a free module over K, we write $[M:K]$ for the rank.

11.5 Theorem. *Let K be a field of characteristic 0, e a nonzero idempotent in $\mathscr{M}_N(KG)$ and P the projective KG-submodule of KG^N generated by the rows of e. If $\mathrm{Tr}(e) = \mathrm{tr}(e)$ then for any finite quotient group B of G, $KB \otimes_{KG} P$ is a nonzero free KB-module, and for any normal subgroup H of finite index in G, $[K \otimes_{KH} P:K] = (G:H)[K \otimes_{KG} P:K]$.*

Proof. To see that the final part follows from the first, observe that if $B = G/H$ then $[K \otimes_{KH} P:K] = [K \otimes_{KH} KG \otimes_{KG} P:K] = [KB \otimes_{KG} P:K] = [KB \otimes_{KG} P:KB][KB:K] = [K \otimes_{KB} (KB \otimes_{KG} P):K]|B| = (G:H)[K \otimes_{KG} P:K]$.

It remains to prove the first part. On replacing K by the subfield generated by the $e_{g,i,j}$ we may assume that K is finitely generated, and here there is an embedding of K in $\mathbb{C}$, so $\operatorname{tr}(e) \neq 0$ by Kaplansky's Theorem 11.3.

Since $\operatorname{Tr}(e) = \operatorname{tr}(e)$ it follows that the image $\bar{e}$ of e in $\mathcal{M}_N(KB)$ satisfies $\operatorname{tr}(\bar{e}) = \operatorname{tr}(e)$ in K, and thus $\operatorname{Tr}(\bar{e}) = \operatorname{tr}(\bar{e}) \neq 0$. Hence, without loss of generality, we may assume that $G = B$ is finite, and it remains to show that P is free.

There is a KG-split surjection $KG \otimes_K P \to P$, $r \otimes p \mapsto rp$, with right inverse $P \mapsto KG \otimes_K P$ given by

$$p \mapsto \frac{1}{|G|} \sum_{g \in G} g \otimes g^{-1} p.$$

The composite

$$\alpha: KG \otimes_K P \to KG \otimes_K P,\ 1 \otimes p \mapsto \frac{1}{|G|} \sum_{g \in G} g \otimes g^{-1} p,$$

is the projection onto the copy of P.

Let $b_1, \ldots, b_m$ be a K-basis of P, and write $gb_i = \sum_{j \in [1,m]} a_{ij}(g) b_j$ with $a_{ij}(g) \in K$, for all $g \in G$, $j \in [1, m]$. Then $1 \otimes b_i$, $i \in [1, m]$, is a KG-basis of $KG \otimes_K P$, and the matrix representing α with respect to this basis is

$$E = \frac{1}{|G|} \sum_{i,j,g} g a_{ij}(g^{-1}) e_{ij} \in \mathcal{M}_m(KG).$$

Thus the KG-module generated by the rows of E is isomorphic to P, so $\operatorname{Tr}(E) = \operatorname{Tr}(e)$, and $\operatorname{tr}(E) = \operatorname{tr}(e)$.

Now

$$\operatorname{Tr}(E) = \frac{1}{|G|} \sum_{i, g \in G} [g] a_{ii}(g^{-1}) = \frac{1}{|G|} \sum_{[g] \in C} [g] \chi(g^{-1})(G : C_G(g)),$$

where $\chi: G \to K$ is the character associated to P, and $C_G(g)$ is the centralizer of g in G. Since $\operatorname{Tr}(E) = \operatorname{tr}(E)$ we see that $\chi(g) = 0$ for all $g \neq 1$. It now follows from character theory that P is a free KG-module.

(In fact, in the only application of interest to us, B will be a finite subgroup of K and hence cyclic generated by a root of unity $\phi = \zeta$ of order $n = |B|$. Here the character theory needed is rather slight. By linear algebra, there is a decomposition $P = P_1 \oplus \cdots \oplus P_n$ with ϕ acting on P_j as ζ^j, and thus, for each $i \in [1, n]$, $\chi(\phi^i) = \sum_j m_j \zeta^{ji}$, where $m_j = [P_j : K]$, so $m_1 + \cdots + m_n = m$. In our situation $\sum_j m_j \zeta^{ji} = \chi(\phi^i) = 0$ for $i = 1, \ldots, n-1$, and it follows that $\sum_j (m_j - (m/n)) \zeta^{ji} = 0$ for $i = 1, \ldots, n$. Hence each $m_j - (m/n)$ is 0, and P is free of rank m/n.) ∎

Recall that the *characteristic* of K is the non-negative integer n which generates the kernel of the ring homomorphism $\mathbb{Z} \to K$.

11.6 Theorem. *Suppose that K is an integral domain, and suppose that for each $g \in G$ and $n \geqslant 2$, if $[g] = [g^n]$ then $g = 1$. Then, for any idempotent $e \in \mathcal{M}_N(KG)$, $\mathrm{Tr}(e) = \mathrm{tr}(e)$.*

Proof. (i) Suppose that the characteristic of K is a prime $p \neq 0$.

For any $h \in G$, and $a \in \mathcal{M}_N(KG)$ let us write $h \in a$ if some $a_{h,i,j} \neq 0$; for any $b \in K[C]$, let us write $[h] \in b$ if $[h]$ occurs with nonzero coefficient in b. Suppose that $[h] \in \mathrm{Tr}(e)$; we wish to show that $h = 1$.

Let q be a nontrivial power of p, that is $q = p^i$ for some $i \geqslant 1$.

We claim that for $a, b \in \mathcal{M}_N(KG)$, $\mathrm{Tr}((a+b)^q) = \mathrm{Tr}(a^q + b^q)$. Here $(a+b)^q$ is the sum of the 2^q words $w_1 \cdots w_q$ in a and b. Since $[w_1 \cdots w_q] = [w_2 \cdots w_q w_1]$, the formal words equivalent under cyclic permutation can be collected together. These equivalence classes are orbits under the action of C_q, so the number of elements in a class is one or a multiple of p. Since it is clear that the only two formal words which are stabilized by C_q correspond to a^q, b^q, we have proved the claim.

Now $[h] \in \mathrm{Tr}(e) = \mathrm{Tr}(e^q) = \mathrm{Tr}\left(\left(\sum_{g,i,j} g e_{i,j} e_{g,i,j}\right)^q\right) = \mathrm{Tr}\left(\sum_{g,i,j} g^q e^q_{i,j} e^q_{g,i,j}\right)$ $= \sum_{g,i} [g^q] e^q_{g,i,i}$, so $[h] = [g^q]$ for some $g \in e$. Since q is an arbitrary nontrivial power of p, and there are only finitely many $g \in e$, we see there must be some $g \in e$, and distinct $m, n \geqslant 1$ such that $[h] = [g^{p^n}] = [g^{p^m}]$. We may assume $m > n$, and then $[h^{p^{m-n}}] = [(g^{p^n})^{p^{m-n}}] = [g^{p^m}] = [h]$, so, by the hypothesis on $G, h = 1$ as desired.

(ii) Suppose now that K has characteristic zero.

On replacing K by the subring generated by the $e_{g,i,j}$ we may assume that K is finitely generated. Suppose that $\mathrm{Tr}(e) \neq \mathrm{tr}(e)$, and let $t \in K$ be a nonzero coefficient occurring in $\mathrm{Tr}(e) - \mathrm{tr}(e)$.

Then t is not nilpotent, so the ring $K[t^{-1}]$ obtained from K by inverting t is nonzero. Let F be a field obtained by factoring out a maximal ideal of $K[t^{-1}]$.

We claim that the characteristic of F is finite. Suppose not. Let Y be a finite transcendence basis for F over $\mathbb{Q}$, and write Y^* for the free commutative monoid on Y, so that F is algebraic over the polynomial ring $\mathbb{Z}[Y^*]$. Let X be a finite set which generates F as a ring. For some nonzero $r \in \mathbb{Z}[Y^*]$, every element of X is integral over $\mathbb{Z}[Y^*][r^{-1}]$, so F itself is integral over $\mathbb{Z}[Y^*][r^{-1}]$. But there is some prime integer p not dividing any power of r in $\mathbb{Z}[Y^*]$, which contradicts p^{-1} being integral over $\mathbb{Z}[Y^*][r^{-1}]$, and the claim is proved. For the natural map $\mathcal{M}_N(KG)\to\mathcal{M}_N(FG)$, $a\mapsto\bar{a}$, we have $\mathrm{Tr}(\bar{e})=\mathrm{tr}(\bar{e})$ in $F[C]$ by (i), contradicting the fact that the image of t in F invertible. ∎

11.7 Lemma. *If G is a torsion-free* PD^2*-group over K then for each $g\in G$ and $n\geqslant 2$, if $[g]=[g^n]$ then $g=1$.*

Proof. Suppose $x^{-1}gx=g^n$ with $g, x\in G$ and $n\geqslant 2$.

For each $i\geqslant 0$, let $g_i=x^igx^{-i}$. Then $g_{i+1}^n=x^{i+1}g^nx^{-i-1}=x^{i+1}(x^{-1}gx)x^{-i-1}=g_i$. Let H be the subgroup of G generated by the g_i, that is, the directed union of the cyclic groups $\langle g_i\rangle$.

We claim that H is finitely generated. This is clear if H has finite index in G, since G is finitely generated by Remark 3.4(iii). If H has infinite index in G then, by Corollary 6.4, H is a free group. As it is abelian, H must be cyclic, so finitely generated.

Thus for some $i, H=\langle g_i\rangle=\langle g_{i+1}\rangle$, so for some $m\in\mathbb{Z}$, $g_{i+1}=g_i^m=g_{i+1}^{nm}$. Since $n\geqslant 2$, g_{i+1} has finite order, and, since G is torsion-free, $g_{i+1}=1$. Thus $H=1$, so $g=1$. ∎

Let us combine the last three results.

11.8 Theorem. *Let K be a field of characteristic* 0 *and G a torsion-free* PD^2*-group over K. For any nonzero finitely generated projective left KG-module P, $K\otimes_{KG}P\neq 0$; moreover, for any normal subgroup H of finite index in G, $[K\otimes_{KH}P:K]=(G:H)[K\otimes_{KG}P:K]$.* ∎

12 PD^2-groups

In this section we bring to fruition the labour of the entire chapter by characterizing the PD^2-groups.

Recall that G is *indicable* if there is a surjective homomorphism $G\to\mathbb{Z}$.

12.1 Theorem. *If G is indicable and* FP$_2$ *over R then $G=A\underset{C}{*}x$ for some finitely generated subgroups $C\leqslant A\leqslant G$, and some $x\in G$ with $C^x\leqslant A$.*

Proof. Suppose that G is FP$_2$ over R, and indicable, say $\sigma:G\to\mathbb{Z}$ is surjective.

Choose an element $\bar{x}_0\in G$ with $\sigma(\bar{x}_0)=1$. Since G is finitely generated by Proposition 3.2(ii), we can extend $\bar{x}_0$ to a finite generating set $\bar{X}=$

$\{\bar{x}_0, \bar{x}_1, \ldots, \bar{x}_n\}$ of G. For each $i = 1, \ldots, n$ let us replace $\bar{x}_i$ with $\bar{x}_i \bar{x}_0^{-\sigma(x_i)}$ so that $\sigma(\bar{x}_i) = 0$.

Let X be a set in bijective correspondence with $\bar{X}$, let F be the free group on X, and denote the kernel of the resulting map $F \to G$ by N. Notice that the composite $F \to G \xrightarrow{\sigma} \mathbb{Z}$ is given by the exponent sum on x_0, so has as kernel the subgroup of F generated by $\{x_0^{-j} x_i x_0^j \mid j \in \mathbb{Z},\ 1 \leqslant i \leqslant n\}$. In particular, each element of N can be expressed in terms of these words.

By Proposition 3.2(iii), $R \otimes_{\mathbb{Z}} N^{ab}$ is finitely generated as left RF-module, so we can choose a finite generating set of the form $\{1 \otimes n_1 N', \ldots, 1 \otimes n_m N'\}$ with $n_1, \ldots, n_m \in N$. Choose a positive integer J sufficiently large so that $n_1, \ldots, n_m$ all lie in the subgroup A of F generated by $\{x_0^{-j} x_i x_0^j \mid -J \leqslant j \leqslant J, 1 \leqslant i \leqslant n\}$. Let C be the subgroup of F generated by $\{x_0^{-j} x_i x_0^j \mid -J \leqslant j < J, 1 \leqslant i \leqslant n\}$. There is a right conjugation map $x_0: C \to A$, and it follows from the universal properties of $A \underset{C}{*} x_0$ and F that there is a natural identification $F = A \underset{C}{*} x_0$.

Let $\bar{A}$, $\bar{C}$ denote the images of A, C in G, respectively, so there is a right conjugation map $\bar{x}_0: \bar{C} \to \bar{A}$. There is then a natural surjective map $\beta: \bar{A} \underset{\bar{C}}{*} \bar{x}_0 \to G$, which we shall prove is injective. Notice that $\operatorname{Ker} \beta$ does not meet $\bar{A}$ so acts freely on the standard tree for the HNN extension, and is therefore free. We now lift this data back to F.

There is a natural surjection $\alpha: F = A \underset{C}{*} x_0 \to \bar{A} \underset{\bar{C}}{*} \bar{x}_0$, so we have maps $F = A \underset{C}{*} X_0 \xrightarrow{\alpha} \bar{A} \underset{\bar{C}}{*} \bar{x}_0 \xrightarrow{\beta} G \xrightarrow{\sigma} \mathbb{Z}$, and it is clear that the composite $\beta\alpha: F \to G$ is the given map.

Write $N_\alpha = \operatorname{Ker} \alpha$, so $N_\alpha \subseteq \operatorname{Ker} \beta\alpha = N$, and $N/N_\alpha \approx \operatorname{Ker} \beta$ which we know is a free group.

The elements $n_1, \ldots, n_m$ lie in the kernel of the map $A \to \bar{A}$, so lie in N_α. Hence the surjective RF-linear map $R \otimes_{\mathbb{Z}} N^{ab} \to R \otimes_{\mathbb{Z}} (N/N_\alpha)^{ab}$ vanishes on the chosen generating set, from which it follows that $R \otimes_{\mathbb{Z}} (N/N_\alpha)^{ab}$ is zero. Since R is nonzero and N/N_α is a free group, we see that $N/N_\alpha = 1$, which proves that β is injective. ∎

12.2 Theorem. *If G is an indicable torsion-free* PD^2*-group over K then G is an infinite surface group.*

Proof. By Definition 3.3(*d*), G is FP_2 over K, and G is indicable by hypothesis, so, by Theorem 12.1, $G = A \underset{C}{*} b$ with C, A finitely generated. Since C is of infinite index, it is free by Corollary 6.4.

It follows that there exists a finitely generated free subgroup C of G of minimal rank such that there exists a G-tree T with $ET = G/C$ as G-set, and with vertex stabilizers finitely generated and different from G.

Choose an edge e_0 of T such that $G_{e_0} = C$, and let $A = G_{\iota e_0}$, $B = G_{\tau e_0}$, $W_A = \operatorname{star}(\iota e_0)$, $W_B = \operatorname{star}(\tau e_0)$, and, if there is only one vertex orbit, choose $b \in G$ with $b\iota e_0 = \tau e_0$. Then one of the following holds:

(1) $G = A \underset{C}{*} B$, $W_A = Ae_0 = A/C$, $W_B = Be_0 = B/C$ and A, B are finitely generated, different from G;

(2) $G = A \underset{C}{*} b$, $W_A = Ae_0 \vee Ab^{-1}e_0 = A/C \vee A/C^b$ and A is finitely generated.

We shall show by contradiction that rank $C = 1$.

Suppose rank $C = 0$. Then $H^1(G, KG) \neq 0$ by Theorem IV.6.10 $(d1) \Rightarrow (b)$, but G is a PD^2-group over K, so $H(G, KG)$ is concentrated in degree 2, a contradiction.

Now suppose rank $C \geqslant 2$. Here we will find a contradiction to the minimality of rank C.

Let $V = VT$, $E = ET$ and consider the exact sequence

(3) $$\operatorname{Ext}^1_{KG}(K, KG) \to \operatorname{Ext}^1_{KG}(KV, KG) \xrightarrow{\partial^*} \operatorname{Ext}^1_{KG}(KE, KG) \to \operatorname{Ext}^2_{KG}(K, KG),$$

which arises from applying $\operatorname{Ext}^1_{KG}(-, KG)$ to the exact sequence $0 \to KE \xrightarrow{\partial} KV \to K \to 0$. Since G is a PD^2-group over K, $\operatorname{Ext}^1_{KG}(K, KG) = 0$, and as K-module $\operatorname{Ext}^2_{KG}(K, KG) \approx K$. Moreover, V and E are G-finite with finitely generated stabilizers, so, by Lemma 7.10(iii), we can rewrite (3) as an exact sequence of K-modules

(4) $$0 \to \bigoplus_V H^1(G_v, KG_v) \xrightarrow{\partial^*} \bigoplus_E H^1(G_e, KG_e) \to K.$$

We leave it as a simple exercise to verify that if $\eta \in H^1(G_v, KG_v)$ then

$$\partial^*(\eta) = \sum_{e \in \tau^{-1}(v)} \eta_e - \sum_{e \in \iota^{-1}(v)} \eta_e,$$

where, for $e \in \operatorname{star}(v)$, $H^1(G_v, KG_v) \to H^1(G_e, KG_e)$, $\eta \mapsto \eta_e$, is the natural map arising from the inclusion $G_e \subseteq G_v$.

For each $e \in E$, G_e is free of rank at least two, so $H^1(G_e, KG_e)$ is a free K-module of infinite rank by Remark 4.3. In particular, the map $H^1(G_e, KG_e) \to K$ cannot be injective so there exists a nonzero element of $H^1(G_e, KG_e)$ which maps to zero in K. By the exactness of (4) at the V-term, this element can be lifted back to some $\sum_V \eta_v \in \bigoplus_V H^1(G_v, KG_v)$.

In summary, we can choose $e_1 \in E$, and $\sum_V \eta_v \in \bigoplus_V H^1(G_v, KG_v)$ such that the E-support of $\sum_V \partial^*(\eta_v)$ consists of e_1. Among the finitely many v such that $\eta_v \neq 0$, choose v so as to maximize the distance to e_1. It follows that $\partial^*(\eta_v) = \pm \eta_{v,e}$, where e is the first edge in the geodesic from v to e_0. Thus e has $(G_v, \operatorname{star}(v))$-weight at most 1. From the exactness of (4) at the V-term the $(G_v, \operatorname{star}(v))$-weight cannot be 0. Thus the $(G_v, \operatorname{star}(v))$-weight of e is 1.

Without loss of generality we may assume that $e = e_0$, $v = \iota e_0$ so $\operatorname{star}(v) = W_A$ and e_0 has W_A-weight 1. Since A is torsion-free and rank $C \geqslant 2$, Corollary 9.4 shows that one of the following holds:

(5) $\quad A = A_1 * A_2$ and $C = C_1 * C_2$ with $C_1 \leqslant A_1$, $C_2 \leqslant A_2$, and $C_1, C_2 \neq 1$;

(6) $\quad A = A_1 * a$ and $C = C_1 * C_2^a$ with $C_1, C_2 \leqslant A_1$, $C_1, C_2 \neq 1$.

Further, in case (2) C^b lies in an A-conjugate of A_1 or A_2, and without loss of generality we assume C^b lies in A_1.

There are four possibilities, each of which leads to a reduction in the rank of C.

For (1), (5), $G = (A_1 \underset{C_1}{*} B) \underset{C_2}{*} A_2$, and if $C_2 = A_2$ then $G = A_1 \underset{C_1}{*} B$.

For (1), (6), $G = (A_1 \underset{C_1}{*} B) \underset{C_2}{*} a$.

For (2), (5), $G = (A_1 \underset{C_1}{*} b) \underset{C_2}{*} A_2$, and if $C_2 = A_2$ then $G = A_1 \underset{C_1}{*} b$.

For (2), (6), $G = (A_1 \underset{C_1}{*} b) \underset{C_2}{*} a$.

This completes the proof that rank $C = 1$.

By Theorem 4.4, C is a PD^1-group over K, and, by Theorem 8.2, (A, W_A), (B, W_B) are PD^2-pairs over K. Let c be a generator of C.

In case (1), by Theorem 10.1, A has a nonempty free generating set $x_1, \ldots, x_n$ and $c = [x_1, x_2] \cdots [x_{n-1}, x_n]$ or $x_1^2 \cdots x_n^2$, while B has a nonempty free generating set $x_{n+1}, \ldots, x_m$ and $c = [x_{n+1}, x_{n+2}] \cdots [x_{m-1}, x_m]$ or $x_{n+1}^2 \cdots x_m^2$. Thus G is generated by $x_1, \ldots, x_m$ with $m \geqslant 2$, and one relation identifying the two expressions for c. It follows from Lemma 8.4 that G is an infinite surface group.

In case (2), by Theorem 10.1, A has a nonempty free generating set t, $x_1, \ldots, x_n$ and $c = tq$, where q is $[x_1, x_2] \cdots [x_{n-1}, x_n]$ or $x_1^2 \cdots x_n^2$ and moreover $c^b = t^{\pm a}$ for some $a \in A$. Hence G is generated by $t, x_1, \ldots, x_n$, $x = ab^{-1}$ with one relation identifying the two expressions for c, that is

$t^{\pm x}tq = 1$. It follows from Lemma 8.4 that G is an infinite surface group. ∎

We now arrive at the main theorem of the chapter.

12.3 Theorem. *If the characteristic of K is zero, then the torsion-free* PD^2*-groups over K are precisely the infinite surface groups.*

Proof. We observed one direction in Corollary 4.10.

Conversely, suppose that G is a torsion-free PD^2-group over K. We shall denote the orientation involution by $^\circ$ and the KG-dual by $*$.

By hypothesis, K contains a copy of $\mathbb{Z}$. Therefore $\mathbb{Q} \otimes_{\mathbb{Z}} K$ contains a copy of $\mathbb{Q}$, and $\mathbb{Q} \otimes_{\mathbb{Z}} K$ modulo a maximal ideal is a field of characteristic zero. Thus there exists a homomorphism from K to a field of characteristic zero, over which G remains a PD^2-group by Proposition 3.7. Hence we may assume that K is itself a field of characteristic zero.

By the preceding theorem, it suffices to show that G is indicable, and since G is finitely generated it suffices to show that G^{ab} is not torsion, or equivalently the K-module $G^{ab} \otimes_{\mathbb{Z}} K$ is nonzero.

Now recall that $G^{ab} \otimes_{\mathbb{Z}} K = H_1(G, K)$, a fact which is proved as follows. Applying $- \otimes_{KG} K$ to the exact sequence $0 \to \omega KG \to KG \to K \to 0$ gives an exact sequence $0 \to H_1(G, K) \to \omega KG \otimes_{KG} K \to K \to K \to 0$, where the middle arrow is zero, so $H_1(G, K) = \omega KG \otimes_{KG} K$. It is straightforward to verify that there is a K-module isomorphism $\omega KG \otimes_{KG} K \approx G^{ab} \otimes_{\mathbb{Z}} K$, with $(g-1) \otimes 1 \in \omega KG \otimes_{KG} K$ corresponding to $gG' \otimes 1 \in G^{ab} \otimes_{\mathbb{Z}} K$, where G' denotes the derived group.

We now calculate $H_1(G, K)$ another way to show that it is nonzero.

Construct an exact sequence of left KG-modules

$$0 \to Q_2 \to Q_1 \to KG \to K \to 0,$$

where Q_1, Q_2 are finitely generated projective left KG-modules; we shall now arrange that Q_2 is free of rank 1.

By Poincaré duality, applying $*^\circ$ gives another exact sequence of left KG-modules, which we write as

$$0 \to KG \to P_1 \xrightarrow{\phi} P_0 \to K \to 0.$$

Choose an element p of P_0 which maps to $1 \in K$, and consider the map $\cdot p: KG \to P_0$, $r \mapsto rp$. Let P be the kernel of the split surjective map $(\phi \quad p)^t: P_1 \oplus KG \to_0$. Then P is a finitely generated projective left KG-

module and we have a commuting diagram with exact rows

$$\begin{array}{ccccccccc} 0 \to & KG & \to & P & \to & KG & \xrightarrow{\varepsilon} & K & \to 0 \\ & \| & & \downarrow & & \downarrow \cdot p & & \| & \\ 0 \to & KG & \to & P_1 & \xrightarrow{\phi} & P_0 & \to & K & \to 0. \end{array}$$

In particular, we get the desired exact sequence of left KG-modules

$$(7) \qquad 0 \to KG \xrightarrow{\alpha} P \xrightarrow{\beta} KG \xrightarrow{\varepsilon} K \to 0.$$

Since this is a projective KG-resolution of K, $H_*(G, K)$ is the homology of the complex

$$0 \to K \xrightarrow{K \otimes \alpha} K \otimes_{KG} P \xrightarrow{K \otimes \beta} K \to 0.$$

Notice that $K \otimes \beta = 0$ since $\beta P = \omega KG$. Hence, $H_1(G, K) =$ Coker $(K \otimes \alpha)$ and it suffices to show that $K \otimes \alpha$ is not surjective.

Consider first the case where G is orientable. Then $H^2(G, KG) = K$ is right annihilated by ωKG, but, from (7), $H^2(G, KG) = KG/\alpha^*(P^*)$. Thus $\alpha^*(P^*) \subseteq \omega KG$, and it follows that $\alpha(1) \in \omega KGP$, and hence $K \otimes \alpha = 0$. But it is clear that $P \neq 0$, so by Theorem 11.8, $K \otimes_{KG} P \neq 0$, and therefore $K \otimes \alpha$ is not surjective. Thus we have proved that if G is an orientable PD^2-group over K then G is an infinite surface group.

Let H be the kernel of the orientation map $\varepsilon^\circ: G \to UK$.

If H has infinite index in G then $\varepsilon^\circ(G)$ is infinite, so G^{ab} is infinite, and G is indicable, and thus G is an infinite surface group by Theorem 12.2.

This leaves the case where H has finite index in G, say $(G:H) = m$. By Theorem 5.5, H is an orientable PD^2-group over K. By the preceding part of the proof, H is an infinite surface group, so, by Corollary 4.10 or Lemma 4.8, there is an exact sequence of left KH-modules $0 \to KH \to KH^n \to KH \to K \to 0$ for some $n \geqslant 2$. Viewing (7) as a KH-resolution, we see, by two applications of Schanuel's Lemma IV.3.3, that there is an isomorphism of KH-modules $KH \oplus P \oplus KH \approx KG \oplus KH^n \oplus KG$. Hence $2 + [K \otimes_{KH} P : K] = 2m + n$. By Theorem 11.8, $m[K \otimes_{KG} P : K] = [K \otimes_{KH} P : K] = (n + 2m - 2) \geqslant 2m$. Hence $[K \otimes_{KG} P : K] \geqslant 2$, so $K \otimes_{KG} \alpha$ is not surjective. ∎

We note the case $K = \mathbb{Z}$.

12.4 Theorem. *The* PD^2*-groups are precisely the infinite surface groups.* ∎

Theorem 12.4, Corollary 6.4 and Theorem 5.5 combine to give two important results.

12.5 Theorem. *Let H be a subgroup of an infinite surface group G. If H has finite index in G then H is an infinite surface group. If H has infinite index in G then H is free.* ∎

12.6 Theorem. *If G is torsion-free and has an infinite surface subgroup of finite index then G is itself an infinite surface group.* ∎

Notes and comments

The study of Poincaré duality groups was initiated by Johnson and Wall (1972) and Bieri (1972), and in particular, they noted Theorem 1.1. Poincaré duality pairs were introduced in Bieri and Eckmann (1978), where the CW-complex form of Theorem 1.3 is given.

Most of Section 2 can be found in any text on homological algebra, although our treatment of the cap-product is not completely standard. Theorem 2.10 is due to Brown (1975). From Theorem 2.21 to the end of Section 2 is extracted from Bieri and Eckmann (1978).

Section 3 requires no comment, and similarly for most of Section 4. Theorem 4.4 arises from the work of Hopf (1943) and Wall (1967). Conjecture 4.6 seems natural.

Lemma 4.8 is a very special case of the Simple Identity Theorem of Lyndon (1950).

The single commutator case of Theorem 4.11 (so $m = 2$) goes back to Nielsen (1917), and the general result is due to Zieschang (1964); the proof given here is new and was worked out in collaboration with E. Formanek.

Theorem 5.3 is from Serre (1971); the reader can find a leisurely treatment with all the details in Passman (1977). Theorem 5.5 was proved independently by Johnson and Wall (1972) and Bieri (1972). Another important result which they proved independently, and which we have not mentioned because the proof seems to require spectral sequences, is that an extension of a PD^n-group by a PD^m-group is a PD^{m+n}-group.

Section 6 is from Strebel (1977), with the treatment here making more explicit use of trees.

Sections 7 and 8 are based on Bieri and Eckmann (1978).

The first part of Theorem 9.1 is due to Swarup (1977), and everything else in Section 9 is due to Müller (1981). We have substantially simplified the proofs.

Section 10 is taken from Eckmann and Müller (1980).

Theorem 11.3 is from Kaplansky (1969); the proof given here, which provides an improved lower bound, is inspired by that of Passman (1971). The Hattori–Stallings Trace was defined independently by Hattori (1965) and Stallings (1965′). Theorems 11.5 and 11.8 were shown to us by P.A. Linnell. Theorem 11.6 is from Formanek (1973) and Lemma 11.7 is from Eckmann and Linnell (1983).

Theorem 12.1 is due to Bieri and Strebel (1978) and Theorem 12.2 is due to Eckmann and Müller (1980). Theorem 12.3 for the case $K = \mathbb{Z}$ is due to Eckmann and Linnell (1983), and the characteristic zero case was shown to us by P.A. Linnell. Theorem 12.4 was first proved in Eckmann and Linnell (1983), Theorem 12.5 is classical, and Theorem 12.6 was proved by Eckmann and Müller (1980). The latter result is a special case of the Nielsen Realization Conjecture proved by Kerckhoff (1983).

VI

Two-dimensional complexes and three-dimensional manifolds

Suppose that a G-graph X occurs as the one-skeleton of a two-dimensional G-complex K. If the underlying polyhedron $|K|$ has certain geometrical properties then it is possible to obtain results about the action of G on X fairly easily. This is the approach adopted in this chapter. Some of the results of Chapter II are duplicated for the G-graphs satisfying these extra conditions. We are then able to deduce some stronger results. Thus it is proved that almost finitely presented groups are accessible.

Of particular interest is the case when K is the two-skeleton of a three-manifold. By applying the techniques in this case we are able to give proofs of the Equivariant Loop and Sphere Theorems.

1 Simplicial complexes

Most of the definitions and results of this section can be found in an appropriate text book. They are included here to establish our terminology and notation. References are given for the results.

A *simplicial complex* K consists of a set VK of vertices of K and a set SK of finite subsets of VK called *simplexes* satisfying the following conditions:

(i) if $v \in VK$ then $\{v\} \in SK$,

(ii) if $\sigma \in SK$ and $\sigma' \subset \sigma$ then $\sigma' \in SK$.

If $\sigma \in SK$ contains $n+1$ vertices then σ is called an *n-simplex*. The set of all n-simplexes of K is denoted $S_n K$. We identify $S_0 K$ with VK in the obvious way. If $\sigma, \tau \in SK$ and $\sigma \subseteq \tau$ then σ is called a face of τ. If K has some n-simplexes but no $(n+1)$-simplexes, then K is called *n-dimensional*. We define the *i-skeleton* K^i, $i = 0, 1, 2, \ldots$, of K to be the simplicial complex such that $VK^i = VK$ and $SK^i = \bigcup_{j=0}^{i} S_j K$. The simplicial complex K is said to be *connected* if K^1 is a connected (unoriented) graph.

We now describe how to associate a topological space $|K|$, called the *polyhedron* of K, with K. Let $\sigma = \{v_0, v_1, \ldots, v_n\}$ be an n-simplex of K. Let $|\sigma|$ be the set of all maps

$$\lambda: \sigma \to I = \{x \in \mathbb{R} \mid 0 \leqslant x \leqslant 1\}$$

which satisfy the condition $\sum_{i=0}^{n} \lambda(v_i) = 1$. We make $|\sigma|$ into a metric space by putting

$$d(\lambda, \mu) = \left(\sum_{v \in \sigma} (\lambda(v) - \mu(v))^2 \right)^{\frac{1}{2}}.$$

Now let $|K|$ be the set of maps $\alpha: VK \to I$ such that the support of α is in SK and also $\sum_{v \in VK} \alpha(v) = 1$. For any $\sigma \in SK$ there is an obvious map $|\sigma| \to |K|$. We identify $|\sigma|$ with its image in $|K|$. Clearly

$$|K| = \bigcup_{\sigma \in SK} |\sigma|.$$

Now define a topology on $|K|$ by requiring that C be closed in $|K|$ if and only if $C \cap |\sigma|$ is closed in $|\sigma|$ for every $\sigma \in SK$.

A complex K is called *locally finite* if every $v \in VK$ belongs to only finitely many $\sigma \in SK$.

1.1 Theorem. *Let $|K|$ be the polyhedron of the simplicial complex K. Then*
(i) *$|K|$ is Hausdorff.*
(ii) *$|K|$ is compact if and only if VK is finite.*
(iii) *$|K|$ is locally compact if and only if K is locally finite.*

Proof. See Hilton and Wylie (1962, pp. 46–8). ∎

By an abuse of notation the subset $|\sigma|$ of $|K|$ is called an n-simplex of $|K|$.

Let $u_0, u_1, \ldots, u_n \in \mathbb{R}^m$. We say that $u_0, u_1, \ldots, u_n$ are *affinely independent* if the equations $\lambda_0 u_0 + \lambda_1 u_1 + \cdots + \lambda_n u_n = 0$ and $\lambda_0 + \lambda_1 + \cdots + \lambda_n = 0$ together imply $\lambda_0 = \lambda_1 = \cdots = \lambda_n = 0$.

If $\sigma = \{v_0, v_1, \ldots, v_n\}$ is an n-simplex and $u_0, u_1, \ldots, u_n$ are affinely independent in $\mathbb{R}^m$, then there is a continuous map $\rho: |\sigma| \to \mathbb{R}^m$

$$\rho(\lambda) = \lambda(v_0) u_0 + \lambda(v_1) u_1 + \cdots + \lambda(v_n) u_n.$$

We see that $\operatorname{Im} \rho$ is the convex hull of $u_0, u_1, \ldots, u_n$, and ρ induces a homeomorphism $|\sigma| \to \operatorname{Im} \rho$. More generally let K be a simplicial complex, and let $\rho: VK \to \mathbb{R}^m$ be an injective map such that if σ is an n-simplex of K then $\rho(\sigma)$ consists of $n+1$ affinely independent points in $\mathbb{R}^m$. Then we can extend ρ to a continuous map $|\rho|:|K| \to \mathbb{R}^m$ such that if $\lambda \in |\sigma|$ then $|\rho|(\lambda) = \lambda(v_0)\rho(v_0) + \cdots + \lambda(v_n)\rho(v_n)$. The map $|\rho|$ will be injective if, for any $\sigma, \tau \in SK$, $\rho(\sigma \cup \tau)$ is an affinely independent set of points of $\mathbb{R}^m$. If $|\rho|$ is injective, the induced map $|K| \to \operatorname{Im} \rho$ is a homeomorphism if the inverse map is continuous. This is the case if K is countable and locally finite. In this situation we say that ρ is a *realization* of K in $\mathbb{R}^m$.

In $\mathbb{R}^{2n+1}$ it is possible to find a countably infinite, nowhere dense, set of points such that any subset of $2n+2$ points is affinely independent. This means that if K is an n-dimensional simplicial complex which is countable and locally finite, then K has a realization in $\mathbb{R}^{2n+1}$. Of course, K may have a realization in $\mathbb{R}^m$ where $n \leqslant m < 2n+1$, though examples exist where no such realization exists. See Hilton and Wylie (1962, p. 43).

Let K be a simplicial complex. We define the *first derived complex* K' of K as follows:

(i) $VK' = SK$.

(ii) If τ is a finite subset of SK, then $\tau \in SK'$ if and only if τ can be ordered $\tau = \{\sigma_0, \sigma_1, \ldots, \sigma_n\}$ so that $\sigma_0 \subset \sigma_1 \subset \sigma_2 \cdots \subset \sigma_n$.

Clearly if τ satisfies (ii) then so does any subset of τ and so K' is a simplicial complex.

1.2 Theorem. *If $\rho: VK \to \mathbb{R}^m$ is a realization of K then $\rho': VK' = SK \to \mathbb{R}^m$ is a realization of K' where if $\sigma = \{v_0, v_1, \ldots, v_n\} \in SK$*

$$\rho'(\sigma) = (\rho v_0 + \rho v_1 + \cdots + \rho v_n)/(n+1).$$

Also $\rho(|K|) = \rho(|K'|)$.

Proof. See Hilton and Wylie (1962, p. 26). ∎

Let K and L be simplicial complexes. A *simplicial map* $\phi: K \to L$ is a map (also denoted) $\phi: VK \to VL$ such that if $\sigma \in SK$ then $\phi(\sigma) \in SL$.

If $\phi: K \to L$ is a simplicial map, then there is a continuous map $|\phi|:|K| \to |L|$, where, if $\alpha: VK \to I$, $\alpha \in |K|$, then for $u \in VL$

$$|\phi|(\alpha)(u) = \sum_{\phi(v)=u} \alpha(v).$$

Let K be a simplicial complex. We turn K into an *oriented* simplicial complex as follows. Let $\sigma = \{v_0, v_1, \ldots, v_n\} \in SK$. There are $(n+1)!$ ways of ordering the vertices of σ. Two orderings are said to determine the same orientation of σ if and only if an even permutation of σ changes one ordering into the other. Otherwise the orderings are said to determine opposite orientations of σ. We write $s = v_0 v_1 \cdots v_n$ to denote the oriented n-simplex s corresponding to the ordering $v_0 < v_1 < v_2 < \cdots < v_n$ of the vertices of σ. We write $-s$ to denote the simplex with the opposite orientation. Thus if $\sigma = \{v_0, v_1, v_2\}$ then $s = v_0 v_1 v_2 = v_1 v_2 v_0 = v_2 v_0 v_1$ and $-s = v_1 v_0 v_2 = v_0 v_2 v_1 = v_2 v_1 v_0$.

The *simplicial chain complex* $C(K)$ of K is the chain complex of free abelian groups $C_n(K)$, where the elements of $C_n(K)$, called *n-chains*, are formal sums with integer coefficients of oriented n-simplexes, e.g. $3v_0 v_1 v_2 - 4v_0 v_1 v_3$ is a two-chain of the complex consisting of all faces of the 3-simplex $\{v_0, v_1, v_2, v_3\}$. The *boundary map* $\partial_n : C_n(K) \to C_{n-1}(K)$ is given by

$$\partial(v_0 v_1 v_2 \cdots v_n) = \sum_{i=0}^{n} (-1)^i v_0 v_1 \cdots \hat{v}_i \cdots v_n,$$

where $v_0 v_1 \cdots \hat{v}_i \cdots v_n$ denotes the oriented $(n-1)$-simplex obtained by omitting v_i from the ordering $v_0 v_1 \cdots v_{i-1} v_i v_{i+1} \cdots v_n$.

Now $\partial_{n-1}\partial_n = 0 : C_n(K) \to C_{n-2}(K)$; see Hilton and Wylie (1962, p. 57). Hence, if $B_n(K) = \operatorname{Im} \partial_{n+1}$ and $Z_n(K) = \operatorname{Ker} \partial_n$, we have $B_n(K) \leqslant Z_n(K)$ and we can define $H_n(K) = Z_n(K)/B_n(K)$. The group $H_n(K)$ is called the nth *homology group* of K. The elements of $B_n(K)$ and $Z_n(K)$ are called *n-boundaries* and *n-cycles*, respectively.

More generally, if G is an abelian group, then the simplicial chain complex $C(K, G)$ is the chain complex of abelian groups $C_n(K, G)$, where the elements of $C_n(K, G)$ are formal sums with coefficients in G of oriented n-simplexes. Thus $C_n(K, G) = C_n(K) \otimes G$. There is a straightforward generalization of the definition of cycle, boundary and homology groups. Thus the homology group $H_n(K, G)$ is defined to be the nth homology group of the chain complex $C(K) \otimes G$.

Let $\operatorname{Hom}(C(K), G)$ be the chain complex

$$0 \to \operatorname{Hom}(C_0(K), G) \to \cdots \to \operatorname{Hom}(C_n(K), G)$$

$$\xrightarrow{\delta_n} \operatorname{Hom}(C_{n+1}(K), G) \to \cdots,$$

where, if $\gamma \in \operatorname{Hom}(C_n(K), G)$, $\delta(\gamma) : C_{n+1}(K) \to G$ is the homomorphism $\delta(\gamma) = \gamma\partial$. The groups $Z^n(K, G) = \operatorname{Ker} \delta_n$ and $B^n(K, G) = \operatorname{Im} \delta_{n-1}$ are called, respectively, the group of *n-cocycles* and the group of *n-coboundaries* of K. The elements of $\operatorname{Hom}(C_n(K), G)$ are called *n-cochains* (with coefficients in G). It is easy to see that $B^n(K, G) \leqslant Z^n(K, G)$, and so we can define

$H^n(K, G) = Z^n(K, G)/B^n(K, G)$ called the nth *cohomology group* of K with coefficients in G.

If $c \in C_n(K)$ and $d \in \mathrm{Hom}(C_n(K), G)$, we can form the *Kronecker product* $(c, d) = d(c)$. This product is linear in both factors.

1.3 Proposition. *An n-cochain is a cocycle if its Kronecker product with every n-boundary is zero.*

Proof. Let d be a cocycle and b a boundary so that $b = \partial c$, then

$$(b, d) = (\partial c, d) = (c, \delta d) = (c, 0) = 0.$$

Conversely, if d is a cochain such that $(b, d) = 0$ for every $b \in B_n(K)$ then for each oriented n-simplex $s, (s, \delta d) = (\partial s, d) = 0$, and so $\delta d = 0$. ∎

1.4 Proposition. *Let K be a connected simplicial complex. For any abelian group G*

$$H^1(K, G) \approx \mathrm{Hom}(H_1(K), G).$$

Proof. Let $d \in Z^1(K, G)$. By Proposition 1.3, $(c, d) = 0$ if $c \in B_1(K)$. Hence, d determines a homomorphism $H_1(K) \to G$. If $d = \delta a$ for $a \in Z^0(K, G)$ then for $c \in Z_1(K)$

$$(c, d) = (c, \delta a) = (\partial c, a) = 0.$$

Hence there is a homomorphism

$$\theta: H^1(K, G) \to \mathrm{Hom}(H_1(K), G).$$

Choose $v_0 \in VK$. For each $v \in VK$ choose an oriented edge path $v_0 v_1, v_1 v_2, \ldots, v_{m-1} v_m$ such that $v_m = v$. Let $c(v)$ be the one-chain

$$c(v) = v_0 v_1 + v_1 v_2 + \cdots + v_{m-1} v_m.$$

Define $\rho: C_1(K) \to Z_1(K)$, $\rho(uv) = c(u) + uv - c(v)$. If z is a closed path, then it is easy to see that $\rho(z) = z$. But $Z_1(K)$ is generated by closed paths so ρ is a retraction onto $Z_1(K)$.

Let $\beta: Z_1(K) \to G$ be a homomorphism which is zero restricted to $B_1(K)$. Let $d: C_1(K) \to G$, $d = \beta\rho$. Then d is zero on $B_1(K)$ and so $d \in Z^1(K; G)$, by Proposition 1.3. Thus we can define a homomorphism

$$\phi: \mathrm{Hom}(H_1(K), G) \to H^1(K, G),$$

which is inverse to θ. ∎

We will mainly be interested in the case when $G = \mathbb{Z}_2$. Now $H_1(K) \otimes \mathbb{Z}_2 = 0$ if and only if $\mathrm{Hom}(H_1(K), \mathbb{Z}_2) = 0$. Thus we have the following.

1.5 Proposition. *If K is a connected simplicial complex, then $H_1(K,\mathbb{Z}_2)=0$ if and only if $H^1(K,\mathbb{Z}_2)=0$.* ∎

We assume that the reader is familiar with the concept of the *fundamental group* of a topological space. If K is a connected simplicial complex, it is possible to write down a presentation for the fundamental group $\pi_1(|K|,v_0)$, where $v_0 \in VK$, in which the generators correspond to the oriented one-simplexes γ of K and the relations are either of the form $\gamma=1$ if γ is in a maximal tree $T\subset K^1$ or of the form $\gamma_0\gamma_1\gamma_2=1$ for each two-simplex $\{v_0,v_1,v_2\}$ of K, where $\gamma_0=v_1v_2$, $\gamma_1=v_2v_0$, $\gamma_2=v_0v_1$. Thus if K is finite $\pi_1(|K|,v_0)$ has a finite presentation. For details of this process, see Hilton and Wylie (1962, chapter 6).

1.6 Theorem. *Let K be a connected simplicial complex. There is a surjective homomorphism $\pi_1(|K|,v_0)\to H_1(K)$, whose kernel is the commutator subgroup of $\pi_1(|K|,v_0)$.*

Proof. See Hilton and Wylie (1962, p. 348). ∎

2 Orbit spaces

Let G be a group and let X be a topological space which is also a left G-set. We say that G acts on X (or X is a G-space) if for each $g\in G$ the map $x\mapsto gx$ is continuous. Put $Y=G\backslash X$. We make Y into a topological space by giving it the weakest topology for which the natural map $\pi:X\to Y$ is continuous, that is U is open in Y if and only if $\pi^{-1}(U)$ is open in X. The space Y is called the *orbit space* for the action of G on X.

2.1 Examples. Put $X=\mathbb{R}$ and let $G=\mathbb{Z}$. Let G act on X by addition, that is the homeomorphism corresponding to $n\in\mathbb{Z}$ is $r\mapsto n+r$. In this case $G\backslash X$ is homeomorphic to S^1, the unit circle. For if we regard S^1 as the subset of the complex plane

$$S^1=\{z\in\mathbb{C}\mid |z|=1\}$$

then there is a map $\theta:\mathbb{R}\to S^1$, $\theta(r)=e^{2\pi ir}$, and it can be seen that θ is surjective and, for each $z\in S^1$, $\theta^{-1}(z)$ is a G-orbit. Thus $S^1\approx G\backslash X$.

Similarly if $X=\mathbb{R}^2$ and $G=\mathbb{Z}\otimes\mathbb{Z}$, then $G\backslash X\approx S^1\times S^1$, the torus. Here the homeomorphism corresponding to the pair (n,m) is

$$(r,s)\to(r+n,s+m).$$

In this case the open square $U=(0,1)\times(0,1)$ is a *fundamental domain* for the action; see Beardon (1983, p. 204). This means that if Gx is a G-orbit, then Gx contains at most one point of U and at least one point of $\bar{U}$ the closure of U in X. Thus X is tesselated by the translates of $\bar{U}$. One can construct $Y = G\backslash X$ by identifying points of ∂U which lie in the same G-orbit. This gives us the familiar construction of a torus in which the opposite edges of the closed unit square are identified.

In both these examples X is simply connected and $\pi: X \to G\backslash X$ is a covering map. A *covering map* $\rho: X \to Y$ is a continuous map such that for each $y \in Y$ there is a path-connected open set $U \subseteq Y$ for which $y \in U$ and ρ maps each component of $\rho^{-1}(U)$ homeomorphically onto U. The map $\pi: X \to G\backslash X$ from a G-space to the orbit space is a covering map if for each $x \in X$ there is a path-connected open set $V \subseteq X$ such that $x \in V$ and $gV \cap V = \varnothing$ if $g \in G - \{1\}$.

If $\rho: X \to Y$ and $\rho': X' \to Y$ are covering maps, Y is path-connected and X is simply connected, then there is a covering map $q: X \to X'$ such that $\rho' q = \rho$; see Massey (1967, p. 159). In particular if X' is simply connected then q is a homeomorphism. We say that X is the *universal cover* of Y and write $X = \tilde{Y}$. Choose $y_0 \in Y$ and put $G = \pi_1(Y, y_0)$. There is an action of G on X so that ρ factors $\rho = \phi\pi$, where ϕ is a homeomorphism. Thus if a path-connected space Y has a universal cover then we can regard Y as an orbit space for the action of $G = \pi_1(Y, y_0)$ on $X = \tilde{Y}$.

Necessary and sufficient conditions are known for a space to have a universal cover. See Massey (1967) for this result and a full discussion of covering space theory. Any space of the form $|K|$, where K is a simplicial complex, has a universal cover. In fact, there is a simplicial complex $\tilde{K}$ such that $|\tilde{K}|$ is the universal cover of $|K|$. If K is one-dimensional, that is if K is a graph, then $\tilde{K}$ is a tree, as in Theorem I.9.2.

If K is a simplicial complex and VK is a G-set for some group G, then we say that K is a *G-complex* if for each $g \in G$ the map $v \mapsto gv$ induces a simplicial map, that is if $\sigma \in SK$ implies $g\sigma \in SK$. If K is a G-complex then we can make $|K|$ into a G-space by extending the action of G on VK linearly over each simplex. Thus we can construct the orbit space $G\backslash|K|$. We can define $G\backslash K$ in the obvious way, but it is not always the case that $|G\backslash K|$ is homeomorphic to $G\backslash|K|$. Thus, let $L = G\backslash K$, where $VL = G\backslash VK$, and, if $\rho: VK \to VL$ is the natural map, then the subsets of VL which are simplexes of L are all the subsets of the form $\rho(\sigma)$, where $\sigma \in SK$. Clearly ρ defines a simplicial map (also denoted ρ) $\rho: K \to L$. If $g \in G$ and $v \in VK$ then $\rho(gv) = \rho(v)$. It follows that there is a map $\theta: G\backslash|K| \to |L|$ such that the map $|\rho|: |K| \to |L|$ factorizes $|\rho| = \theta\pi$. Now $|\rho|$ is surjective and so θ is surjective. It may be the case that

θ is not injective. For example, let $VK=\mathbb{Z}$ and $SK=\{\{i\},\{i,i+1\}|i\in\mathbb{Z}\}$. Let $G=\mathbb{Z}$ and let G act on K by addition. Now $L=G\backslash K$ consists of a single vertex. But $G\backslash|K|\approx G\backslash\mathbb{R}$ is homeomorphic to S^1. Thus θ cannot be injective. Again if $G=2\mathbb{Z}$ acting on K by addition, then $G\backslash K$ has two vertices and a single one-simplex. Thus $|G\backslash K|$ is homeomorphic to $I=[0,1]$, while again $G\backslash|K|$ is homeomorphic to S^1. In fact, if we identify $|K|$ with $\mathbb{R}$ in the obvious way, then $|\rho|$ takes all points $n+\frac{1}{2}, n\in\mathbb{Z}$ to the same point of $|L|$. However, $\pi(\frac{1}{2})$ and $\pi(\frac{3}{2})$ are distinct points of $G\backslash|K|$.

We now find necessary and sufficient conditions for θ to be injective. We will see that this means that θ is a homeomorphism.

Suppose θ is injective. For each one-simplex γ of K, the two vertices u, v of γ must lie in distinct G-orbits. For otherwise $\rho(u)=\rho(v)$ and so $\pi(u)=\pi(v)$ and every point of $|\gamma|$ would be mapped to $\rho(u)$.

Now let $\sigma=\{u_0,u_1,\ldots,u_n\}$ and $\tau=\{v_0,v_1,\ldots,v_n\}$ be n-simplexes of K. Suppose $\rho(u_i)=\rho(v_i), i\in[0,n]$. Since θ is injective, $\pi(u_i)=\pi(v_i)$. Let $\alpha,\beta\in|K|$ be such that $\alpha(u_i)=\beta(v_i)$, $\alpha(u)=\beta(v)=0$ if $u\notin\sigma$, $v\notin\tau$. Then $|\rho|(\alpha)=|\rho|(\beta)$ and so $\pi(\alpha)=\pi(\beta)$. Choose α so that its values on $u_0,u_1,\ldots,u_n$ are all distinct. There exists $g\in G$ such that $g\alpha=\beta$. Hence $gu_i=v_i$, $i\in[0,n]$. In particular, such a $g\in G$ exists if $u_i=v_i$, $i\in[0,n-1]$ and u_n and v_n are in the same G-orbit. This means that we have proved the 'only if' part of the following theorem.

2.2 Theorem. *The map $\theta: G\backslash|K|\to|G\backslash K|$ defined above is a homeomorphism if and only if*

(*a*) *no one-simplex of K has both its vertices in the same G-orbit, and*

(*b*) *If $\sigma=\{u_0,u_1,\ldots,u_n\}$ and $\tau=\{v_0,v_1,\ldots,v_n\}$ are n-simplexes of G, where $u_i=v_i$, $i\in[0,n-1]$ and u_n and v_n are in the same G-orbit, then there exists $g\in G$ such that $gu_i=v_i$, $i\in[0,n]$.*

Proof. It remains to show that (*a*) and (*b*) are sufficient to ensure that θ is a homeomorphism. Suppose (*a*) and (*b*) are satisfied. Let $\alpha,\beta\in|K|$ be such that $|\rho|(\alpha)=|\rho|(\beta)$. Let $\sigma=\{u_0,u_1,\ldots,u_n\}$ be the support of α, and $\tau=\{v_0,v_1,\ldots,v_m\}$ be the support of β. Thus σ and τ are simplexes. Now ρ is a simplicial map, and, by (*a*), ρ does not map distinct vertices of the same simplex of K to the same vertex of $G\backslash K$. Hence if $|\rho|(\alpha)=|\rho|(\beta)$ then $m=n$ and we can number the vertices of τ so that $\rho(u_i)=\rho(v_i)$, $i\in[0,n]$. Now u_i is in the same G-orbit as v_i. We want to show that there exists $x\in G$ such that $xu_i=v_i$, $i\in[0,n]$. Choose g_0 so that $g_0u_0=v_0$. Let $k\in[0,n-1]$. Suppose we have chosen g_k so that $g_ku_j=v_j$ if $j\leqslant k$. Consider the $(k+1)$-simplexes $\{v_0,v_1,\ldots,v_{k+1}\}$ and $\{v_0,v_1,\ldots,g_ku_{k+1}\}$. Now, by (*b*), there exists $g\in G$ such that $gu_i=v_i$, $i=0,1,\ldots,k$ and $gg_ku_{k+1}=v_{k+1}$.

Put $g_{k+1} = gg_k$. Thus we can successively define $g_1, g_2, \ldots, g_n$ and take $x = g_n$. It follows that $x\alpha = \beta$ and so $\pi(\alpha) = \pi(\beta)$, that is θ is injective. We remarked earlier that θ is surjective since $|\rho|$ is surjective. Thus θ is bijective and there is an inverse map $\theta^{-1}:|G\backslash K| \to G\backslash|K|$. Now θ^{-1} is continuous if it is continuous on each simplex of $|G\backslash K|$. Clearly this is the case. Thus θ is a homeomorphism. ∎

Let K be a G-complex. Let K' be the derived complex of K. We make K' into a G-complex in the obvious way. Similarly $K'' = (K')'$ becomes a G-complex.

2.3 Theorem. *For any G-complex K, the map $\theta:G\backslash|K''| \to |G\backslash K''|$ is a homeomorphism.*

Proof. We have to show that conditions (*a*) and (*b*) of Theorem 2.2 are satisfied for the G-complex K''.

First note a property of the G-complex K'. Let σ be a simplex of K' and let u, v be vertices of σ. If $gu = v$ for some $g \in G$ then $u = v$. For since u, v are vertices of σ, u and v are simplexes of K and either $u \subseteq v$ or $v \subseteq u$ and since inclusion is preserved under the action of g, equality follows immediately.

Condition (*a*) of Theorem 2.2 is true for K' and so, *a fortiori*, for K''. To prove condition (*b*) is satisfied, let $\tau = \{u_0, u_1, \ldots, u_n\}$, $\tau' = \{v_0, v_1, \ldots, v_n\}$, where $\tau, \tau' \in SK''$. Suppose $u_i = v_i$ for $i \in [0, n-1]$ and $gu_n = v_n$ for some $g \in G$. The vertices $u_0, u_1, \ldots, u_n$ of τ can be arranged so that they are ordered by inclusion. If u_n is not the largest, then $u_n \subset u_i$ for some $i \neq n$ and $v_n = gu_n \subset v_i = u_i$. But now the remark above shows that $u_n = v_n$ and so $u_i = v_i$, $i \in [0, n]$.

If u_n is the largest vertex of τ then we can assume $u_0 \subset u_1 \subset \cdots \subset u_n$ and $v_0 \subset v_1 \subset \cdots \subset v_n$. But now $gu_n = v_n$ implies $gu_i = u_i = v_i$ for $i \in [0, n-1]$. For note that $gu_i \subset v_n$ and $v_i = u_i \subset v_n$ and again the remark above shows that $gu_i = v_i$. ∎

It follows from Theorem 2.3 that, if K is G-complex, then there is no real loss of generality in assuming that $G\backslash|K|$ and $|G\backslash K|$ are homeomorphic. For, if K does not have this property, then we can replace K by K''.

3 Patterns and tracks

Throughout this section K and L will denote connected two-dimensional simplicial complexes. We write $S_0K = VK, S_1K = EK$ *and*

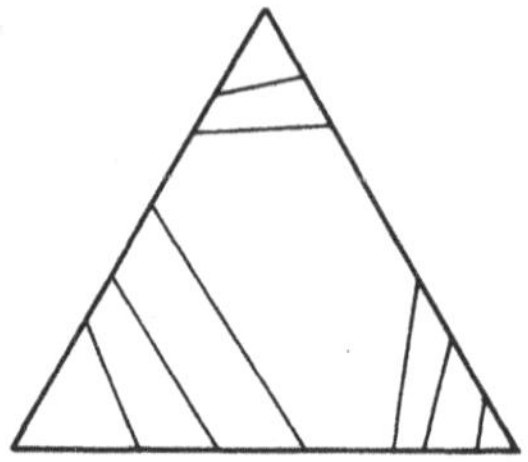

Fig. VI.1

$S_2K = FK$. Clearly the one-skeleton K^1 of K can be regarded as an unoriented graph which will be denoted X.

3.1 Definition. A *pattern* P is a subset of $|K|$ with the following properties:

(i) For each two-simplex σ of $K, P \cap |\sigma|$ is a union of finitely many disjoint straight lines joining distinct faces of s.

(ii) If γ is a one-simplex of K then $P \cap |\gamma|$ consists of finitely many points in the interior of $|\gamma|$.

Thus if P is a pattern in $|K|$ and $\sigma \in SK$, then $|\sigma| \cap P$ is as in Fig. VI.1.

A component of a pattern is called a *track*.

It is important to note that the definition requires that if γ is a one-simplex of K and $x \in |\gamma| \cap P$ then, in *every* two-simplex that has γ as a face, there is a straight line with x as one of the end points.

The case when every one-simplex of K is contained in exactly two two-simplexes is of particular interest. This is the case if $|K|$ is a two-dimensional manifold (without boundary). In this case, for any point $u \in P \cap |K|$, there is a neighbourhood U of u in $|K|$ such that $P \cap U$ is an open interval. Thus P is a one-manifold. If K is finite, P will consist of finitely many simple closed curves.

Let $P \subset |K|$ be a pattern. Let $j_P : EK \mapsto \mathbb{Z}^+ = \{0, 1, \dots\}$, be the map such that $j_P(\gamma)$ is the number of points in $|\gamma| \cap P$. We show that this map determines the pattern P in an almost unique way. This is because P is determined uniquely by $P \cap |K^1|$. For, suppose $\sigma \in FK$ and σ has faces $\gamma_1, \gamma_2, \gamma_3$. There are $j(\gamma_i)$ points of $P \cap \partial|\sigma|$ in $|\gamma_i|$. Suppose $P \cap |\sigma|$ contains α_1 lines joining γ_2 and γ_3, α_2 lines joining γ_1 and γ_3, and α_3 lines joining γ_1 and γ_2. Then $j_P(\gamma_1) = \alpha_2 + \alpha_3, j_P(\gamma_2) = \alpha_1 + \alpha_3, j_P(\gamma_3) = \alpha_1 + \alpha_2$. Thus, if $m = \alpha_1 + \alpha_2 + \alpha_3$, then $j_P(\gamma_1) + j_P(\gamma_2) + j_P(\gamma_3) = 2m$ and $\alpha_i = m - j_P(\gamma_i)$. Hence $P \cap \partial|\sigma|$ determines $P \cap |\sigma|$.

3.2 Theorem. *Let $f : EK \to \mathbb{Z}^+$ be a map such that for each two-simplex $\sigma \in FK$ with faces $\gamma_1, \gamma_2, \gamma_3$ we have $f(\gamma_1) + f(\gamma_2) + f(\gamma_3) = 2m_\sigma$ for some*

$m_\sigma \in \mathbb{Z}^+$ and $f(\gamma_i) \leqslant m_\sigma$ for $i = 1, 2, 3$. Then there is a pattern P such that $j_P = f$.

If P, Q are patterns such that $j_P = j_Q$ then there is a homeomorphism $h{:}|K| \to |K|$ such that $h|\sigma| = |\sigma|$ for every $\sigma \in SK$, and $h(P) = Q$.

Proof. It follows from the remarks above that, if we choose $f(\gamma)$ points in the interior of each 1-simplex $|\gamma|$ of $|K|$, we can find a unique pattern P such that $P \cap |K^1|$ is the set of points chosen.

If $j_P = j_Q$ then there is a homeomorphism $h_1{:}|K^1| \to |K^1|$ such that $h_1(P \cap |K^1|) = Q \cap |K^1|$. This follows from the fact that, if S, T are subsets of $[0, 1]$ each containing the same finite number of elements, then there is a homeomorphism $[0, 1] \to [0, 1]$ mapping S to T. It is possible to extend h^1 to a homeomorphism $h{:}|K| \to |K|$ such that $h(P) = Q$. ∎

3.3 Definition. Two patterns P and Q are said to be *equivalent*, written $P \equiv Q$, if $j_P = j_Q$. ∎

Let P be a pattern. Let D_P be the graph defined as follows:

VD_P is the set of components of $|K| - P$,

ED_P is the set of components of P.

If $e \in ED_P$ then the vertices of e are the components of $|K| - P$ whose closures contain e. Clearly there are at most two such components. There may be only one component. We allow the possibility in the unoriented graph D_P that the vertices of a particular edge are identified. Think of such an edge as a based loop.

Let $\mathrm{mod}_2{:}\mathbb{Z}^+ \to \mathbb{Z}_2$ be the natural map. Let P be a pattern in $|K|$. Consider the map $z_P = \mathrm{mod}_2\, j_P{:}EK \to \mathbb{Z}_2$. We regard z_P as a 1-cochain, i.e. an element of $\mathrm{Hom}(C_1(K), \mathbb{Z}_2)$. Note that we do not have to orient the simplexes of K since we are using $\mathbb{Z}_2$ coefficients. Now if $\sigma \in FK$, $\partial|\sigma| \cap P$ has an even number of points. Hence $\delta z_P = 0$, that is $z_P \in Z^1(K, \mathbb{Z}_2)$. It is easy to give a necessary and sufficient condition for z_P to belong to $B^1(K, \mathbb{Z}_2)$.

3.4 Proposition. *Let P be a pattern in $|K|$. The following are equivalent:*

(i) $z_P \in B^1(K, \mathbb{Z}_2)$;

(ii) *$|K| - P$ can be coloured with two colours (black and white) so that each component has one colour and each track of P is in the closure of a white region and a black region;*

(iii) *the vertices of the graph D_P can be coloured so that adjacent vertices have distinct colours.*

Proof. Clearly (ii) and (iii) are equivalent.

Assume (i) is true, so that $z_P = \delta c$, where c is a zero-cochain. If $v \in VK$ and $c(v) = 1$, colour v black. Colour v white if $c(v) = 0$. Clearly two vertices of a two-simplex σ have the same colour if and only if the edge joining them intersects P an even number of times. It is possible to extend the colouring to $|\sigma| - P$ so that each component has one colour and adjacent regions have distinct colours. Do this on each two-simplex of $|K|$ to obtain a colouring as in (ii).

Conversely, suppose $|K| - P$ can be coloured as in (ii). Let $c: VK \to \{0, 1\}$ be such that $c(v) = 1$ if v is black, $c(v) = 0$ if v is white. Clearly $\delta c = z_P$ as required. ∎

3.5 Corollary. *A track t separates $|K|$ if and only if $z_t \in B^1(K, \mathbb{Z}_2)$.* ∎

If γ is an oriented one-simplex of K then the tracks of P which intersect $|\gamma|$, when ordered in the obvious way, give an edge path in D_P. Let $\alpha_1(\gamma)$ be the one-chain of D_P corresponding to this path, that is $\alpha_1(\gamma)$ is the sum of the oriented edges of D_P which comprise the path. We then have a commuting diagram

$$\begin{array}{ccccc} C_2(K) & \xrightarrow{\partial} & C_1(K) & \xrightarrow{\partial} & C_0(K) \\ \downarrow & & \downarrow{\scriptstyle \alpha_1} & & \downarrow{\scriptstyle \alpha_0} \\ 0 & \longrightarrow & C_1(D_P) & \longrightarrow & C_0(D_P), \end{array}$$

where, if $v \in VK$, $\alpha_0(v)$ is the component of $|K| - P$ containing v. Applying $\operatorname{Hom}(-, \mathbb{Z}_2)$ to the above diagram we obtain

$$\begin{array}{ccccc} \operatorname{Hom}(C_0(K), \mathbb{Z}_2) & \to & \operatorname{Hom}(C_1(K), \mathbb{Z}_2) & \to & \operatorname{Hom}(C_2(K), \mathbb{Z}_2) \\ \uparrow{\scriptstyle \alpha^0} & & \uparrow{\scriptstyle \alpha^1} & & \uparrow \\ \operatorname{Hom}(C_0(D_P), \mathbb{Z}_2) & \to & \operatorname{Hom}(C_1(D_P), \mathbb{Z}_2) & \to & 0. \end{array}$$

If t is a track of P with corresponding element $z_t \in \operatorname{Hom}(C_1(K), \mathbb{Z}_2)$ then $z_t \in \operatorname{Im} \alpha^1$, since $\alpha^1(\zeta_t) = z_t$, where $\zeta_t(t) = 1$, $\zeta_t(s) = 0$ if s is a track, $s \subset P$, $s \neq t$.

Since the above diagram is commutative, we have an induced map $H^1(D_P, \mathbb{Z}_2) \to H^1(K, \mathbb{Z}_2)$. This map is injective. In order to prove this, we have to show that, if $x \in \operatorname{Hom}(C_1(D_P), \mathbb{Z}_2)$ and $\alpha^1(x) = \delta a$ for some $a \in \operatorname{Hom}(C_0(K), \mathbb{Z}_2)$, then $x = \delta b$ for $b \in \operatorname{Hom}(C_0(D_P), \mathbb{Z}_2)$. Now $\operatorname{Hom}(C_1(D_P), \mathbb{Z}_2) \approx (ED_P, \mathbb{Z}_2)$ and so x can be regarded as a subpattern P_1 of P and $\alpha^1(x) = z_{P_1}$. Since $z_{P_1} = \delta a$, by Proposition 3.4, $|K| - P_1$ can be two-coloured. This colouring determines a 'colouring' of $|K| - P$ (in this colouring, adjacent regions may have the same colour), and a 'colouring' of D_P, which can be regarded as an element b of $\mathrm{Hom}(C_0(D_P), \mathbb{Z}_2)$. Hence, $x = \delta b$ as required.

Let $\chi(D_P)$ be the Euler characteristic of D_P. Then

$$\chi(D_P) = |VD_P| - |ED_P| = 1 - \operatorname{rank} H^1(D_P, \mathbb{Z}_2).$$

Let $\beta = \operatorname{rank} H^1(K, \mathbb{Z}_2)$. By the above argument, rank $H^1(D_P, \mathbb{Z}_2) \leqslant \beta$ and so $\chi(D_P) \geqslant 1 - \beta$. Thus $|VD_P| \geqslant |ED_P| + 1 - \beta$, and we have the following result.

3.6 Theorem. *Let P be a pattern in $|K|$ with a finite number α of tracks. If $\beta = \operatorname{rank} H^1(K, \mathbb{Z}_2)$, then $|K| - P$ has at least $\alpha - \beta + 1$ components.* ∎

Let t be a track in $|K|$ with corresponding function $j_t: EK \to \mathbb{Z}^+$. Now $2j_t$ satisfies the conditions of Theorem 3.2 and so there is a pattern P in $|K|$ such that $j_P = 2j_t$. In Fig. VI.2 we illustrate the situation in a typical two-simplex σ. Here t is shown by dotted lines and P by continuous lines. Clearly there is a two-sheeted covering map $\rho: P \to t$ in which the two lines adjacent to a dotted line are mapped to the dotted line. It follows that P has either one or two component tracks. If P has one track, we say that t is *twisted*. If P has two tracks, we say that t is *untwisted*. In either case there is a connected closed subset B of $|L|$ such that $\partial B = P$. Such a region is called a *band*. The band B is shown shaded in Fig. VI.2. The track t associated with B is written $t(B)$. If $t(B)$ is untwisted, B is homeomorphic to $t(B) \times [0, 1]$ and we say B is *untwisted*. Otherwise, B is said to be *twisted*. In a closed surface a twisted track is an orientation reversing simple closed curve; a twisted band is a Möbius band and an untwisted band is an annulus.

Let $\alpha: K \to L$ be a simplicial map and let Q be a pattern in $|L|$. Then α induces a continuous map $|\alpha|: |K| \to |L|$ and $|\alpha|^{-1} Q$ is a pattern P in $|K|$. If $\gamma = \{v_0, v_1\} \in EK$ then $\alpha(\gamma) = \{\alpha v_0, \alpha v_1\}$ is either a zero-simplex or a

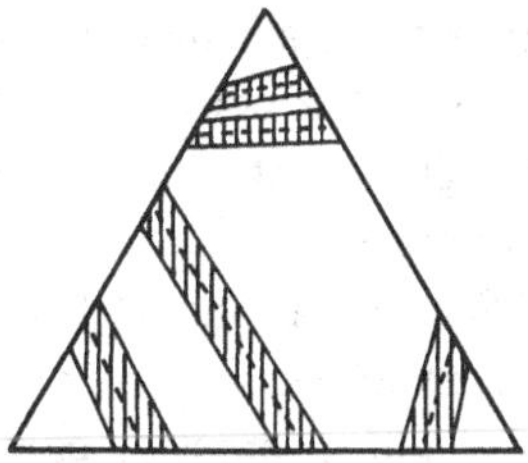

Fig. VI.2

one-simplex of L. If we define j_Q to be zero on zero-simplexes of L, then $j_P(\gamma) = j_Q(\alpha(\gamma))$.

If Q' is a pattern such that $2j_{Q'} = j_Q$, then $|\alpha|^{-1}Q' = P'$, where $2j_{P'} = j_P$. Thus if Q' is a union of bands then so is $|\alpha|^{-1}Q'$. This is also clear by examining $|\alpha|^{-1}(Q' \cap |\sigma|)$ for each individual two-simplex σ.

3.7 Proposition. *Let $\beta = \text{rank}\, H^1(L, \mathbb{Z}_2)$. If $B_1, B_2, \ldots, B_k$ are disjoint twisted bands in $|L|$ then $k \leqslant \beta$.*

Proof. Let $t_i = t(B_i)$ and $z_i = z_{t_i}, i \in [1, k]$. We show that $z_1, z_2, \ldots, z_k$ represent linearly independent elements of $H^1(L, \mathbb{Z}_2)$. For suppose P is a pattern consisting of a subset of $\{t_1, t_2, \ldots, t_k\}$. Now $\sum(z_t : t \in P)$ belongs to $B^1(L, \mathbb{Z}_2)$ if and only if $|L| - P$ can be two-coloured as described in Proposition 3.4. But $B_i - t_i$ is connected and so $|L| - P$ is connected and it is not possible to two-colour $|L| - P$. ∎

A pattern $Q \subset |L|$ is called *reduced* if no two distinct tracks of Q are equivalent. A pattern Q is not reduced if and only if there is a component C of $|L| - Q$ whose closure $\bar{C}$ is an untwisted band.

We now show that if L is finite and Q is reduced then there is an upper bound on the number of components of Q.

Suppose then that L is a connected finite two-complex; let v_L be the number of vertices and f_L the number of two-simplexes of L. Let $\beta = \text{rank}\, H^1(L, \mathbb{Z}_2)$. Put $n(L) = 2\beta + v_L + f_L$.

3.8 Theorem. *Let Q be a reduced pattern in $|L|$. There are less than $n(L)$ component tracks in Q.*

Proof. Let Q be a pattern in $|L|$ with $\alpha \geqslant n(L)$ components. We show that $|L| - Q$ has a component C such that $\bar{C}$ is an untwisted band.

If σ is a two-simplex of L and U is the closure of a component of $|\sigma| - Q$, then U is a disc. We say that U is *good* if $\partial U \cap \partial|\sigma|$ consists of two components in distinct faces of σ. For any σ there are at most three U's which contain a vertex of σ and at most one other component which is not good. Thus in Fig. VI.3 the shaded region is the only component which is not good and does not contain a vertex.

By Theorem 3.6, $|L| - Q$ has at least $\alpha - \beta + 1$ components. But $\alpha \geqslant n(L)$ and so $|L| - Q$ has more than $\beta + v_L + f_L$ components. There are at most v_L components which contain a vertex of L and at most f_L components which contain a region of a two-simplex which is not good and does not

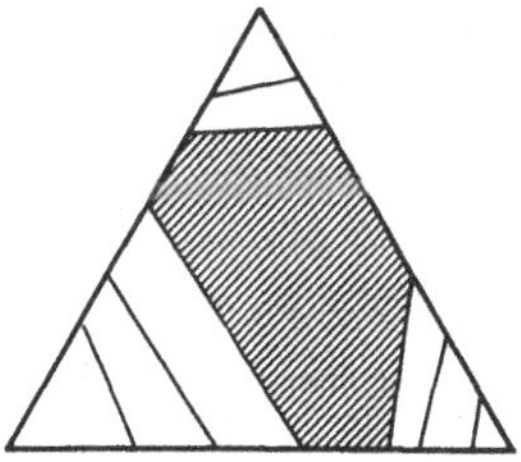

Fig. VI.3

contain a vertex. Thus there are at least $\beta+1$ components which are bands. By Proposition 3.7 there is at least one component which is an untwisted band. ∎

4 Almost finitely presented groups

4.1 Definition. A group G is said to be *almost finitely presented* (afp) if it is FP_2 over $\mathbb{Z}_2$. ∎

By Proposition V.3.2(ii), if G is afp then G is finitely generated. If $G = F/N$, where N is a normal subgroup of the finitely generated free group F, then, by Proposition V.3.2(iii), G is afp if $\mathbb{Z}_2 \otimes_{\mathbb{Z}} N^{ab}$ is a finitely generated $\mathbb{Z}_2 G$-module. If G is finitely presented then we can choose F and N so that N is the normal closure in F of a finite number of elements. This means that N will be a finitely generated $\mathbb{Z}G$-module. Thus if G is finitely presented then G is afp.

Bieri and Strebel (1980) have given an example of a group which is afp but not finitely presented.

We first characterize afp groups geometrically.

4.2 Theorem. *A group G is afp if and only if there is a connected G-finite simplicial two-complex K, on which G acts freely, such that $H^1(K, \mathbb{Z}_2) = 0$.*

Proof. Let S be a finite generating set for G. We have an exact sequence corresponding to the Cayley graph of G with respect to S

$$\mathbb{Z}_2[G \times X] \xrightarrow{\partial_1} \mathbb{Z}_2 G \to \mathbb{Z}_2 \to 0.$$

Now $\operatorname{Ker} \partial_1$ is generated by closed paths in the Cayley graph. Thus we can obtain a G-finite two-dimensional cell complex C with $H_1(C) = 0$ by attaching two-cells to finitely many G-orbits of closed paths in the Cayley graph. Since the G-action on the Cayley graph is free, we can arrange that the G-action on C is free also. Now triangulate C to obtain the two-complex K. Since $H_1(K)=0$, we see from Proposition 1.4 that $H^1(K,\mathbb{Z}_2)=0$.

Conversely if there is a connected two-complex on which G acts freely and $H^1(K,\mathbb{Z}_2)=0$, then $H_1(K,\mathbb{Z}_2)=0$ by Proposition 1.5. Hence if $C(K)$ is the chain complex associated with K, then $\mathbb{Z}_2\otimes C(K)$ gives an exact sequence

$$C_2 \to C_1 \to \mathbb{Z}_2 \to 0$$

of free $\mathbb{Z}_2 G$-modules. Thus G is FP_2 over $\mathbb{Z}_2$ as required. ∎

If G is a finitely presented group, there is a finite two-complex L such that $\pi_1(L)\approx G$. In this case $K=\tilde{L}$, the universal cover of L, satisfies the conditions of Theorem 4.2.

The reason why it is useful to consider afp groups is provided by the following proposition (using the geometric characterization of afp groups provided by Theorem 4.2).

4.3 Proposition. *Let K be a connected two-complex for which $H^1(K,\mathbb{Z}_2)=0$. If P is a pattern in $|K|$ then D_P is a tree.*

Proof. By Corollary 3.5, if t is a track and $t\subseteq P$, then t separates $|K|$. Thus removing any edge from D_P disconnects the graph. The only graphs with this property are trees. ∎

Let $\alpha: K\to L$ be a simplicial map. We have seen that if Q is a pattern in $|L|$, then $|\alpha|^{-1}Q=P$ is a pattern in $|K|$. Clearly there is an induced map $D_\alpha: D_P\to D_Q$. The case of special interest to us is when $H^1(K,\mathbb{Z}_2)=0$ and the group G acts on K so that $L=G\backslash K$, where $|L|$ is homeomorphic to $G\backslash|K|$. In this case D_P is a G-tree and $D_Q\approx G\backslash D_P$.

Let G be an afp group. By Theorem 4.2 there is a connected two-complex K such that G acts freely on K, $G\backslash K$ is finite and $H^1(K,\mathbb{Z}_2)=0$. Let T be a G-tree. Because G acts freely on VK, there is a G-map $\alpha: VK\to VT$.

Define a map $f: EK\to\mathbb{Z}^+$ as follows:

$$\text{for each } \gamma\in EK \text{ let } f(\gamma)=d(\alpha v_0,\alpha v_1),$$

where $\gamma=\{v_0,v_1\}$ and, for $x,y\in VT$, $d(x,y)$ is the number of edges in the unique reduced edge path in T joining x and y. We show that f satisfies the conditions of Theorem 3.2.

Let σ be a two-simplex of K with vertices v_0, v_1, v_2. Let T_σ be the minimal subtree of T containing $\alpha v_0, \alpha v_1$ and αv_2. Up to reordering of v_0, v_1 and v_2 there are four possible configurations for T_σ which are illustrated in Fig. VI.4.

Let $\gamma_0 = \{v_1, v_2\}$, $\gamma_1 = \{v_0, v_1\}$, $\gamma_2 = \{v_0, v_1\}$. It is easy to see that $f(\gamma_0) + f(\gamma_1) + f(\gamma_2)$ is even. In fact, $f(\gamma_0) + f(\gamma_1) + f(\gamma_2) = 2m$, where m is the number of edges in T_σ. Since $\alpha v_0, \alpha v_1$ and αv_2 are in T_σ, the distance between any two of them is at most m. By Theorem 3.2 $f = j_P$, where P is a G-pattern. Also D_P is a G-tree since $H^1(K, \mathbb{Z}_2) = 0$. We now show that there is a G-map $\beta: D_P \to T$ which is a morphism of graphs.

Let σ be the two-simplex of K, as above. Mark off $j_P(\gamma_i)$ points on the edge γ_i and join by straight lines in $|\sigma|$ in the usual way. Now, for each pair of adjacent points of the pattern in the interior of an edge, put in another point and join these points by straight lines (shown dotted in Fig. VI.5).

In the case corresponding to Fig. V.4(i) there will be a triangular region (shown shaded in Fig. VI.5) bounded by three dotted lines. There is a continuous map extending α restricted to $\{v_0, v_1, v_2\}$ from $|\sigma|$ to the polyhedron of T_σ in which the shaded regions and the dotted lines are

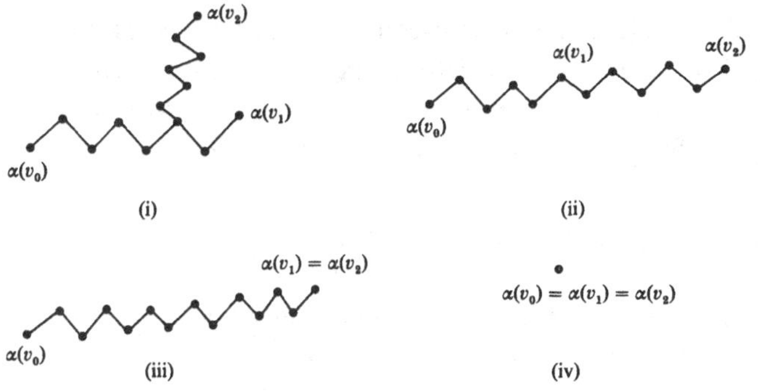

Fig. VI.4

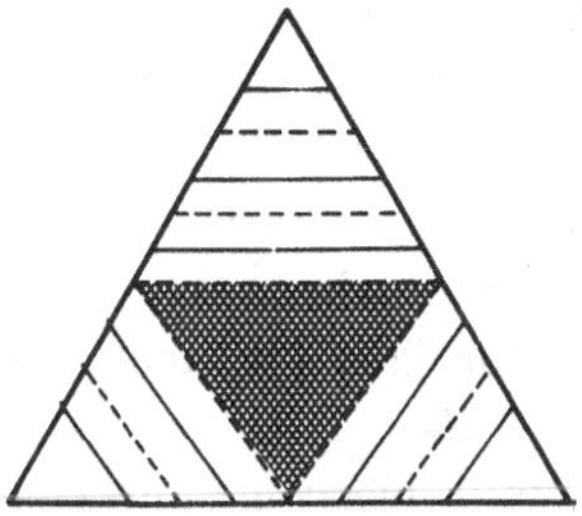

Fig. VI.5

mapped to vertices and the lines of the pattern are mapped to the midpoints of edges. The maps for any two two-simplexes with a common face can be made to match up on that face and so we obtain a continuous map $|\alpha|:|K|\to|T|$ which extends $\alpha: VK\to VT$. Clearly each track of P is mapped to the midpoint of an edge of T. If $v\in VT$ then $|\alpha|^{-1}(v)$ will consist of complete simplexes or subsets of two-simplexes which are either 'dotted lines' or 'triangular regions' as described above. Let C be a component of $|K|-P$. If $C\cap VK=\varnothing$ then the union of the 'dotted lines' and 'triangular regions' contained in C form a deformation retract R of C. Since C is connected, so is R. If $C\cap VK\neq\varnothing$ then let R be the union of all simplexes $|\sigma|$ completely contained in C, that is

$$R=\cup\{|\sigma|,\sigma\in SK\,|\,|\sigma|\subset C\}.$$

Again it is not hard to see that R is a deformation retract of C and so R is connected. This means that $|\alpha|(R)$ is a single vertex of T and so we have a well-defined map $\beta: VD_P\to ET$. This map, together with the previously defined $\beta: ED_P\to ET$, gives a morphism of graphs $\beta: D_P\to ET$.

Let T' be a G-tree. A vertex $v\in VT'$ is called *inessential* if it is incident with just two edges e_1 and e_2 and $G_{e_1}=G_v=G_{e_2}$. Any other vertex of T' is called *essential.*

As in Section 3, for a finite two-complex L, put $n(L)=2\,\text{rank}\,H^1(L,\mathbb{Z}_2)+v_L+f_L$. For an afp group G let $n(G)$ be the minimum value $n(L)$ takes as K ranges over all G-complexes on which G acts freely so that $H^1(K,\mathbb{Z}_2)=0$, $L=G\backslash K$ is finite and $|L|$ is homeomorphic to $G\backslash|K|$.

4.4 Theorem. *Let G be an afp group and let T be a G-tree. There is a G-tree T' and a G-map $\beta: T'\to T$ such that*

(i) *T' has less than $n(G)$ orbits of essential vertices,*
(ii) *G_v is finitely generated for each $v\in VT'$,*
(iii) *G_e is finitely generated for each $e\in ET'$.*

Proof. Let K be a two-complex on which G acts freely so that $H^1(K,\mathbb{Z}_2)=0$ and $L=G\backslash K$ is such that $n(L)=n(G)$. As above, we construct from a G-map $\alpha: VK\to VT$, a G-pattern $P\subset|K|$. Put $T'=D_P$. We have a G-morphism $\beta: T'\to T$. If $t\in ET'$, then t is a track of P. Now $G_t\backslash t$ maps injectively into $G\backslash|K|=|L|$. Thus $G_t\backslash t$ is the polyhedron of a finite graph. Since G_t acts freely on t it follows from Corollary I.9.3 that G_t is finitely generated. If C is a component of $|K|-P$ then C determines a vertex v of T' and if $\bar{C}$ is the closure of $C, G_v\backslash\bar{C}$ maps injectively into $G\backslash K$ and so $G_v\backslash\bar{C}$ is compact. Again it follows from Corollary I.9.3 that G_v is finitely generated since it acts freely on $\bar{C}$. Thus we have verified (ii) and (iii).

To prove (i) consider the covering map $\pi:|K|\to|L|$. The G-pattern P in $|K|$ will map to a pattern Q in $|L|$. Let C be a component of $|L|-Q$ whose closure is an untwisted band B. Then $\pi^{-1}B$ is a G-orbit of untwisted bands in $|K|$. Corresponding to $\pi^{-1}C$ is a G-orbit of inessential vertices of $D_P = T'$. But we saw in the proof of Theorem 3.8 that there are less than $n(L)$ components of $|L|-Q$ which are not untwisted bands. ∎

We are now able to give a geometric proof of Theorem V.12.1 in the case when $R=\mathbb{Z}_2$.

4.5 Theorem. *If G is an afp indicable group then $G = A\underset{C}{*}x$, where A and C are finitely generated subgroups of G and $x\in G$.*

Proof. A group is indicable if and only if it acts on the tree $T=\mathbb{R}$ with one edge orbit. For we can identify VT with $\mathbb{Z}$ and if $\phi:G\to\mathbb{Z}$ is a surjective homomorphism, then the map $n\mapsto\phi(g)+n$ can be extended to the required action.

Suppose then that G acts on $T=\mathbb{R}$ as described above. By Theorem 4.4 there is a G-tree T' and a G-morphism $\beta:T'\to T$ so that the stabilizers of edges and vertices of T' are finitely generated. Now, if $v\in VT$, $G_v = \operatorname{Ker}\phi$. This means that $G\backslash T'$ cannot be a tree. For, if $G\backslash T'$ is a tree, G is generated by vertex stabilizers by Proposition I.4.4 and so $\operatorname{Ker}\phi = G$. If we choose $e\in ET'$ so that Ge is a nonseparating edge in $G\backslash T'$, then by collapsing all other edges of T' apart from those in Ge we obtain a G-tree T'' such that T'' has one orbit of edges and one orbit of vertices. Thus $G=A\underset{C}{*}x$ as required, where $C=G_e$. ∎

5 Addition of patterns

Let P,Q be patterns in $|K|$ with associated functions $j_P, j_Q: EK\to\mathbb{Z}^+$. Now j_P+j_Q satisfies the conditions of Theorem 3.2 and so there is a pattern R such that $j_R = j_P + j_Q$. We say R is the *sum* of P and Q and write $R=P+Q$. Generally R is only defined up to equivalence. However, if $P\cap Q\cap|K^1|=\varnothing$, we can take R to be the unique pattern such that $R\cap|K^1|=(P\cup Q)\cap|K^1|$.

5.1 Example. Let P,Q each consist of two tracks as is illustrated in Fig. VI.6. The pattern P is shown intersecting the one-skeleton of K with

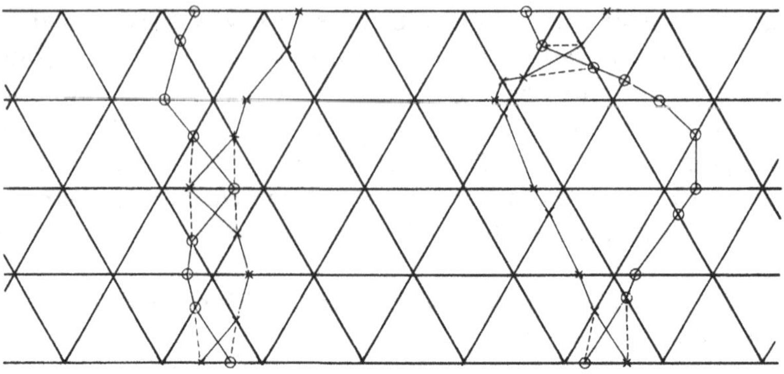

Fig. VI.6

o's and pattern Q is shown intersecting $|K^1|$ with ×'s. The pattern $P+Q$ is the same as $P \cup Q$ except in those two-simplexes where the patterns intersect. In these two-simplexes $P+Q$ is indicated by dotted lines. ∎

We are particularly interested in tracks t which separate $|K|$, that is so that $|K|-t$ has two components C_0 and C_1. If t is such a track, let $b_t: VK \to \mathbb{Z}_2$, $b_t(v)=i$ if $v \in C_i$. Clearly we could have labelled C_0 as C_1 and vice versa, in which case, instead of b_t, we would have obtained $b_t^* = 1 - b_t$. Note that if we regard b_t as an element of $\mathrm{Hom}(C_0(K), \mathbb{Z}_2)$ then $\delta b_t = \delta b_t^* = \mathrm{mod}_2\, j_t$, where $j_t: EK \to \mathbb{Z}^+$ is the map associated with t and $\mathrm{mod}_2: \mathbb{Z}^+ \to \mathbb{Z}_2$ is the natural map.

For any track t, let $\|t\|$ denote $|t \cap |K^1||$ if $t \cap |K^1|$ is finite; otherwise put $\|t\| = \infty$. If $\|t\|$ is finite, δb_t is finite and so $b_t \in \mathscr{B}X$ where X is the graph K^1. Note that $|\delta b_t| \leqslant \|t\|$ with equality if and only if t intersects each 1-simplex of K^1 at most once, that is if $\mathrm{Im}\, j_t \subseteq \{0,1\}$.

5.2 Definition. The track $t \subset |K|$ is said to be *thin* if t separates $|K|$, $\|t\| = m$ is finite, and the corresponding element $b_t \notin \mathscr{B}_m X$, the subring of $\mathscr{B}X$ generated by $\{a \mid |\delta a| < m\}$. ∎

If t is thin, then clearly $|\delta b_t| = m$ and b_t is a thin element of $\mathscr{B}X$ in the sense described in Definition II.2.1. Conversely we have the following proposition.

5.3 Proposition. *Let $H^1(K, \mathbb{Z}_2) = 0$. If $b \in \mathscr{B}X$ is thin as in Definition II.2.1, that is $b \notin \mathscr{B}_m X$ where $m = |\delta b|$, then there exists a thin track t with $\|t\| = m$ such that $b_t = b$.*

Proof. Let $i:\{0,1\}\to\mathbb{Z}^+$ be the inclusion map. For any $b\in\mathscr{B}X$ the map $i\delta b:EK\to\mathbb{Z}^+$ satisfies the conditions of Theorem 3.2. Thus there is a pattern P such that $j_P=i\delta b$. Since $H^1(K,\mathbb{Z}_2)=0$, it follows from Proposition 4.3 that D_P is a tree. Let E be the finite set consisting of all b_t, b_t^* as t ranges over all the tracks of P. Thus E is a tree subset of $\mathscr{B}X$. Now b is constant on the vertices of a component of $|K|-P$ and so b is in the Boolean ring generated by E. Now $P\cap|K^1|$ has at most $|\delta b|$ elements. Hence $b_t\in\mathscr{B}_mX$ unless P consists of a single track, and this track is thin in the sense of Definition 5.2. ∎

5.4 Definition. Let t_1, t_2 be thin tracks in $|K|$. We write $t_1\sim t_2$ if $\|t_1\|=\|t_2\|=m$ and $b_{t_1}-b_{t_2}\in\mathscr{B}_mX$. ∎

Note that in Definition 5.4 the choice of b_{t_1} and b_{t_2} is not crucial, since $b_{t_1}-b_{t_2}\in\mathscr{B}_mX$ if and only if $b_{t_1}^*-b_{t_2}=1-b_{t_1}-b_{t_2}\in\mathscr{B}_mX$.

Let P, Q be patterns in $|K|$. We write $P\leqslant Q$ if P is equivalent to a subpattern of Q, that is if there exists a pattern $P'\subseteq Q$ such that $P\equiv P'$. If $P\leqslant Q$ then there exists $R\leqslant Q$ such that $Q=P+R$; we write $Q=P\oplus R$. We say two patterns P and R are *compatible* if $P+R=P\oplus R$. Thus, two patterns P and R are compatible if there exist patterns $P'\equiv P$ and $R'\equiv R$ such that $P'\cap R'=\varnothing$. For any pattern P with a finite number of component tracks $t_1, t_2, \ldots, t_n$, we have $P=t_1\oplus t_2\oplus\cdots\oplus t_n$.

5.5 Proposition. *Let* $H^1(K,\mathbb{Z}_2)=0$. *Let* $t_1, t_2\subset|K|$ *be thin tracks. If* $P=t_1+t_2$ *then* P *consists of two thin tracks, that is* $P=t_1+t_2=s_1\oplus s_2$, *where* s_1 *and* s_2 *are thin. The tracks of* P *can be numbered so that* $t_1\sim s_1$ *and* $t_2\sim s_2$. *The pattern* P *is compatible with both* t_1 *and* t_2.

Proof. Let $b_1=b_{t_1}$ and $b_2=b_{t_2}$. If b_1 and b_2 are nested, then t_1 and t_2 are compatible. For suppose, say, $b_1\subseteq b_2$. We construct disjoint tracks t_1', t_2' equivalent to t_1 and t_2, respectively, as follows. Let γ be an oriented edge of $X=K^1$. If $\gamma\in\delta b_1$, $\gamma\notin\delta b_2$ then take $|\gamma|\cap t_1'$ to be the midpoint of $|\gamma|$. Similarly if $\gamma\in\delta b_2$ and $\gamma\notin\delta b_1$, take $|\gamma|\cap t_2'$ to be the midpoint of $|\gamma|$. If $\gamma\in\delta b_1\cap\delta b_2$ then take $|\gamma|\cap t_1'$ to be nearer than $|\gamma|\cap t_2'$ to $\iota\gamma$. By considering what happens on an arbitrary 2-simplex, one can see that t_1' and t_2' are disjoint.

Proposition 5.5 is clearly true if t_1 and t_2 are compatible, since $t_1+t_2=t_1\oplus t_2$. Thus we can assume that t_1 and t_2 are not disjoint and b_1 and b_2 are not nested. Thus each of the four elements of $b_1\square b_2$ are nonempty sets.

In Fig. VI.7 the greek letters represent the number of edges of X which join the relevant pairs of sets. Thus there are ρ edges with one vertex in $b_1^* \cap b_2$ and one vertex in $b_1 \cap b_2^*$. By relabelling b_1 as b_1^* and (or) b_2 as b_2^*, if necessary, we may assume that for $c \in b_1 \square b_2$, $|\delta c|$ takes its smallest value when $c = b_1 \cap b_2$. Since b_1 is thin and $b_1 = b_1 \cap b_2 + b_1 \cap b_2^*$, we must have $|\delta(b_1 \cap b_2^*)| \geqslant |\delta b_1|$, that is $\alpha + \rho + \delta \geqslant \beta + \gamma + \rho + \delta$. Hence, $\alpha \geqslant \beta + \gamma$. Similarly, using the fact that b_2 is thin and $b_2 = b_1 \cap b_2 + b_1^* \cap b_2$ we have $|\delta(b_1^* \cap b_2)| \geqslant |\delta b_2|$. This gives $\beta \geqslant \alpha + \gamma$. Hence, $\gamma = 0$ and $\alpha = \beta$. It is now clear that $|\delta(b_1 \cap b_2^*)| = |\delta b_1|$ and $|\delta(b_1^* \cap b_2)| = |\delta b_2|$.

We now show that $\rho \neq 0$. To see this recall that we know that $t_1 \cap t_2 \neq \varnothing$. Let σ be a 2-simplex of K containing a point of $t_1 \cap t_2$. There must be at least one face e of σ which contains a point of t_1 and a point of t_2; see Fig. VI.8. This is because $\delta|\sigma| \cap (t_1 \cup t_2)$ has four points and σ has only three sides. Thus $e \in \delta b_1 \cap \delta b_2$, so that either γ or ρ is nonzero. Since $\gamma = 0$, $\rho \neq 0$ and $|\delta(b_1 \cap b_2)| < |\delta b_1|$ and $|\delta(b_1 \cap b_2)| < |\delta b_2|$. Since $b_1 = b_1 \cap b_2^* + b_1 \cap b_2$ and $|\delta(b_1 \cap b_2^*)| = |\delta b_1|$ this must mean that $b_1 \cap b_2^*$ is thin. Similarly $b_1^* \cap b_2$ is thin. Hence, by Proposition 5.3, there are thin tracks s_1 and s_2 such that $b_{s_1} = b_1 \cap b_2^*$ and $b_{s_2} = b_1^* \cap b_2$. Since $b_1 \cap b_2 \in \mathscr{B}_m X$, where

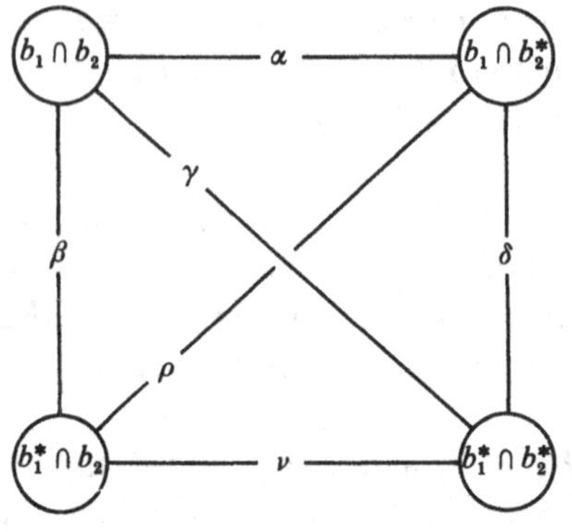

Fig. VI.7

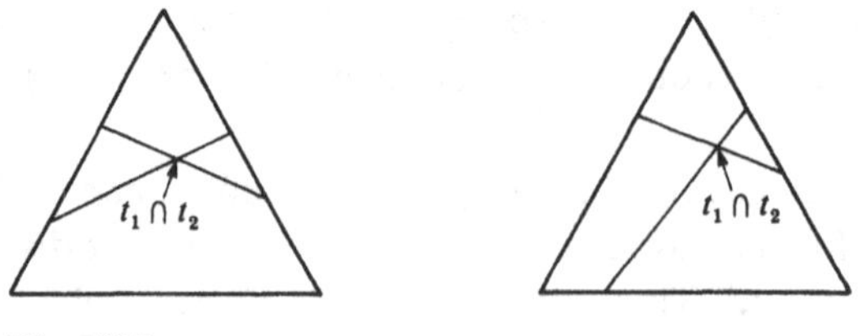

Fig. VI.8

$m = \min(\|t_1\|, \|t_2\|)$, we see that $s_1 \sim t_1$ and $s_2 \sim t_2$. Since $b_1 \cap b_2^*$ and $b_1^* \cap b_2$ are nested, s_1 and s_2 are compatible and we may as well assume they are disjoint. Let $f, g: EK \to \mathbb{Z}^+$, $f = j_{s_1} + j_{s_2}$, $g = j_{t_1} + j_{t_2}$. We show that $f = g$. Let $e \in EK$. Consider Fig. VI.7. If e is in the ρ position of Fig. VI.7, that is e has one vertex in $b_1^* \cap b_2$ and one vertex in $b_1 \cap b_2^*$ then $f(e) = g(e) = 2$. If e is in one of the $\alpha, \beta, \delta, \nu$ positions then $f(e) = g(e) = 1$. Recall that there are no edges in the γ position. If $e \notin \delta b_1 \cup \delta b_2$ then $f(e) = g(e) = 0$. Thus $f = g$ as required. Hence $j_{t_1} + j_{t_2} = j_P$, where $P = s_1 \oplus s_2$.

Finally note that $\{b_1, b_1^* \cap b_2, b_1 \cap b_2^*\}$ and $\{b_2, b_1^* \cap b_2, b_1 \cap b_2^*\}$ are nested sets. This means that P is compatible with either t_1 or t_2. In fact, there is a component C of $|K| - P$ whose closure contains both s_1 and s_2, and C contains tracks t_1' and t_2' such that $t_1' \equiv t_1$ and $t_2' \equiv t_2$. ∎

We illustrate Proposition 5.5 in an important case. Suppose K is as in Proposition 5.5 and also K is locally finite and $X = K^1$ has more than one end. A track $t \subset |K|$ is said to be *minimal* if the two components of $|K| - t$ each have infinitely many vertices and $\|t\|$ takes its smallest value consistent with this condition. If t is minimal then t is thin. To see this note that

$$\mathscr{B}_{\text{fin}} X = \{b \in \mathscr{B} X \mid \text{either } b \text{ or } b^* \text{ finite}\}$$

is a subring of $\mathscr{B}X$. Also $b_s \in \mathscr{B}_{\text{fin}} X$ if $\|s\| < \|t\|$. Thus if $m = \|t\|$ then $\mathscr{B}_m X \subseteq \mathscr{B}_{\text{fin}} X$. However t minimal implies $b_t \notin \mathscr{B}_{\text{fin}} X$. Hence $b_t \notin \mathscr{B}_m X$ and so t is thin.

By Proposition 5.5 if t_1 and t_2 are minimal tracks then $P = t_1 + t_2 = s_1 \oplus s_2$, where $s_1 \sim t_1$ and $s_2 \sim t_2$. But $s_i \sim t_i$ implies s_i is minimal since $b_{s_i} - b_{t_i} \in \mathscr{B}_{\text{fin}} X$. This is illustrated on the left hand side of Fig. VI.6, p. 234. Note that the right hand side of Fig. VI.6 shows that the minimality of $\|t\|$ is important. Two tracks t_1, t_2 are shown and neither b_{t_1} nor b_{t_2} is in $\mathscr{B}_{\text{fin}} X$. However, $t_1 + t_2 = s_1 \oplus s_2$, where both b_{s_1} and b_{s_2} are in $\mathscr{B}_{\text{fin}} X$.

5.6 Proposition. *If P is a pattern in $|K|$ and $t_1, t_2, \ldots, t_n$ are tracks compatible with P, then the pattern $Q = t_1 + t_2 + \cdots + t_n$ is compatible with P.*

Proof. By replacing each t_i by an equivalent track – if necessary – we can assume $P \cap t_i = \varnothing$ for every i, and also $t_i \cap t_j \cap |K^1| = \varnothing$ if $i \neq j$. Thus we can take Q to be the unique pattern such that

$$\left(\bigcup_{i=1}^{n} t_i\right) \cap |K^1| = Q \cap |K^1|.$$

Let $\sigma \in FK$. Then $|\sigma| \cap P$ will consist of disjoint straight lines which are disjoint from $|\sigma| \cap t_i$ for each $i \in [1, n]$. This is illustrated in Fig. VI.9, where $P \cap |\sigma|$ is shown in continuous lines and $\left(\bigcup_{i=1}^{n} t_i \right) \cap |\sigma|$ is shown dotted.

For each component c of $|\sigma| - P$, $\partial |\sigma| \cap c \cap \left(\bigcup_{i=1}^{n} t_i \right)$ can be joined in a unique way to provide disjoint lines contained in Q. Thus Q is disjoint from P and P and Q are compatible as required. ∎

5.7 Theorem. *Let $H^1(K, \mathbb{Z}_2) = 0$. Let $S = \{t_1, t_2, \ldots, t_n\}$ be a finite set of thin tracks in $|K|$. Let $P = t_1 + t_2 + \cdots + t_n$. Then P consists of n thin tracks $P = s_1 \oplus s_2 \oplus \cdots \oplus s_n$ which can be numbered so that $s_i \sim t_i$, $i \in [1, n]$.*

Proof. By replacing the t_i's by equivalent tracks, if necessary, we can assume $t_i \cap t_j \cap |K^1| = \varnothing$ if $i \neq j$. For $i, j \in [1, n]$, $i < j$, let $N_{ij} = |t_i \cap t_j|$. Let $N = \sum_{i,j} N_{ij}$. The theorem is proved by induction on N.

If $N = 0$, the t_i's are disjoint and the theorem is true. Suppose $N \neq 0$. Then $N_{pq} \neq 0$ for some p, q. By Proposition 5.5 there are disjoint tracks

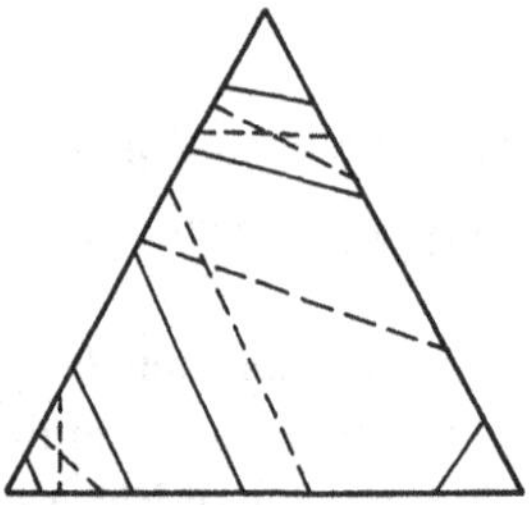

Fig. VI.9

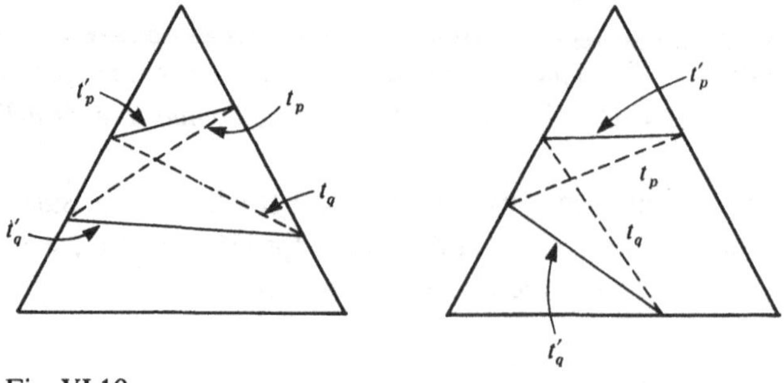

Fig. VI.10

t'_p, t'_q such that $(t'_p \cup t'_q) \cap |K^1| = (t_p \cup t_q) \cap |K^1|$ and $t'_p \sim t_p, t'_q \sim t_q$. Put $t'_i = t_i$ if $i \neq p$ and $i \neq q$. Let $N'_{ij} = |t'_i \cap t'_j|$ and $N' = \sum_{i,j} N'_{ij}$. Suppose $i \in [1, n] - \{p, q\}$. By considering Fig. VI.10 it will be seen that if t_i intersects $t'_p \cup t'_q$ exactly once inside the two-simplex $|\sigma|$ then it must intersect t_p or t_q inside $|\sigma|$. Also if t_i intersects both t'_p and t'_q inside $|\sigma|$ then it must intersect both t_p and t_q. Hence (if, say, $i < p < q$), $N'_{ip} + N'_{iq} \leqslant N_{ip} + N_{iq}$ and $N' < N$. Since $\left(\bigcup_{i=1}^{n} t'_i\right) \cap |K^1| = \left(\bigcup_{i=1}^{n} t_i\right) \cap |K^1|$, the theorem follows by induction. ∎

5.8 Proposition. *Let $H^1(K, \mathbb{Z}_2) = 0$. Let P be a pattern of thin tracks in $|K|$ and let s be any thin track, $s \subset |K|$. There exists a thin track s' such that $s' \sim s$ and s' is disjoint from P.*

Proof. Let $\{t_1, t_2, \ldots, t_n\}$ be the set of tracks of P which intersect s. This set is finite since s is compact. We use induction on n. If $n = 0$, there is nothing to prove. Suppose $n \geqslant 1$. Let $P_1 = t_1 \cup t_2 \cup \cdots \cup t_n$. Now D_{P_1} is a finite tree. Thus we may assume t_1 is a 'twig' of D_{P_1}, that is there is a vertex of t_1 which is incident with no other edge. Consider $s + t_1$. By Proposition 5.5, $s + t_1 = \hat{s} + \hat{t}_1$, where $\hat{s} \sim s$ and $\hat{t}_1 \sim t_1$. Also it can be seen from the proof of Proposition 5.5 that the pair $\hat{s}, \hat{t}_1$ can be chosen so that they lie in the two different components of $|K| - t_1$. Now one of these components contains all the tracks $t_2, t_3, \ldots, t_n$. Hence either $\hat{s}$ or $\hat{t}_1$ is compatible with P. If $\hat{s}$ is compatible with P then take $s' = \hat{s}$ and we are done. If $\hat{t}_1$ is compatible with P, then $\hat{s}$ is compatible with t_1 and $\hat{s}$ is compatible with any track $t \subseteq P - P_1$ by Proposition 5.6. Hence $\hat{s}$ is equivalent to a track which intersects less than n tracks of P. The result follows by induction. ∎

For the rest of this section we assume that the two-complex K satisfies $H^1(K, \mathbb{Z}_2) = 0$; that the group G acts on K so that G_e is finite for each $e \in EX$ (where $X = K^1$). It is also assumed that if $L = G \backslash K$ then $|L|$ is homeomorphic to $G \backslash |K|$: there is no real loss of generality in assuming this as explained in Section 2.

Let $t \subset |K|$ be a track with $\|t\|$ finite. We can then define $P = \sum_{g \in G} gt$, where, for $e \in EK$,

$$j_P(e) = \sum_{g \in G} j_t(g^{-1}e).$$

5.9 Theorem. *Let $t \subset |K|$ be a thin track. Then $P = \sum_{g \in G} gt$ is equivalent to a G-pattern of thin tracks. There exists a track $s \subseteq P$ such that $s \sim t$.*

Proof. If G is finite, the theorem follows immediately from Theorem 5.7. Suppose G is not finite.

Let γ_0 be a fixed one-simplex of K, such that $|\gamma_0| \cap P \neq \emptyset$. We show first that there is a finite subcomplex K_i of K such that $|K_i|$ contains any track s such that $s \cap |\gamma_0| \neq \emptyset$ and $\|s\| = i$. To see this, note that if $\gamma \in EK$ and $|\gamma| \cap s \neq \emptyset$, where $\|s\|$ is finite, then γ is the face of only finitely many two-simplexes. Thus we can successively construct $K_1, K_2, \ldots$ so that each K_i is finite and so that:

(i) K_1 contains the fixed one-simplex γ_0,

(ii) K_{i+1} contains K_i and if γ is a one-simplex of K_i and γ is contained in only finitely many two-simplexes of K, then all of these two-simplexes are in K_{i+1}.

Let $m = \|t\|$ and let $S = \{gt \mid gt \cap |K_m| \neq \emptyset\}$. Since K_m is finite, S is a finite set of thin tracks. By Theorem 5.7, $\sum_{gt \in S} gt = P_1$, where P_1 is a pattern of thin tracks s' and $s' \sim gt$ for some $g \in G$. Now if $s' \cap |\gamma_0| \neq \emptyset$ then $s' \subset |K_m|$. But if γ is a one-simplex of K_m, $j_P(\gamma) = j_{P_1}(\gamma)$. Thus $P \cap |K_m|$ and $P_1 \cap |K_m|$ are equivalent patterns of $|K_m|$. We know that any track of P_1 which intersects $|\gamma_0|$ is a thin track contained in $|K_m|$. Hence any track of P which intersects $|\gamma_0|$ is a thin track in $|K_m|$. But γ_0 was arbitrary so that any track of P is thin. Also if s is a track of $P, s \sim gt$ for some $g \in G$. Since $j_P(\gamma) = j_P(x\gamma)$ for every $x \in G$, we can choose P so that $P \cap |K^1| = xP \cap |K^1|$. By Theorem 3.2 this means that $P = xP$ for every $x \in G$ and so P is a G-pattern. Therefore we can choose s so that $s \sim t$. ∎

5.10 Example. Theorem 5.9 is illustrated in Figs. VI.11 and VI.12. The polyhedron $|K|$ is $\mathbb{R} \times [-1, 1]$ triangulated as shown. The group G, the automorphism group of K, is generated by ρ_x, ρ_y and t where the respective transformations of $|K|$ induced by these elements are reflection in the x-axis, reflection in the y-axis, and the translation $(x, y) \mapsto (x+1, y)$.

In Fig. VI.11, a minimal track t is shown together with all its translates under G. The G-pattern P is shown in Fig. VI.12. Since some one-simplexes of K have nontrivial stabilizers we cannot take $P \cap |K^1|$ to be the intersection of $|K^1|$ with the union of all translates of t. Instead we have counted the number $n_\gamma = j_P(\gamma)$ of translates of t which intersect a given one-simplex γ and then divided $|\gamma|$ into $n_\gamma + 1$ equal segments and taken $P \cap |\gamma|$ to be the interior division points. ∎

5.11 Theorem. *If $G \backslash K = L$ is finite, then there is a G-finite tree G-subset of $\mathscr{B}X$ which generates $\mathscr{B}X$ as a Boolean ring.*

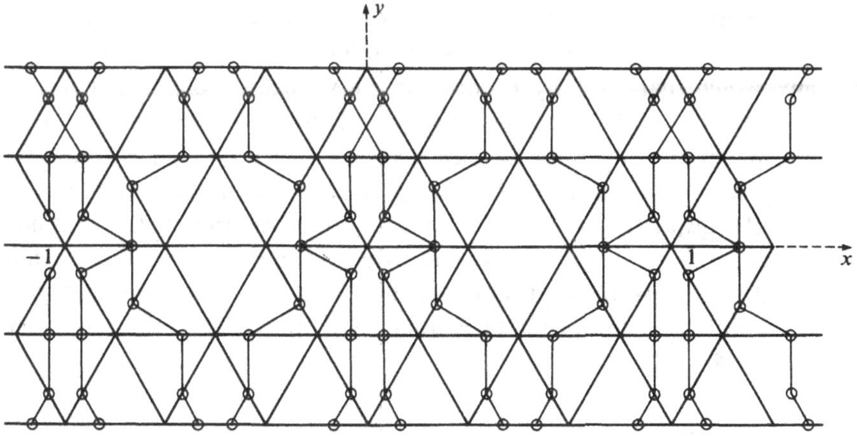

Fig. VI.11

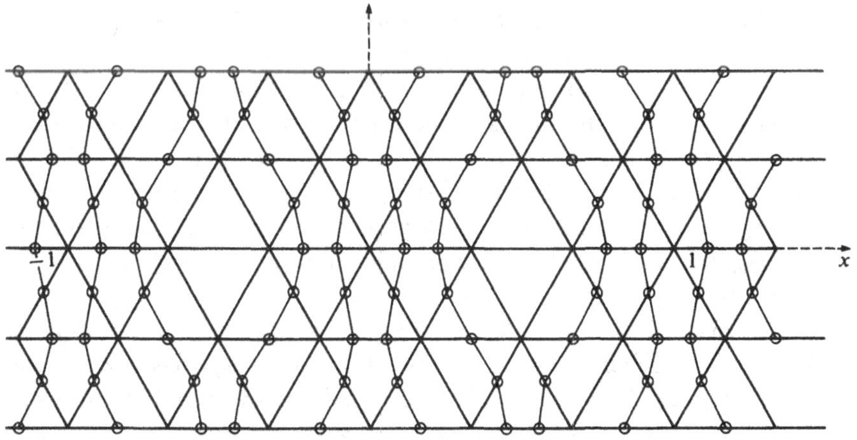

Fig. VI.12

Proof. Consider those reduced patterns $Q \subset |L|$ consisting of tracks t such that if $\pi:|K| \to |L|$ is the natural map, then $\pi^{-1}Q$ consists only of compact tracks. By Theorem 3.8 we can choose Q so that it is maximal, that is so that there is no reduced $Q' \subset |L|$ with similar properties that properly contains Q. Let $P = \pi^{-1}Q$. Let

$$E = \{b_t, b_t^* \mid t \text{ is a track of } P\},$$

and let Z be the subring of $\mathscr{B}X$ generated by E. We have to show that $Z = \mathscr{B}X$. Suppose $\operatorname{Im} \mathscr{B}\phi \neq Z$. Choose $b \in \mathscr{B}X - Z$ so that $|\delta b|$ has the smallest number m of elements. Thus $\mathscr{B}_m X \subseteq Z$ and b is thin in the sense of Definition II.2.1. Thus by Proposition 5.3 there is a thin track t such that $b_t = b$. Since $\mathscr{B}_m X \subseteq Z$, if $s \sim t$ then $b_s \in Z$. Thus we can change t to any s such that $s \sim t$. By Proposition 5.8 we can choose s so that it is disjoint from P. By Theorem 5.9 we can change s so that it is a track of a G-pattern. Also it can be seen from the proof of Theorem 5.9 and Proposition 5.6 that s can be chosen so that it is disjoint from P. Let $P' = P \cup \{gs \mid g \in G\}$. Let $Q' = \pi P'$. Clearly Q' is a pattern in $|L|$. If Q' is not reduced then πs is equivalent ($\equiv$) to a track of Q. But if $B \subset |L|$ is a band, $\pi^{-1} B$ is a union of bands (see Section 3). Thus if πs is equivalent to a track of Q, s is equivalent to a track of P. But this would contradict $b_s \notin Z$. ∎

6 The accessibility of almost finitely presented groups

6.1 Proposition. *Let U, W be connected G-graphs and suppose G_e is finite if $e \in EW$. If $\alpha: VU \to VW$ is a G-map, then α induces a G-map $\mathscr{B}\alpha: \mathscr{B}W \to \mathscr{B}U$.*

Proof. Let $b \in \mathscr{B}W$ so that $b: VW \to \mathbb{Z}_2$. Now $\mathscr{B}\alpha(b) = b\alpha: VU \to \mathbb{Z}_2$. In order to show that $\operatorname{Im} \mathscr{B}\alpha \subseteq \mathscr{B}U$ we have to show that $\delta(b\alpha)$ is finite. Since $G \backslash U$ is finite it suffices to show that for any $e \in EU$ there are any finitely many $g \in G$ such that $ge \in \delta(b\alpha)$. Let $e_1, e_2, \ldots, e_n$ be an edge path in W joining $\alpha(\iota e)$ and $\alpha(\tau e)$. Now if $ge \in \delta(b\alpha)$, $ge_i \in \delta b$ for some i. Since δb is finite and G_{e_i} is finite for each i, there are only finitely many such g. ∎

6.2 Proposition. *Let G be a finitely generated group and let X be a G-graph such that G_e is finite if $e \in EX$. If $\alpha: G \to VX$ is a G-map (regarding G as a left G-set under left multiplication) then α induces a G-map $\mathscr{B}\alpha: \mathscr{B}X \to \mathscr{B}G$. If $G \backslash X$ is finite and G acts freely on X then $\mathscr{B}\alpha$* is surjective.

Proof. If U is the Cayley graph of G with respect to a finite generating set then $\mathscr{B}G = \mathscr{B}U$ by Proposition IV.6.7. Hence the first part of the proposition follows from Proposition 6.1. If G acts freely on X, then we can find a G-map $\beta: VX \to G$ such that $\beta\alpha = 1_G$. If also $G \backslash X$ is finite, then, by Proposition 6.1, β induces $\mathscr{B}\beta: \mathscr{B}G \to \mathscr{B}X$. Clearly $\mathscr{B}\alpha \mathscr{B}\beta = 1_{\mathscr{B}G}$ and so $\mathscr{B}\alpha$ is surjective. ∎

6.3 Theorem. *Let G be an afp group. Then G is accessible.*

Proof. By Theorem 4.2, there is a two-dimensional simplicial complex K such that $H^1(K, \mathbb{Z}_2) = 0$, G acts freely on K and $G \backslash K = L$ is finite. We may also assume that $G\backslash|K|$ is homeomorphic to $|L|$ by Theorem 2.3. By Theorem 5.11 if $X = K^1$ then there is a G-finite tree G-subset E of $\mathscr{B}X$ which generates $\mathscr{B}X$ as a Boolean ring. Since E is nested it follows that E generates $\mathscr{B}X$ as a $\mathbb{Z}_2$-module. Clearly there is a G-map $\alpha: G \to VX$ and by Proposition 6.2 α induces a surjective G-map $\mathscr{B}\alpha: \mathscr{B}X \to \mathscr{B}G$. Hence, $\mathscr{B}G$ is finitely generated as a $\mathbb{Z}_2 G$-module. It follows from Corollary IV.7.6 that G is accessible. In fact the G-tree $T = T[E]$ is terminal. ∎

7 Two-manifolds

In this section $M = |K|$ is a connected two-manifold, and K is a simplicial complex.

Let N be a one-dimensional submanifold of M. We assume that $N = |L|$, where L is a subcomplex of some subdivision of K, though it will not usually be a subcomplex of K itself. We are particularly interested in the cases when N is connected and compact, so that N is either a simple closed curve (scc) in $M - \partial M$ or it is a line segment (homeomorphic to $I = [0, 1]$) properly embedded in M ($\partial N = \partial M \cap N$). We assume that N is in general position with respect to K. This is the case if and only if for each two-simplex $\sigma, |\sigma| \cap N$ is a one-submanifold of $|\sigma|$ properly embedded in $|\sigma|$ and $N \cap VK = \emptyset$. Thus $|\sigma|$ will appear as in Fig. VI.13, $N \cap |\sigma|$ will consist of finitely many scc's and line segments joining pairs of points in $\partial|\sigma| - VK$.

A closed compact subset of N homeomorphic to I is called an *arc*. If N is compact then let $\|N\| = |N \cap |K^1||$. Also if d is an arc of N let $\|d\| = |d \cap |K^1||$.

Two one-submanifolds $N, N' \subset M$ are said to be *disc equivalent* if there is a finite sequence $N_1, N_2, \ldots, N_k$ of one-submanifolds such that $N_1 = N$, $N_k = N'$ and for $i \in [1, n-1]$ there are arcs $d_i \subset N_i$ and $d'_{i+1} \subset N_{i+1}$ such that $N_i - d_i = N_{i+1} - d'_{i+1}$ and $d_i \cup d'_{i+1}$ bounds a disc D_i in M where $D_i \cap N_i = d_i$ and $D_i \cap N_{i+1} = d_{i+1}$.

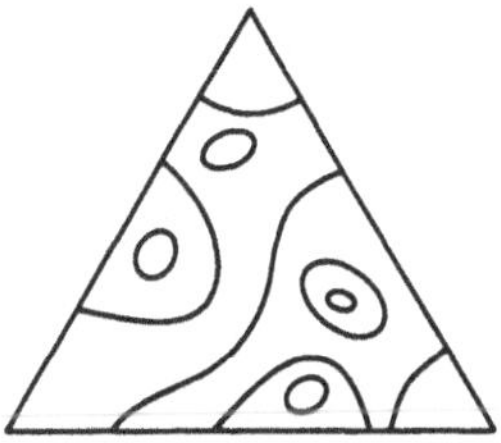

Fig. VI.13

7.1 Proposition. *Let l be an scc in the two-manifold M (in general position as above). Then l is disc equivalent to an scc in the interior of a two-simplex or l is disc equivalent to a track t such that $\|t\| \leqslant \|l\|$.*

Proof. If $\|l\| = 0$ then l is an scc in the interior of a two-simplex. If $\|l\| \neq 0$ then, for each two-simplex σ, $|\sigma| \cap l$ will consist of finitely many arcs joining pairs of points in $\partial|\sigma|$. Suppose there is an arc d joining two points x, y in a face $|\gamma|$ of $|\sigma|$. We can choose d so that it is 'innermost' that is so that there is no arc of $|\sigma| \cap l$ joining points of $|\gamma|$ lying between x and y in $|\gamma|$. Now $\gamma \not\subset \partial M$ and so there is a unique two-simplex $\sigma_1 \neq \sigma$ such that $|\gamma| \subset \partial|\sigma_1|$. There is a disc equivalence which replaces l by l', where $\|l'\| = \|l\| - 2$ in which a neighbourhood of d in l is replaced by an arc d' in l', where $d' \subset |\sigma_1|$ as in Fig. VI.14. If we repeat this process we eventually obtain an scc l such that $\|l\| = 0$ or l contains no arc contained in a two-simplex $|\sigma|$ and joining two points in the same face of $|\sigma|$. Thus $|\sigma| \cap l$ will be as in Fig. VI.15. It is how not hard to verify that we can find a disc equivalence which straightens the line segments in each two-simplex. ∎

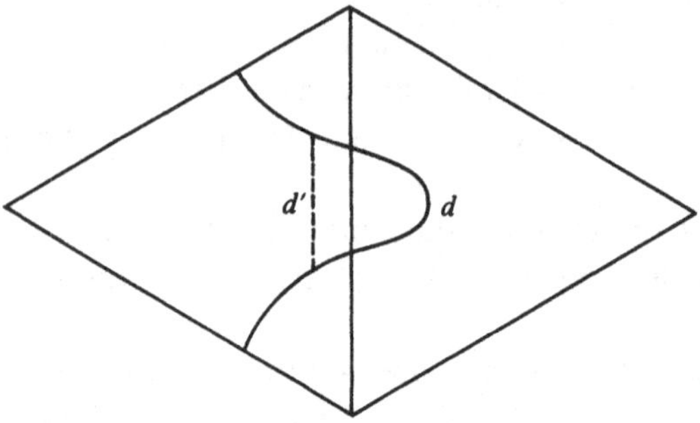

Fig. VI.14

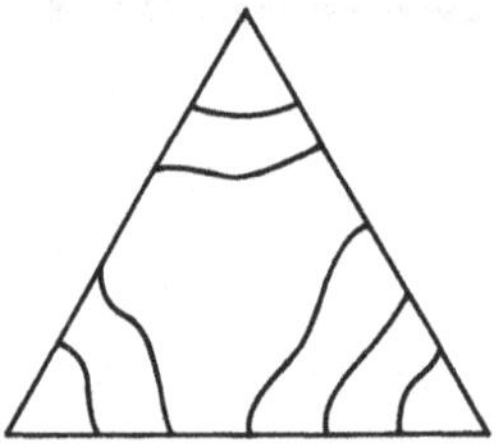

Fig. VI.15

Now let $M = |K|$ be a two-manifold for which $\partial M = \emptyset$ and $H^1(K, \mathbb{Z}_2) = 0$. By Corollary 3.5 every track separates M. If an scc l separates M then so does any scc disc equivalent to l. It follows from Proposition 7.1 that every scc in M separates M. It is known (Richards, 1963) that if M is a two-manifold with $\partial M = \emptyset$ and every scc separates M then $M \approx S^2 - Z$, where Z is a totally disconnected subset of S^2. The set Z can be regarded as the set of ends of M. Thus M has more than one end unless $Z = \emptyset$ or $|Z| = 1$, that is unless $M \approx S^2$ or $\mathbb{R}^2$. Suppose there is a group G which acts on K and thereby induces an action on M. If we assume that G acts faithfully on K then the stabilizer of each simplex $\sigma \in SK$ must be finite. In fact, if $v \in VK$ then G_v must be either finite cyclic or dihedral. Also if $e \in EK$, $|G_e| = 1$ or 2. Suppose M has more than one end. This is the case if and only if there is an scc in M which does not bound a disc. By Proposition 7.1 there is a track t in M that does not bound a disc. Since t separates M the two component s of $M - t$ must each contain infinitely many vertices. By Theorem 5.9 there is an scc l such that $gl = l$ or $gl \cap l = \emptyset$ for every $g \in G$ and l does not bound a closed disc in M. In fact we could take l to be a minimal track as in Section 5.

Let P be a G-pattern in M, where the component tracks of P are scc's. Let C be the closure in M of a component of $M - P$. Thus C is a two-manifold acted on by G_C, and ∂C consists of a G_C-set of scc's. We obtain a two-manifold $\hat{C}$ with $\partial \hat{C} = \emptyset$ by attaching discs to each boundary curve of C. Now C has more than one end if and only if $\hat{C}$ has more than one end. If $\hat{C}$ has more than one end then there exists an scc l in $\hat{C}$ which does not bound a disc in $\hat{C}$. In fact, l can be chosen to lie in C, and l does not bound a disc in $\hat{C}$ if and only if it does not bound a compact region of C. Again as in the proof of Theorem 5.9, we can choose l so that it is a track, disjoint from P and either $gl = l$ or $gl \cap l = \emptyset$ for every $g \in G$. Thus we can replace P by $P \cup (\cup \{gl | g \in G\})$. If $G \backslash K = L$ is finite then we cannot repeat this process more than $n(L)$ times, and so we eventually find a G-pattern P in M such that, if C is the closure of a component of $M - P$, then $\hat{C} \approx \mathbb{R}^2$ or S^2.

We have proved the following.

7.2 Theorem. *Let $M = |K|$ be a two-manifold such that $\partial M = \emptyset$ and $H^1(K, \mathbb{Z}_2) = 0$. Suppose the group G acts on K so that $G \backslash K$ is finite. There is a G-pattern P of scc's in M such that if C is the closure in M of a component of $M - P$ then $\hat{C} \approx \mathbb{R}^2$ or S^2.* ▮

8 Patterns and surfaces in three-manifolds

It is assumed that the reader is familiar with the basic definitions of three-manifolds (in the PL-category) as can be found in Hempel (1976).

Let M be a connected three-manifold. Let $M=|K|$ where the three-dimensional simplicial complex K is a combinatorial three-manifold. We do not assume that M is compact or that $\partial M=\varnothing$. A surface $S\subset M$ is said to be properly embedded if $S=|Y|$, where Y is a subcomplex of some subdivision of K, and $\partial S=\partial M\cap S$. Let M^i denote the i-skeleton $|K^i|, i=0,1,2$.

A properly embedded surface $S\subset M$ is called *patterned* if $S\cap M^2=P$ is a pattern in M^2 and for each three-simplex $\rho\in S_3K$, $|\rho|\cap S$ is a union of disjoint properly embedded discs in ρ. We write $P=P_S$.

8.1 Theorem. *For any pattern $P\subset M^2$ there is a patterned surface S such that $P=P_S$.* If *R,S are patterned surfaces such that $P_R=P_S$, then there is a homeomorphism $h:M\to M$ such that $h|\sigma|=|\sigma|$ for every simplex $\sigma\in K$ and $h(R)=S$.*

Proof. Let ρ be a three-simplex. Now $\partial|\rho|$ is a two-sphere and $P\cap\partial|\rho|$ is a union of finitely many disjoint scc's $c_1,c_2,\ldots,c_r$. We can find disjoint discs $d_1,d_2,\ldots,d_r$ which are properly embedded in ρ such that $\partial d_i=c_i$. Clearly the choice of discs is not unique. However, if $d_i', i\in[1,r]$ is another such set of discs, then there is a homeomorphism $h_\rho:|\rho|\to|\rho|$ which fixes $\partial|\rho|$ such that $h_\rho(d_i)=(d_i')$, $i\in[1,r]$. Carry out this choice of discs for each $\rho\in S_3K$. The union of all the discs will be a patterned surface such that $P_S=P$. If S' is another surface such that $P_{S'}=P_S$, then in each three-simplex ρ we can find a homeomorphism $h_\rho:|\rho|\to|\rho|$ as above so that $h_\rho(|\rho|\cap S)=|\rho|\cap S'$. By combining the h_ρ we obtain a homeomorphism $h':M\to M$ such that $h'S=S'$, h' leaves M^2 fixed and $h'|\rho|=|\rho|$ for every $\rho\in S_3K$. If P and Q are equivalent patterns in M^2 then, by Theorem 3.2, there is a homeomorphism $k':M^2\to M^2$ such that $k'P=Q$ and $k'|\sigma|\subseteq|\sigma|$ for every $\sigma\in S_0K\cup S_1K\cup S_2K$. We can find a homeomorphism $k:M\to M$ such that k restricted to M^2 is k' and $k|\sigma|=|\sigma|$ for every $\sigma\in S_3K$. Thus if $P\equiv Q$ then there exists $k:M\to M$ such that $kP=Q$. But now by the remarks above there exists $h':M\to M$ such that $h'kP=Q$. The second part of the theorem follows by putting $h=h'k$. ∎

We say two patterned surfaces A,B are equivalent, written $A\equiv B$, if $P_A\equiv P_B$. An equivalence class of patterns P in M^2 is determined uniquely by the function $j_P:S_1K\to\mathbb{Z}^+$, where $j_P(\gamma)=|P\cap|\gamma||$. Thus an equivalence class of patterned surfaces S is determined uniquely by the map $j_S: S_1 K \to \mathbb{Z}^+, j_S(\gamma) = |S \cap |\gamma||$.

As for patterns we can define an addition of patterned surfaces. If A, B are patterned surfaces in M then $A + B = C$, where $j_C = j_A + j_B$. Here C is only defined up to equivalence. It would be more precise to define addition on the set of equivalence classes of patterned surfaces. However, we persist with the rather sloppy (but convenient) practice of Section 3 and do not distinguish a patterned surface from its equivalence class.

We carry over the terminology and notation of Section 5 used for patterns to patterned surfaces. Thus if A, B are patterned surfaces we write $A \leqslant B$ if there exists $A' \subseteq B$ such that $A \equiv A'$. We write $A \oplus B = C$ if $A + B = C$ and $A \leqslant C, B \leqslant C$. We say A and B are compatible if $A + B = A \oplus B$. Two patterned surfaces A, B are compatible if and only if there exist $A' \equiv A, B' \equiv B$ such that $A' \cap B' = \varnothing$.

Let $\mathscr{P}M$ denote the set of all patterned surfaces in M. We will be investigating subsets $\mathscr{C}$ of $\mathscr{P}M$ which satisfy the following conditions:

($\mathscr{C}1$) If $A \in \mathscr{C}$, A is connected and compact.

($\mathscr{C}2$) If $A \in \mathscr{C}$ and $A' \equiv A$, then $A' \in \mathscr{C}$.

($\mathscr{C}3$) If $A_1, A_2 \in \mathscr{C}$ then $A_1 + A_2 = B_1 \oplus B_2$, where $B_1, B_2 \in \mathscr{C}$.

8.2 Theorem. *Let $\mathscr{C}$ be a G-subset of $\mathscr{P}M$ satisfying $\mathscr{C}1, \mathscr{C}2$ and $\mathscr{C}3$. If $A_1, A_2, \ldots, A_n \in \mathscr{C}$ then $A_1 + A_2 + \cdots + A_n = B_1 \oplus B_2 \oplus \cdots \oplus B_n$, where $B_i \in \mathscr{C}, i \in [1, n]$.*

Proof. The argument is essentially a repeat of the argument of Theorem 5.7. By replacing the A_i's by equivalent patterned surfaces if necessary, we may assume that $A_i \cap A_j \cap M^1 = \varnothing$ if $i \neq j$. For $i, j \in [1, n]$, $i < j$, let $N_{ij} = |A_i \cap A_j \cap M^2|$. Let $N = \sum_{ij} N_{ij}$. We use induction on N. If $N = 0$ then the patterns $P_{A_1}, P_{A_2}, \ldots, P_{A_n}$ are all disjoint and we can construct disjoint patterned surfaces $B_1, B_2, \ldots, B_n$ such that $B_i \equiv A_i$. If $N \neq 0$ then choose p, q so that $N_{pq} \neq 0$. By ($\mathscr{C}3$) $A_p + A_q = A'_p \oplus A'_q$, where $A'_p, A'_q \in \mathscr{C}$. Let σ be a two-simplex. The position is slightly different from that of Theorem 5.7 in that $A_i \cap |\sigma|$ may have more than one component. Thus, replacing $(A_p \cup A_q) \cap \sigma$ by $(A'_p \cup A'_q) \cap \sigma$ may involve replacing more than one pair of crossing lines. However, the change can be effected by a finite number of changes as illustrated in Fig. VI.10, page 238. Each such change can only reduce the total number of intersections in σ. Thus N is reduced and we can conclude from the induction hypothesis as in the proof of Theorem 5.7 that $A_1 + A_2 + \cdots + A_n = B_1 \oplus B_2 \oplus \cdots \oplus B_n$, where $B_i \in \mathscr{C}$, $i \in [1, n]$. ∎

If $S \in \mathscr{P}M$ let $\|S\| = |M^1 \cap S|$. Thus $\|S\|$ is finite if S is compact.

Suppose now that K is a G-complex, where $G\backslash K = L$ and $G\backslash|K| \approx |L|$. We suppose that the G-action is faithful. Since M is connected this means that $G_\rho = 1$ for every $\rho \in S_3 K$, $|G_\sigma| = 1$ or 2 for every $\sigma \in S_2 K$ and G_ρ is finite if $\rho \in S_1 K$.

If $A \in \mathscr{P}M$ and A is compact, we can define $S = \sum_{g \in G} gA$, where $j_S(e) = \sum_{g \in G} j_A(g^{-1}e)$.

8.3 Theorem. *Let $\mathscr{C} \subseteq \mathscr{P}M$ be a nonempty G-set satisfying $(\mathscr{C}1), (\mathscr{C}2)$ and $(\mathscr{C}3)$. Let $A \in \mathscr{C}$ and let $S = \sum_{g \in G} gA$. The patterned surface S can be chosen so that $gS = S$ for every $g \in G$. If C is a component of S then $C \in \mathscr{C}$ and for every $g \in G$ either $gC = C$ or $gC \cap C = \varnothing$.*

Proof. This argument is mainly a repeat of the argument of Theorem 5.9. It is a bit simpler here because we know that K is locally finite.

Let γ_0 be a fixed one-simplex of K such that $|\gamma_0| \cap S \neq \varnothing$. Let $m = \|A\|$ and let K_m be a finite subcomplex of K such that $|K_m|$ contains every patterned surface B for which $\|B\| = m$ and $B \cap |\gamma_0| \neq \varnothing$. One could construct such a K_m inductively by putting $|K_1| = |\gamma_0|$ and taking K_{i+1} to be the union of K_i together with all the faces of every simplex which has a face in K_i. The set $\mathscr{S}' = \{gA \mid gA \cap |K_m| \neq \varnothing\}$ is finite. Hence, by Theorem 8.2, each component of $S' = \sum_{gA \in \mathscr{S}} gA$ is in $\mathscr{C}$. For any $\gamma \in S_1 K_m$, $j_S(\gamma) = j_{S'}(\gamma)$. Hence, $S \cap |K_m| = S' \cap |K_m|$. Now any component of S' which intersects $|\gamma_0|$ is in $|K_m|$. Hence any component of S which intersects $|\gamma_0|$ belongs to $\mathscr{C}$. However the choice of γ_0 was arbitrary, and so every component of S must be in $\mathscr{C}$.

As in the proof of Theorem 5.9 it is possible to arrange that $S \cap |K^1| = gS \cap |K^1|$ for every $g \in G$. This forces $S \cap |K^2|$ and $gS \cap |K^2|$ to be the same and then we can choose the discs in the three-simplexes of K so that $gS = S$ for every $g \in G$. ∎

8.4 Remark. The concept of a patterned surface is closely related to that of a *normal surface* introduced by Haken (1961) in his algorithm for determining the genus of a knot. A patterned surface S in the triangulated three-manifold M is a normal surface if the intersection of S with each three-simplex consist of discs whose boundaries intersect each one-simplex at most once. Two such discs in a given three-simplex are said to have the same type if they intersect the same one-simplexes. In normal surface theory a normal surface is specified by the number of discs of each type within each three-simplex. The sum of two normal surfaces is not always defined in that there may be no normal surface for which the number of discs of each type is the sum of the corresponding numbers for each summand. However, if the sum is defined then the surface obtained is the same as the surface obtained by regarding the surfaces as patterned surfaces and summing them as patterned surfaces. Thus the set $\mathcal{N}M$ of normal surfaces is proper subset of the set $\mathcal{P}M$ of patterned surfaces, and $\mathcal{P}M$ is closed under addition but $\mathcal{N}M$ is not closed under addition. Clearly then it is easier to work in $\mathcal{P}M$ rather than $\mathcal{N}M$. ∎

9 Simplifying surface maps

Let S be a compact two-manifold and let $f:(S, \partial S) \to (M, \partial M)$ be a general position map; see Hempel (1976) for a full discussion of general position. It is also assumed that f is in general position with respect to K. We define an *i-piece* of f to be a component of $f^{-1}|\sigma|$, where σ is an $(i+1)$-simplex of K. Thus the set of zero-pieces of f is a finite set of points of S. The one-pieces are either scc's or arcs joining two zero-pieces. The two-pieces are surfaces with boundary where the boundary is a union of one-pieces.

We describe a way of changing the map f to a new map $\hat{f}:(\hat{S}, \partial\hat{S}) \to (M, \partial M)$, where the pieces of $\hat{f}$ have a particularly simple form. In fact, they will give a regular cell decomposition of $\hat{S}$. The change from f to $\hat{f}$ is effected in a finite number of steps. It is also shown that if f is an embedding then it can be arranged that $\hat{f}$ is an embedding.

Let p be a one-piece of f. Let σ be the two-simplex of K such that $f(p) \subset |\sigma|$. Let V be a regular neighbourhood of p in S. (We are assuming that f is simplicial with respect to some triangulation of S and some subdivision of K.) We can thereby choose V so that if τ is a simplex of K and $|\tau| \cap |\sigma| = \varnothing$ then $f(V) \cap |\tau| = \varnothing$. Let $\mathring{V}$ denote the interior of V regarded as a subspace of S. Thus $p \subset \mathring{V}$ even if $p \subset \partial S$. Similarly let βV denote the boundary of V regarded as a subspace of S, so that $\beta V = V - \mathring{V}$. If ∂V is the boundary of V regarded as a two-manifold, then $\partial V = \beta V$ if and only if $p \cap \partial S = \varnothing$.

We first show that if f has any one-pieces that are scc's then we can change f to f', where f' has one less one-piece. Let p be a one-piece as above such that p is an scc. First we consider the case when $|\sigma| \not\subset \partial M$ or, equivalently, $p \cap \partial S = \varnothing$. In this case V is an annulus containing p in its interior. Let S' be the surface obtained from S by surgery along p. Thus

S' is obtained from S by removing the interior of V and attaching discs Δ_1, Δ_2 to the two components l_1, l_2 of ∂V. Note that $f(l_1)$ and $f(l_2)$ will lie in the interiors of the two three-simplexes $|\rho_1|$ and $|\rho_2|$, respectively, which contain $|\sigma|$. It is easy to see that there is a continuous map $f': S' \to M$ such that f' and f are equal when restricted to $S - \mathring{V}$ and $f'(\Delta_i) \subset |\rho_i|$ for $i = 1, 2$.

If f is an embedding, then $f(S) \cap |\sigma|$ will consist of finitely many disjoint scc's and arcs as in Fig. VI.13, page 243. We choose a one-piece p so that $f(p)$ is innermost in $|\sigma|$, that is so that if Δ is the disc in $|\sigma|$ such that $\partial\Delta = f(p)$ then $\Delta \cap f(S) = f(p)$. With this choice of p it is not difficult to see that f' can be chosen to be an embedding.

Suppose now that p is an scc and $\sigma \subset \partial M$. This means that $p \subset \partial S$, V is an annulus, $\beta V = l$, where l is an scc, and $f(l)$ is in the interior of the unique three-simplex $|\rho|$ such that $|\sigma| \subset |\rho|$. Let S' be the surface obtained by removing $\mathring{V}$ from S and attaching a disc D to l. Choose f' so that f' is f when restricted to $S - V$ and $f'(D) \subset |\rho|$. If f is an embedding, then we choose p so that $f(p)$ is innermost in $|\sigma|$ and then f' can be chosen to be an embedding. Changes of the above two types each of which reduce by one the number of 1-pieces of f which are scc's are called *changes of Type A*. These changes are illustrated in Fig. VI.16.

Suppose then that every one-piece of f is an arc. A one-piece p is called *returning* if the two zero-pieces u, v of f joined by p are both mapped by

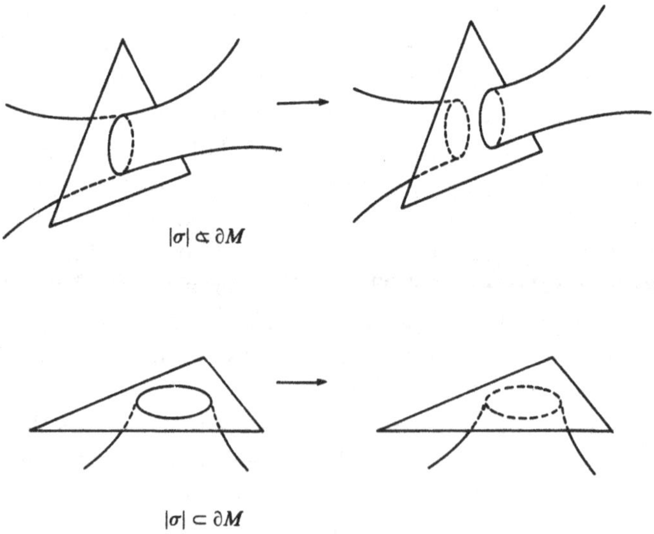

Fig. VI.16

f into the same one-simplex $|\gamma|$ of K. We now describe changes which will eventually remove all returning one-pieces.

Let p be a returning arc. There are three cases to consider: (i) $|\gamma| \not\subset \partial M$, (ii) $|\gamma| \subset \partial M$, $|\sigma| \not\subset \partial M$, (iii) $|\sigma| \subset \partial M$. In Case (i) the regular neighbourhood V is a disc and $\beta V = \partial V$ is an scc in $S - \partial S$. The union of all the three-simplexes of K which contain $|\gamma|$ is a closed ball B and $f(\beta V) \subset \mathring{B} - |\sigma|$. Note that $\mathring{B} - |\sigma|$ is contractible. Define $f':(S, \partial S) \to (M, \partial M)$ so that f' is continuous, f' and f are the same when restricted to $S - \mathring{V}$, and $f'(V) \subset \mathring{B} - |\sigma|$. In Case (ii), βV consists of two arcs l_1 and l_2. Let S' be the surface obtained by removing $\mathring{V}$ from S and attaching two discs D_1 and D_2 by identifying l_1 and l_2 with connected closed subsets of $\partial D_i, i = 1, 2$. Let $f':(S', \partial S') \to (M, \partial M)$ be a map such that f and f' are the same on $S - \mathring{V}$. Also $f'(\partial D_i - l_i) \subset |\sigma_i|$ for $i = 1, 2$, where σ_1 and σ_2 are the two-simplexes of ∂M such that $|\gamma| = |\sigma_1| \cap |\sigma_2|$, and $f'(D_1)$ and $f'(D_2)$ lie in the two distinct components of $\mathring{B} - |\sigma|$. In Case (iii) βV consists of a single arc. Let σ' be the two-simplex of ∂M such that $\sigma' \neq \sigma$ and $|\gamma| = |\sigma| \cap |\sigma'|$. Define $f':(S, \partial S) \to (M, \partial M)$ so that f' is the same as f on $S - \mathring{V}$ and $f'(p) \subset |\sigma'|$.

Note that it is not clear that f' will have fewer returning one-pieces than f. However, it is always the case that $\|f'\| = \|f\| - 2$. (Here $\|f\| = |f(S) \cap M^1|$.) Changes of the above type are called *changes of Type B*. If f has returning one-pieces then we carry out a change of Type B and reduce $\|f\|$. If the new map has returning one-pieces then repeat the process. Since $\|f\|$ is a positive integer this process cannot be repeated indefinitely, and so we eventually get a map with no returning one-pieces. Note that a Type B change may introduce a map which has one-pieces which are scc's. These, however, can be removed using Type A changes. A Type A change does not alter $\|f\|$. Changes of Type B are illustrated in Fig. VI.17.

If f is an embedding and has returning one-pieces then we can choose p so that $f(p)$ is innermost in $|\sigma|$, that is so that the disc D bounded by p and the arc $[u, v]$ of $|\gamma|$ satisfies $f(p) \cap D = f(p)$. Here we are assuming that we have removed one-pieces which are scc's by Type A changes. With such a choice of p we can arrange that f' is also an embedding.

Note that if $S = S'$, then the maps f and f' in a Type B change are homotopic. If f is an embedding and $S = S'$ then f and f' are isotopic.

Now we show how to change f to simplify the two-pieces. If S has a connected component C which is a two-piece, that is there is a three-simplex ρ of K such that $f(C)$ is contained in $|\rho|$, then put $S' = S - C$ and change f to $f':(S', \partial S') \to (M, \partial M)$, where f' is f restricted to S'. Every other

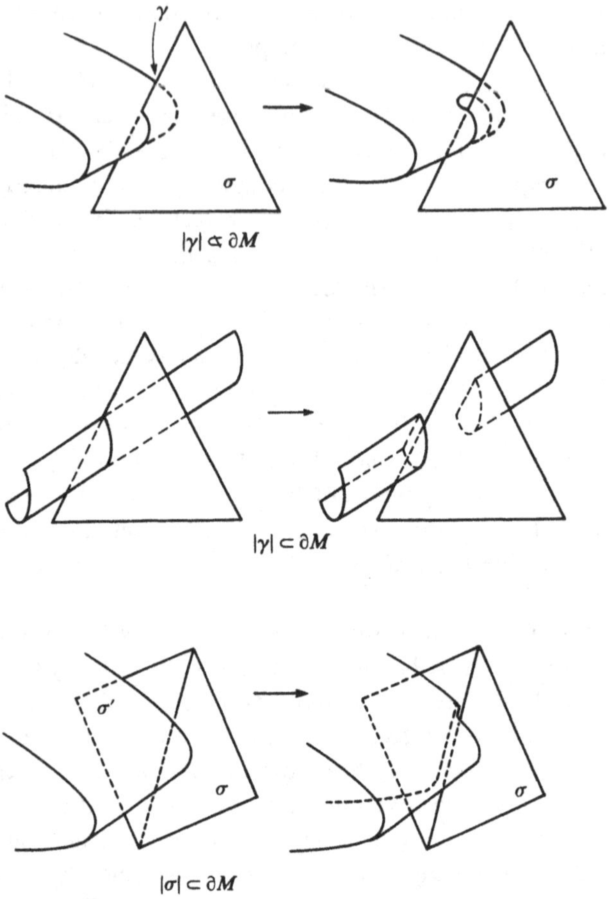

Fig. VI.17

two-piece of f must be a two-manifold with nonempty boundary. Suppose there is a two-piece P which is not a disc. Then there is an scc l in P which does not bound a disc in P. Let S' be the surface obtained from S by surgery along l. Thus S' is obtained from S by removing the interior of a regular neighbourhood V of l and attaching discs to the boundary curves thereby created (two discs Δ_1, Δ_2 if l is orientation preserving, one disc Δ_1 if l is orientation reversing). Let f' be the same as f on $S - \mathring{V}$ and map the discs Δ_1, Δ_2 into $|\rho|$. Now surgery increases $\chi(P)$ (by two if l is orientation preserving and by one if l is orientation reversing). Clearly surgery may separate P. However, the number b of boundary components of P is an upper bound on the number of two-pieces that can be created

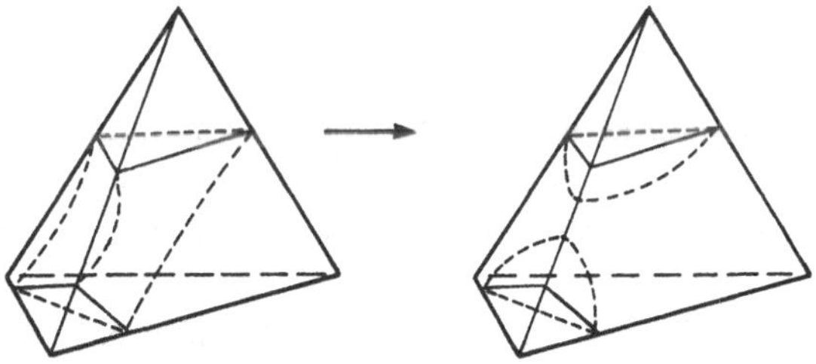

Fig. VI.18

by successive surgeries. Thus eventually we will obtain b discs by this process. Changes of this sort are called changes of Type C. A Type C change is illustrated in Fig. VI.18.

If f is an embedding, then f restricted to l is an embedding and there is a disc $D \subset |\rho|$ such that $\partial D = l$. We can assume that D meets $f(S)$ transversely. Thus $f(S) \cap D$ is a finite set of scc's. Choose l' to be an scc in $f(S) \cap D$ which is innermost in D. Thus l' bounds a disc D' in D and $D' \cap f(S) = \partial D'$. If l' bounds a disc D_1 in $f(P)$, then we can alter D by swapping D' for D_1 and making a small alteration in a neighbourhood of D_1 obtaining a new disc D'' such that $\partial D'' = l$ and $D'' \cap f(S)$ has one less component than $D \cap f(S)$. By repeating this process we eventually obtain a disc Δ such that $\Delta \cap f(S) = \partial \Delta$ and $\partial \Delta$ does not bound a disc in $f(P)$. If we now replace l by $f^{-1}(\partial \Delta)$ we can carry out the Type C change above so that f' is also an embedding.

Finally we show that we can alter f so that, if P is a two-piece of the changed map, then ∂P contains either three or four zero-pieces, and no two of these zero-pieces are mapped by f to the same one-simplex of K. Suppose, then, that P is a two-piece of f and $\{u, v\} \subset \partial P$, where $u \neq v$ and $\{f(u), f(v)\} \subset |\gamma|$, where γ is a one-simplex of K. There is an arc $a \subset P$ such that $\partial a = \{u, v\}$. Let V be a regular neighbourhood of a in S. There are two cases to consider: $|\gamma| \not\subset \partial M$ and $|\gamma| \subset \partial M$. If $|\gamma| \not\subset \partial M$ then we alter f as in a Type B(i) change. If $|\gamma| \subset \partial M$ then we alter f as in a Type B(ii) change.

We see then that, if σ is a two-face of ρ, where ρ is the three-simplex of K such that $f(P) \subset |\rho|$, then $f^{-1}|\sigma| \cap P$ is either empty or a single arc. Thus ∂P consists of at most four one-pieces. It is not possible for ∂P to consist of either one or two one-pieces. Hence ∂P must contain either three or four one-pieces. It therefore contains either three or four zero-pieces. A change of this sort is called a Type D change. As in the discussion for Type B changes, if f is an embedding then we can carry out a Type D change so that f' is an embedding.

We now record some of the properties of the map $\hat{f}:(\hat{S},\partial\hat{S})\to(M,\partial M)$ resulting from carrying out all the changes described above. The changes that have been carried out here on surfaces in three-manifolds are similar to the changes made on scc's in two-manifolds in Proposition 7.1. There is a corresponding notion of disc equivalence. Two surfaces $S, S' \subset M$ are said to be *disc equivalent* if there is a sequence $S = S_1, S_2, \ldots, S_n = S'$ of surfaces $S_i \subset M, i \in [1,n]$, such that, for each $j \in [2,n]$, there exist discs $D_j \subset S_j, D_{j'} \subset S_{j-1}$ such that $D_j \cup D_{j'}$ bounds a three-ball $B_j \subset M, S_j - D_j = S_{j-1} - D'_j, B_j \cap S_j = D_j$ and $B_j \cap S_{j-1} = D'_j$.

9.1 Theorem. *If f is an embedding and $f(S)$ has a component two-sphere C which does not bound a three-ball then $\hat{f}(\hat{S})$ has a component two-sphere $\hat{C}$ which does not bound a three-ball.*

Proof. If we examine the above changes, the ones which affect C are either disc equivalence or of the following type. There is a disc D such that $D \cap C = \partial D$, and the change is by surgery along D. This creates two embedded and disjoint two-spheres C_1 and C_2 in M. If $C_1 = \partial B_1$ and $C_2 = \partial B_2$, where both B_1 and B_2 are three-balls, then we shall show that $C = \partial B$, where B is a three-ball which contradicts our hypothesis. If $B_1 \cap B_2 = \varnothing$ then $B = B_1 \cup B_2 \cup b$, where b is a three-ball bounded by $D_1 \cup D_2 \cup A$ with A a regular neighbourhood of ∂D in C and $D_1 \subset C_1$ and $D_2 \subset C_2$ are disjoint discs lying close to and on either side of D. It is clear that B is a three-ball since attaching two three-balls by a common boundary disc creates a three-ball. If $C_1 \subset B_2$ then we may assume $B_1 \subset B_2$ by the Schoenflies Theorem; see M. Brown (1960). But now the closure of $B_2 - b \cup B_1$ is a three-ball B such that $\partial B = C$. If $C_2 \subset B_1$ then the symmetric argument works. If neither $C_1 \subset B_2$ nor $C_2 \subset B_1$ then it is easy to see that B_1 and B_2 are disjoint, which is the case with which we have already dealt.

It is easy to see that if C does not bound a three-ball, then neither will any two-sphere disc equivalent to C. ∎

9.2 Theorem. *If S contains a disc Δ such that $f|\partial\Delta \not\simeq 0$ in ∂M then $\hat{S}$ contains a disc $\hat{\Delta}$ such that $\hat{f}|\partial\hat{\Delta} \not\simeq 0$ in ∂M.*

Proof. It is clear that the changes, except a Type B(ii) change, can only change $f|\partial\Delta$ by a homotopy. A Type B(ii) change on Δ creates two discs Δ_1 and Δ_2 such that $f|\partial\Delta \simeq f'|\partial\Delta_1 + f'|\partial\Delta_2$. Also, surgery on a disc can only create a disc and a two-sphere. ∎

10 The equivariant sphere and loop theorems

We retain the notation of the previous section.

Let $\mathscr{S} \subseteq \mathscr{P}M$ be the set of patterned surfaces S satisfying the following conditions:

($\mathscr{S}_1$) $S \approx S^2$ (the two-sphere).

($\mathscr{S}2$) S does not bound a three-ball in M, that is there is no closed subset $B \subset M$ such that $B \approx B^3$ and $\partial B = S$.

($\mathscr{S}3$) If $A \in \mathscr{P}M$ and A satisfies ($\mathscr{S}1$) and ($\mathscr{S}2$) then $\|S\| \leqslant \|A\|$.

Thus there is a constant k such that $\|S\| = k$ for every $S \in \mathscr{S}$.

We see from Theorem 9.1 that, if $S \subseteq M$ is a two-sphere which does not bound a three-ball and which meets M^2 transversely, then there is a patterned surface S' with the same properties as S, that is S' is a two-sphere and S' does not bound a three-ball, such that $S' \cap M^1 \subseteq S \cap M^1$. Thus if $\|S\| = k$, then $S' \in \mathscr{S}$ and $S' \cap M^1 = S \cap M^1$.

10.1 Theorem. *The subset $\mathscr{S}$ of $\mathscr{P}M$ satisfies ($\mathscr{C}1$), ($\mathscr{C}2$) and ($\mathscr{C}3$).*

Proof. Clearly $\mathscr{S}$ satisfies ($\mathscr{C}1$) and ($\mathscr{C}2$).

We show that $\mathscr{S}$ satisfies ($\mathscr{C}3$). Suppose $A_1, A_2 \in \mathscr{S}$. Replacing A_1, A_2 by equivalent surfaces if necessary, we can assume that A_1, A_2 meet transversely if at all. Thus $A_1 \cap A_2$ consists of finitely many scc's and $A_1 \cap A_2 \cap M^2$ is a finite set of points. Let $\alpha = |A_1 \cap A_2 \cap M^2|$ and let β be the number of components of $A_1 \cap A_2$. Choose A_1, A_2 from their respective equivalence classes so that (α, β) is the smallest possible in the lexicographic ordering of pairs of non-negative integers ($(n, m) \leqslant (n', m')$ if $n < n'$ or $n = n'$ and $m \leqslant m'$).

If $(\alpha, \beta) = (0, 0)$ then A_1 and A_2 are disjoint and so ($\mathscr{C}3$) is trivially satisfied. Assume then that $(\alpha, \beta) \neq (0, 0)$. Two curves $l_1, l_2 \subset A_1 \cap A_2$ are said to be *parallel* in A_i ($i = 1$ or 2) if there is an annulus $a \subseteq A_i$ such that $\|a\| = 0$ and $\partial a = l_1 \cup l_2$. A disc $D \subset A_i$, $i = 1$ or 2, is called *minimal* if $\partial D = D \cap A_1 \cap A_2$ and $\|D\| \leqslant \|D'\|$ for any other disc D' contained in either A_1 or A_2 such that $\partial D \subset A_1 \cap A_2$. There is a constant d such that if D is a minimal disc then $\|D\| = d$. Notice that both $A_1 - A_1 \cap A_2$ and $A_2 - A_1 \cap A_2$ have components which are open discs, so that minimal discs exist. By relabelling A_2 as A_1 if necessary we can assume that there is a minimal disc $D \subset A_1$. (In fact, we will soon see that there are minimal discs in both A_1 and A_2 so that no relabelling is required.) Let $l = \partial D$. Now l bounds two discs in A_2 whose union is all of A_2. Thus we can choose $D' \subset A_2$ so that $\partial D' = l$ and $\|D'\| \leqslant \frac{1}{2}k$. If $d < \|D'\|$ then $D \cup D'$ is an embedded two sphere in M and $\|D \cup D'\| < k$. Hence $D \cup D' = \partial B$, where

B is a three-ball. Thus A_2 is disc equivalent to the two-sphere S obtained from A_1 by replacing D' by D. But $\|S\| = k - \|D'\| + d < k$ and so we have a contradiction. Hence $\|D'\| = d$. However, note that D' may not be minimal in that there may be components of $A_1 \cap A_2$ other than l in D'. Note that all these components will have to be parallel to l in A_2 and D' will contain a minimal disc Δ.

Our object is to show that there is a minimal disc D such that the corresponding D' as described above is also minimal and $D \cup D' = \partial B$, where B is a three-ball. Let D, D' be as above, so that D is a minimal disc, $D \subset A_1$, $D' \subset A_2$, $\partial D = \partial D'$ and $\|D\| = \|D'\| = d$. The first case we consider is when $d < \frac{1}{2}k$. In this case $\|D \cup D'\| < k$ and so $D \cup D' = \partial B$, where B is a three-ball. We may suppose that D is the only component of $B \cap A_1$ which is a minimal disc. For, if there are other components which are minimal discs, then we alter our choice of D so that it is the minimal disc component of $B \cap A_1$ for which ∂D is innermost in D'. It is also the case that $B \cap A_1$ has no component which is an annulus a such that $\|a\| = 0$. For should such a component exist then ∂a bounds an annulus $a' \subset A_2$ such that $\|a'\| = 0$. If we now form the two-sphere $(A_2 - a') \cup a$ and make a small alteration in a neighbourhood of a' we obtain a two-sphere A_2' such that $A_1 \cap A_2'$ has at least two fewer components than $A_1 \cap A_2$. Now $(\partial B - a') \cup a$ is a two-sphere in B and hence it bounds a ball $B' \subset B$. Thus A_2' is disc equivalent to $(A_2 - D') \cup D$ which is disc equivalent to A_2. Since A_2' is disc equivalent to A_2 it follows that A_2' does not bound a three-ball in M. Now $A_2' \cap M^1 = A_2 \cap M^1$. If σ is a two-simplex of K then $A_2 \cap |\sigma|$ consists of disjoint straight lines joining a certain set of points in $\partial|\sigma|$. Clearly $A_2' \cap |\sigma|$ will consist of disjoint arcs (not necessarily straight lines) joining the same set of points. Note that there will be no returning arcs (see Section 9) since, if there were, A_2' would be disc equivalent to a patterned surface S such that $\|S\| < k$ and so A_2' would bound a three-ball. It follows that the end points of each arc of $A_2' \cap |\sigma|$ are the end point of an arc of $A_2 \cap |\sigma|$. This means that $A_1 \cap A_2' \cap |\sigma|$ will contain at least as many points as $A_1 \cap A_2 \cap |\sigma|$, since an arc joining two points will have at least as many intersections with a fixed set of straight lines as the straight line joining the two points. But $A_1 \cap A_2$ is a subset of $A_1 \cap A_2'$ and so the components that have been lost cannot intersect M^2. We now see that A_2' is a patterned surface equivalent to A_2. In fact $A_2' \cap M^2 = A_2 \cap M^2$, and so $A_1 \cap A_2 \cap M^2 = A_1 \cap A_2' \cap M^2$. But $A_1 \cap A_2'$ has fewer components than $A_1 \cap A_2$ which contradicts the choice of A_1 and A_2.

Now consider the minimal disc Δ contained in D'. The argument above shows that $\partial\Delta$ bounds a disc Δ' in A_1 such that $\|\Delta'\| = d$ and A_1 is disc equivalent to $A=(A_1-\Delta')\cup\Delta$. We can alter A in a neighbourhood of Δ so that we replace Δ by a disc lying in the interior of B and 'parallel' to Δ and we obtain a two-sphere $A'\in\mathscr{S}$ such that $A'\cap\partial B$ has fewer components than $A_1\cap\partial B$. Since $A_1\cap B$ contains no annulus for which $\|a\|=0$, we do not create any new minimal disc in B by the change from A_1 to A'. We now repeat this process, and eventually we obtain a two-sphere S such that $S\in\mathscr{S}$ and $S\subset B$. Plainly this is absurd, since any two-sphere embedded in a three-ball must bound a three-ball. Thus if $d<\frac{1}{2}k$ there are minimal discs $D\subset A_1$ and $D'\subset A_2$ such that $\partial D=\partial D'$.

It remains to treat the case when $d=\frac{1}{2}k$. In this case $A_1\cap A_2$ will consist of a set of scc's which are parallel in both A_1 and A_2. Let σ be a two-simplex of K which contains a point of $A_1\cap A_2$. It is always possible to choose $x\in A_1\cap A_2\cap|\sigma|$ as in Fig. VI.19 so that there are points p, r in the same face of σ with $p\in A_2$, $r\in A_1$ and x is the only point of $A_1\cap A_2$ in the line segments xp and xr. In the present situation every point of $M^1\cap(A_1\cup A_2)$ is contained in a minimal disc. Let D be the minimal disc containing r and let D' be the minimal disc containing p. With this choice of D and D' it can be seen that $D\cup D'$ has a returning arc and so $D\cup D'$ is isotopic to an embedded two-sphere S such that $\|S\|=k-2$. Thus $D\cup D'=\partial B$, where B is a three-ball.

Let $P_1=(A_1-D)\cup D'$ and $P_2=(A_2-D')\cup D$. If $\partial D\cap M^2=\varnothing$ then P_1 and P_2 are patterned surfaces and P_1 and P_2 can be altered in a neighbourhood of ∂D so that $P_1\cap P_2$ has one fewer component than $A_1\cap A_2$. This would contradict the choice of A_1 and A_2. Assume then that there is a two-simplex σ which contains a point of ∂D. Since $\|P_i\|=k$ and P_i is disc equivalent to $A_i, i=1,2$, there are patterned surfaces $P'_i, i=1,2$, such that $P'_i\cap M^1=P_i\cap M^1$. In $|\sigma|$ the situation will be as in Fig. VI.19. Clearly $P'_1\cap P'_2\cap|\sigma|$ will contain fewer points than $P_1\cap P_2\cap|\sigma|$. Now $(A_1\cup A_2)\cap M^1=(P_1\cup P_2)\cap M^1=(P'_1\cup P'_2)\cap M^1$, and

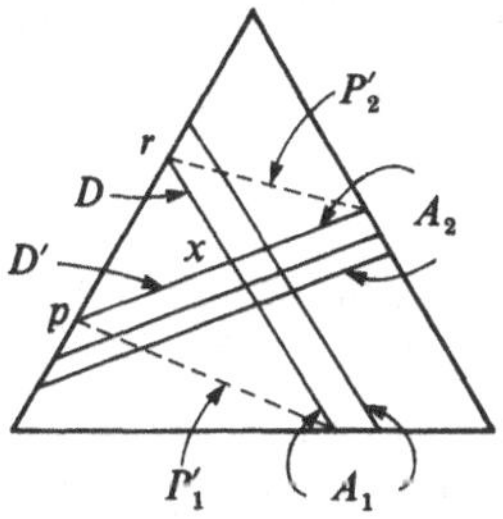

Fig. VI.19

so $A_1 + A_2 = P'_1 + P'_2$. The Theorem now follows by induction on (α, β) since $|P'_1 \cap P'_2 \cap M^2| < |A_1 \cap A_2 \cap M^2| = \alpha$. ∎

Suppose now that we are in the situation of Theorem 8.3, so that there is a G-action on M. If $\mathscr{S} \neq \varnothing$ then it is easy to see that $\mathscr{S}$ is a G-set. Thus combining Theorems 8.3 and 10.1 we have the following result.

10.2 Theorem. *If $\mathscr{S} \neq \varnothing$, then there exists $S \in \mathscr{S}$ such that $gS = S$ or $gS \cap S = \varnothing$ for every $g \in G$.* ∎

A three-manifold M is called *irreducible* if each two-sphere $S \subset M$ bounds a three-ball; M is called *boundary irreducible* if for each properly embedded disc $(D, \partial D) \subset (M, \partial M)$ there is a disc $D' \subset \partial M$ such that $\partial D' = \partial D$.

Let Σ be a set of disjoint two-spheres in M. Let $M[\Sigma]$ be the three-manifold (not usually connected) obtained from M by cutting along the two-spheres of Σ and attaching a three-ball to each boundary two-sphere that has been created by the cutting. Thus if Σ consists of a single two-sphere S then cutting along S creates two copies of S in the new manifold and $M[\Sigma]$ is obtained by attaching three-balls to each of these. In this case $M[\Sigma]$ will be connected if and only if S does not separate M. (We assume throughout that M is connected.)

10.3 Theorem. *Let M, K, G be as in Theorem 8.3. Suppose $G \backslash K = L$ is finite. There is a G-finite G-set Σ of disjoint two-spheres in M such that $M[\Sigma]$ is irreducible.*

Proof. Let Σ be a G-finite G-set of disjoint two-spheres in M. We show that if $M[\Sigma]$ is not irreducible then we can enlarge Σ in a nontrivial way. However, an argument similar to that of Theorem 3.8 shows that there is a bound on the number of orbits of Σ.

Let S be a two-sphere in M, and suppose $S \cap A = \varnothing$ for every $A \in \Sigma$. Clearly S becomes an embedded two-sphere S' in $M[\Sigma]$. Let $\mathscr{S}[\Sigma]$ denote the set of all two-spheres $S \subset M$ satisfying the following conditions:

(*a*) $S \in \mathscr{P}M$,

(*b*) $S \cap A = \varnothing$ for every $A \in \Sigma$,

(*c*) the corresponding two-sphere S' does not bound a three-ball in $M[\Sigma]$,

(*d*) If $T \subset M$ is a two-sphere satisfying (*a*), (*b*) and (*c*) then $\|S\| \leqslant \|T\|$.

First we show that if $M[\Sigma]$ is not irreducible then $\mathscr{S}[\Sigma] \neq \varnothing$. Assume then that there is a two-sphere $S' \subset M[\Sigma]$ that does not bound a three-ball.

We can assume that S' meets each $A \in \Sigma$ transversely. Let $c(S')$ denote the number of components of $S' \cap (\cup A \mid A \in \Sigma)$. Choose S' so that $c(S')$ takes a minimal value. Suppose $c(S') \neq 0$. Choose $A \in \Sigma$ so that $A \cap S' \neq \emptyset$. Choose an scc l in $A \cap S'$ so that l is innermost in A. Thus l bounds a disc D in A such that $D \cap S' = \partial D$. Now do surgery on S' along l. By this we mean the following. Remove a small open neighbourhood of l from S'. This creates two boundary curves l_1 and l_2. Now attach two discs D_1 and D_2 to l_1 and l_2, respectively, where D_1 and D_2 are discs which are close to but disjoint from D. Instead of S' we now have two two-spheres S'' and S''', where $c(S'') + c(S''') = c(S') - 1$. It follows from the argument of Theorem 9.1 that if both S'' and S''' bound three-balls then so does S'. Hence either S'' or S''' does not bound a ball. Clearly this contradicts the minimality of $c(S')$. Hence $c(S') = 0$. Any two-sphere S' which does not bound a three-ball and for which $c(S') = 0$ must lie in the image of the inclusion map $M - (\cup A \mid A \in \Sigma) \to M[\Sigma]$ and so there is a two-sphere $S \subset M$ which becomes S' under this inclusion map.

We have shown that we can choose a two-sphere S so that it satisfies conditions (*b*) and (*c*) above. If we choose such an S for which $\|S\|$ is minimal then it will also satisfy (*d*). However, we know from Theorem 9.1 that S can be assumed to be a patterned surface, since the changes described in Section 9 can all be carried out while preserving the empty intersection with the two-spheres of Σ. Thus $\mathscr{S}[\Sigma] \neq \emptyset$.

We can now repeat the argument of Theorem 10.1 and show that $\mathscr{S}[\Sigma]$ satisfies ($\mathscr{C}1$), ($\mathscr{C}2$) and ($\mathscr{C}3$). By Theorem 8.3 there is an $S \in \mathscr{S}[\Sigma]$ such that for every $g \in G$ either $gS = S$ or $gS \cap S = \emptyset$. Thus we can form a new G-set Σ' of disjoint two-spheres where $\Sigma' = \Sigma \cup \{gS \mid g \in G\}$. Let $\pi: M \to |L|$ be the natural map. Let $Q = |L^2| \cap (\cup(\pi(A) \mid A \in \Sigma))$ and let $Q' = |L^2| \cap (\cup(\pi(A) \mid A \in \Sigma'))$. Then Q and Q' are patterns in $|L^2|$, and Q' is the union of Q and an extra track $t = \pi(S) \cap |L^2|$. If t is equivalent to a track t' of Q then S will be equivalent to some $A \in \Sigma$. Thus $S \cup A = \partial R$, where $R \approx S^2 \times I$. But this would mean that the two-sphere S' in $M[\Sigma]$ bounds a three-ball, contradicting (*c*). Thus if Q is a reduced pattern then so is Q'. We know from Theorem 3.8 that a reduced pattern in $|L^2|$ has at most $n(L^2)$ component tracks. If we start with $\Sigma = \emptyset$ we can use the above process to enlarge Σ at most $n(L^2)$ times. Thus we must eventually obtain a G-set Σ for which $M[\Sigma]$ is irreducible. ∎

Suppose that $\partial M \neq \emptyset$. Let $2M$ be the three-manifold obtained by taking two copies of M and identifying their boundaries. Clearly the cyclic group $C_2 = \{1, x\}$ acts on $2M$; the element x transposes the two copies of M, which will be denoted M and xM, and x acts trivially on $\partial M = M \cap xM$. We can now make $H = G \times C_2$ act on $2M$ by requiring that $gxm = xgm$, for every $m \in M$ and $g \in G$.

A disc D properly embedded in M (so that $(D, \partial D) \subset (M, \partial M)$ and $D \cap \partial M = \partial D$) is called a *compressing disc* if ∂D does not bound a disc in ∂D. Thus M is boundary incompressible if and only if M contains no compressing discs.

10.4 Proposition. *Let $D \subset M$ be a compressing disc. Then the two-sphere $S = D \cup xD$ does not bound a three-ball in $2M$.*

Proof. Suppose D is a compressing disc and $S = \partial B$, where B is a ball. Now $xS = S$ and S separates $2M$. Hence either $xB = B$ or $xB \cap B = \varnothing$. But x acts trivially on ∂M and so $xB = B$. Let $C = M \cap B$. Let $i: C \to B$ be the inclusion map and $j: B \to C$ be the map such that $j(c) = j(xc) = c$ for every $c \in C$. Then $ji = 1_C$. It follows that $H_1(C) = 0$ since the identity map on $H_1(C)$ factors through $H_1(B) = 0$. Consider the Mayer–Vietoris sequence

$$H_2(B) \to H_1(B \cap \partial M) \to H_1(C) \oplus H_1(xC) \to H_1(B).$$

We see that $H_1(B \cap \partial M) = 0$ since all the other groups in the sequence are trivial. It follows that the component of $B \cap \partial M$ containing ∂D must be a disc, contradicting the hypothesis that D is a compressing disc. ∎

10.5 Theorem. *Let M, K, G be as in Theorem 8.3 and suppose that M contains a compressing disc. Then there is a compressing disc D such that $gD = D$ or $gD \cap D = \varnothing$ for every $g \in G$.*

Proof. Consider the manifold $2M$ acted on by the group $H = G \times C_2$. By Theorem 10.3 there is an H-set Ω of disjoint two-spheres in $2M$ such that $2M[\Omega]$ is irreducible. It can also be seen from the proof of Theorem 10.3 that we can construct Ω so that it contains a G-set Σ of two-spheres in M for which $M[\Sigma]$ is irreducible. We show that $\Omega \neq \Sigma \cup x\Sigma$ and that for some $A \in \Omega - \Sigma \cup x\Sigma$, $D = A \cap M$ is a compressing disc with the required properties.

We know that M contains a compressing disc Δ. By modifying Δ by appropriate surgery we can assume that $\Delta \cap A = \varnothing$ if $A \in \Sigma$. This means that Δ corresponds to a compressing disc Δ' in $M[\Sigma]$. Clearly $2(M[\Sigma]) \approx 2M[\Sigma \cup x\Sigma]$. Now $\Delta' \cup x\Delta'$ does not bound a ball in $2(M[\Sigma])$ by Proposition 10.4. Hence $2M[\Sigma \cup x\Sigma]$ is not irreducible and so $\Omega \neq \Sigma \cup x\Sigma$.

If $A \in \Omega$ and $A \cap \partial M \neq \emptyset$ then $xA \cap A \neq \emptyset$ and so $xA = A$. Thus $A = D \cup xD$ where $D = M \cap A$. If we calculate Euler characteristics, we see that $2 = \chi(A) = 2\chi(D)$ and so D is a disc. Suppose $A \cap \partial M = \partial D'$, where D' is a disc in ∂M. Then $D \cup D'$ is a two-sphere in M which bounds a ball B' when regarded as a two-sphere in $M[\Sigma]$. Now $B' \cup xB'$ is a three-ball in $2(M[\Sigma])$ and $D \cup xD = \partial(B' \cup xB')$. Thus if, for every $A \in \Omega$, either $A \cap \partial M$ is empty or it bounds a disc in ∂M, then $M[\Sigma \cup x\Sigma]$ is irreducible and we have seen that this is not the case. Thus, for some $A \in \Omega$, $A \cap M$ is a compressing disc in M. ∎

Let Ψ be a set of compressing discs in M. The manifold obtained from M by cutting along each element of Ψ is denoted $M[\Psi]$.

10.6 Theorem. *Let M, K, L and G be as in Theorem 8.3. There is a G-set Ψ of disjoint compressing discs in M such that $M[\Psi]$ is boundary incompressible.*

Proof. Let Ω be as in the proof of Theorem 10.5. Let Ψ be the set of compressing discs D for which $D = A \cap M$ for some $A \in \Omega$. Clearly Ψ is a G-set of disjoint compressing discs. Using arguments similar to those of Theorem 10.3 it is not hard to show that if $M[\Psi]$ contains a compressing disc then $2M[\Omega]$ contains a two-sphere which does not bound a ball. ∎

Let $\mathscr{D} \subseteq \mathscr{P}M$ be the set of patterned surfaces D satisfying the following conditions:

($\mathscr{D}1$) D is a compressing disc.

($\mathscr{D}2$) If $D' \in \mathscr{P}M$ is a compressing disc, then $\|D'\| \geqslant \|D\|$.

There is a constant k such that $\|D\| = k$ if $D \in \mathscr{D}$.

It follows from Theorem 9.2 that if there is a compressing disc in M, then $\mathscr{D} \neq \emptyset$. If we could show that $\mathscr{D}$ satisfies ($\mathscr{C}1$), ($\mathscr{C}2$) and ($\mathscr{C}3$), then Theorem 8.3 would provide an alternative proof of Theorem 10.5. In fact we prove that $\mathscr{D}$ satisfies ($\mathscr{C}1$), ($\mathscr{C}2$) and ($\mathscr{C}3$) for the case when two-spheres in M separate.

10.7 Theorem. *Let M be a three-manifold in which every two-sphere separates. Then $\mathscr{D}$ satisfies ($\mathscr{C}1$), ($\mathscr{C}2$) and ($\mathscr{C}3$).*

Proof. Clearly $\mathscr{D}$ satisfies ($\mathscr{C}1$) and ($\mathscr{C}2$).

We show that $\mathscr{D}$ satisfies ($\mathscr{C}3$). The proof is very similar to that of Theorem 10.1. Let $A_1, A_2 \in \mathscr{D}$. As in the proof of Theorem 10.1 replace A_1

and A_2 by equivalent surfaces so that A_1 and A_2 meet transversely and so that the pair (α, β) is minimized, where $\alpha = |A_1 \cap A_2 \cap M^2|$ and β is the number of components of $A_1 \cap A_2$. Note that a component of $A_1 \cap A_2$ may be either an scc or an arc properly embedded in M. Suppose that $A_1 \cap A_2$ contains an scc. We prove that there is an scc l in $A_1 \cap A_2$ such that l bounds minimal discs in both A_1 and A_2.

Let D be a minimal disc in A_1. As in the proof of Theorem 10.1 there is a disc $D' \subset A_2$ such that $\|D'\| = \|D\|$ and $\partial D' = \partial D$. Clearly $D \cup D'$ is a two-sphere which, by hypothesis, separates M and so there is a closed subspace B of M such that $B \cap \overline{(M - B)} = D \cup D'$ and $B \cap \partial A_1 = \emptyset$. By changing the choice of D if necessary we can arrange that D is the only component of $B \cap A_1$ which is a minimal disc. Suppose D' is not a minimal disc. There is then a minimal disc Δ properly contained in D'. Now $\partial\Delta$ bounds a disc Δ' in A_1 such that $\|\Delta'\| = \|D\|$, and $A = (A_1 - \Delta') \cup \Delta$ is a compressing disc such that $\|A\| = k$. By moving Δ inside B and straightening $A \cap M^2$ so that it becomes a pattern we obtain $A' \in \mathscr{D}$ such that $A' \cap \partial B$ has fewer components than $A_1 \cap \partial B$. We do not create any new minimal disc in B by this process, since (as in the proof of Theorem 10.1) there is no component of $B \cap A_1$ which is an annulus a such that $\|a\| = 0$. If we repeat this process we must eventually create $A'' \in \mathscr{D}$ such that $A'' \cap B$ has a component C which is a disc such that $\|C\| > \|D\|$ and ∂C bounds a disc $D'' \subset D'$ such that $\|D''\| = \|D\|$. This is impossible. Hence D' is a minimal disc. The argument of the last paragraph of the proof of Theorem 10.1 shows that $A_1 + A_2 = P'_1 \oplus P'_2$, where $P'_1, P'_2 \in \mathscr{D}$.

It remains to treat the case when $A_1 \cap A_2$ contains no scc's. Suppose then that every component of $A_1 \cap A_2$ is an arc. Consider the discs $D \subset A_i, i = 1$ or 2, such that $\partial D \subseteq (A_1 \cap A_2) \cup \partial A_i$. In this new context a disc is called *minimal* if $D \cap (A_1 \cap A_2) = \partial D \cap (A_1 \cap A_2)$, and $\|D\|$ is smallest amongst such discs. We show that there are minimal discs $D_1 \subset A_1$, $D_2 \subset A_2$ such that $D_1 \cap A_2 = D_2 \cap A_1$ and $\partial(D_1 \cup D_2)$ bounds a disc in ∂M. Let $D_1 \subset A_1$ be a minimal disc. Then $l = D_1 \cap A_2$ is an arc. It is not hard to adapt the argument in the proof of Theorem 10.1 to show that there is a disc $\Delta \subset A_2$ such that $\partial\Delta \subset l \cup \partial A_2$ and $\|\Delta\| = \|D_1\|$:Δ is uniquely determined unless $\|\Delta\| = \frac{1}{2}k$. If $\|\Delta\| < \frac{1}{2}k$ then $\|\Delta \cup D_1\| < k$ and so $\partial(\Delta \cup D_1)$ bounds a disc in ∂M. It follows from the fact that two-spheres separate that $D_1 \cup \Delta$ separates M. Let B be the closed subset of M such that $B \cap \overline{(M - B)} = D_1 \cup \Delta$ and B contains the disc in ∂M bounded by $\partial(D_1 \cup \Delta)$. As in the proof of Theorem 10.1 it can be assumed that D_1 is the only component of $A_1 \cap B$ which is a minimal disc, and there is no component that is a disc d such that $\|d\| = 0$, which is bounded by two arcs in $A_1 \cap A_2$

joined by two arcs in ∂M. Again, an argument very similar to that given in the proof of Theorem 10.1 shows that D_1 is the only component of $A_1 \cap B$, and so Δ is a minimal disc. Putting $D_2 = \Delta$ we have minimal discs $D_1 \subset A_1$ and $D_2 \subset A_2$ such that $A_1 \cap D_2 = A_2 \cap D_1$ and $\partial(D_1 \cup D_2)$ bounds a disc in ∂M. If $\|D_1\| = \frac{1}{2}k$ then an argument similar to that of the corresponding case in the proof of Theorem 10.1 shows that, after a possible change in the choice of the minimal disc D_1, we can find a minimal disc $D_2 \subset A_2$ such that $A_1 \cap D_2 = A_2 \cap D_1$, and, for some two-simplex $\sigma, |\sigma| \cap (D_1 \cup D_2)$ has a component which is a returning arc. This means that $(D_1 \cup D_2, \partial(D_1 \cup D_2))$ is isotopic in $(M, \partial M)$ to a disc $(D', \partial D')$ such that $\|D'\| < k$. Since D' cannot be a compressing disc, it follows that $D_1 \cup D_2$ cannot be a compressing disc, and so $\partial(D_1 \cup D_2)$ bounds a disc in ∂M. In either case the now familiar argument, in which one swaps D_1 and D_2, gives two new compressing discs A_1', A_2' in $\mathscr{D}$ whose intersection has fewer components. These discs are in $\mathscr{D}$ since $\partial A_i'$ is disc equivalent to ∂A_i in $\partial M, i = 1, 2$, with disc equivalence as defined in Section 7. The theorem follows by induction. ∎

11 The Loop Theorem

11.1 Proposition. *Let M be a compact three-manifold and suppose some component of ∂M is not a two-sphere, then $\pi_1(M) \neq 1$.*

Proof. By Propositions 1.4 and 1.5 and Theorem 1.6 it suffices to show that $H_1(M, \mathbb{Z}_2) \neq 0$. Suppose $H_1(M, \mathbb{Z}_2) = 0$. By Poincaré duality $H_2((M, \partial M), \mathbb{Z}_2) \approx H^1(M, \mathbb{Z}_2) = 0$. The exact sequence

$$H_2((M, \partial M), \mathbb{Z}_2) \to H_1(\partial M, \mathbb{Z}_2) \to H_1(M, \mathbb{Z}_2)$$

shows that $H_1(\partial M, \mathbb{Z}_2) = 0$. However, the two-sphere is the only connected closed surface with trivial first homology group. ∎

In this section we need to extend the concept of a pattern and patterned surface to two-complexes and three-manifolds which have a regular cell decomposition rather than just a simplicial decomposition. Thus we assume that the three-manifold M is a CW-complex where every attaching map is injective. We assume that the boundary of each i-cell consists of finitely many $(i-1)$-cells. Also if σ_1 and σ_2 are distinct i-cells, $i = 1, 2, 3$, then $\sigma_1 \cap \sigma_2$ contains at most one $(i-1)$-cell. We further assume that every two-cell has a fixed map to a convex polygonal disc in $\mathbb{R}^2$; each one-cell in the boundary of the two-cell being mapped to an edge of the disc. This allows us to talk about straight lines joining points on the boundary of a two-cell. Let C_i denote the set of i-cells, $i = 0, 1, 2, 3$. Let M^i denote the i-skeleton of M. Thus M^i is the union of all j-cells, $j \leqslant i$.

Let Φ be a subset of C_1 satisfying the following condition:

(Φ1) For each $\sigma \in C_2$ there are at most three one-cells $\gamma \in \Phi$ such that $\gamma \subset \partial\sigma$.

Note that a simplicial decomposition for M satisfies all the above conditions for the cell decomposition and, if we take Φ to be the set of all one-simplexes of the simplicial decomposition, then Φ satisfies (Φ1).

A *pattern* (or *Φ-pattern*) in M^2 is a subset $P \subset M^2$ such that, for each two-cell $\sigma \in C_2, \sigma \cap P$ consists of disjoint straight lines joining distinct one-cells in $\partial\sigma \cap \Phi$. A patterned surface is an embedded surface $(S, \partial S) \subset (M, \partial M)$ such that $S \cap M^2$ is a pattern and, for every three-cell $\rho \in C_3$, $S \cap \rho$ is a union of disjoint discs. The whole theory of patterns and patterned surfaces described previously for simplicial two-complexes and three-manifolds with simplicial decompositions works equally well in the slightly more general situation described above. In particular an equivalence class of patterns or patterned surfaces corresponds to a map $j: \Phi \to \mathbb{Z}^+$ for which if $\sigma \in C_2$ then the sum of $j(\gamma)$ taken over all $\gamma \in \Phi \cap \partial\sigma$ is an even integer $2m_\sigma$, and $j(\gamma) \leqslant m_\sigma$ if $\gamma \in \Phi \cap \partial\sigma$.

Let B^2 denote the unit disc $B^2 = \{z \in \mathbb{C} \mid |z| \leqslant 1\}$. Let $\mathscr{F}$ denote the set of all maps $f: (B^2, \partial B^2) \to (M, \partial M)$ such that $f|\partial B^2 \not\simeq 0$ in ∂M and such that if $\gamma \in C_1$ and $\gamma \cap f(B^2) \neq \varnothing$ then $\gamma \in \Phi$. Our aim is to show that if $\mathscr{F} \neq \varnothing$ then $\mathscr{F}$ contains an embedding. Note that if f is an embedding then $f(B^2)$ is a compressing disc in M. We say that $f \in \mathscr{F}$ is *minimal* if $\|f\| \leqslant \|g\|$ for every $g \in \mathscr{F}$. Here $\|f\| = |M^1 \cap f(B^2)|$. If $\mathscr{F} \neq \varnothing$ then we see from Theorem 9.2 that there exists a minimal $f \in \mathscr{F}$ such that the one-pieces and two-pieces of f are cells and there are no returning one-pieces. Of course in Section 9 M had a simplicial decomposition. Now we assume that M has a cell decomposition as above and the i-pieces are defined in the obvious way. We see that we can choose f so that, if p is a one-piece of f, then $f(p)$ is a straight line joining distinct one-cells in $\partial\sigma \cap \Phi$. For the two-cell $\sigma, \sigma \cap f(B^2)$ will be as in Fig. VI.20.

Let

$$j: \Phi \to \mathbb{Z}^+, \quad j(\gamma) = |\gamma \cap f(B^2)|.$$

It is not hard to see that the pattern conditions are satisfied and so there is a patterned surface S such that $j = j_S$.

11.2 The Loop Theorem. *With f and M as above, there exists an embedding $g \in \mathscr{F}$ such that $g(B^2) = S$.*

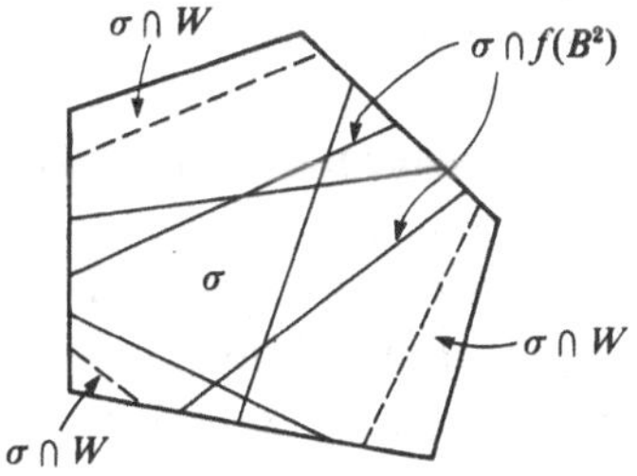

Fig. VI.20

Proof. We first prove the Theorem in the case when M is simply connected. Let

$$c_1(f) = |\{\gamma \in \Phi \mid \gamma \cap f(B^2) \neq \varnothing\}|$$

and let

$$c(f) = \|f\| - c_1(f).$$

Clearly $c(f) \geqslant 0$ and $c(f) = 0$ if and only if $f(B^2)$ intersects each one-cell of Φ at most once. We use induction on $c(f)$.

If $c(f) = 0$ then for each i-cell σ there is at most one $(i-1)$-piece p such that $f(p) \subset \sigma$. Thus f must be an embedding. Assume then that $c(f) > 0$ and the Theorem is true for any map

$$f' : (B^2, \partial B^2) \to (M', \partial M')$$

for which $c(f') < c(f)$. Note that we do not assume that $M' = M$. However, M' must be simply connected and have a regular cell decomposition with a distinguished set Φ' of one-cells with similar properties to those of M and Φ.

Let $k : \Phi \to \mathbb{Z}^+$, $k(\gamma) = 2$ if $\gamma \cap f(B^2) \neq \varnothing$ and $k(\gamma) = 0$ if $\gamma \cap f(B^2) = \varnothing$. It is clear that for any two-cell σ, the number of one-cells γ in $\partial\sigma \cap \Phi$ for which $\gamma \cap f(B^2) \neq \varnothing$ is either $0, 2$ or 3. This means that the function k satisfies the pattern conditions and so there is a patterned surface W such that $j_W = k$. We can arrange that if γ is a one-cell for which $\gamma \cap f(B^2) \neq \varnothing$ then all the points of $\gamma \cap f(B^2)$ lie between the two points of $\gamma \cap W$. If we do this then for any two-cell σ the lines of $\sigma \cap W$ will not intersect $\sigma \cap f(B^2)$ (see Fig. VI.20) and we can choose the discs of $\rho \cap W$ for each three-cell ρ so that $W \cap f(B^2) = \varnothing$. By further considering the situation in each three-cell we see that there is a compact connected submanifold V of M such that $W \subset \partial V$ and $f(B^2) \subset V$. Note that V inherits a regular cell decomposition from that of M. There are two-cells in this decomposition of two types. They are either of the form $\sigma \cap V$, where σ is a two-cell of

M, or they are discs of $W \cap \rho$, where ρ is a three-cell of M. Each two-cell of M comes with a fixed map to a convex polygonal disc in $\mathbb{R}^2$, and the restriction of this map to $\sigma \cap V$ is taken to be the fixed map for the two-cell $\sigma \cap V$ of V. The map f factors through the inclusion $i: V \to M$.

Let C be a component of W. We show that, for each one-cell γ, $|\gamma \cap C| \leqslant |\gamma \cap f(B^2)|$. For if $|\gamma \cap C| > |\gamma \cap f(B^2)|$ we must have $|\gamma \cap C| = 2$ and $|\gamma \cap f(B^2)| = 1$. But this means that there is a loop in M that intersects both $f(B^2)$ and S transversely just once. However, $H^1(M, \mathbb{Z}_2) = 0$ and so, by Proposition 3.4, $M^2 - S$ can be two-coloured. This is a contradiction. Thus for each one-cell $|\gamma \cap C| \leqslant |\gamma \cap f(B^2)|$ and hence $\|C\| \leqslant \|f\|$. If C is a compressing disc in M then there is a map f' in $\mathscr{F}$ such that $f'(B^2) = C$. Thus by the minimality of $\|f\|$ we have $\|C\| \geqslant \|f\|$. Hence if C is a compressing disc $\|C\| = \|f\|$ and $|C \cap \gamma| = |\gamma \cap f(B^2)|$ for every one-cell γ. Thus in this situation C and S are equivalent patterned surfaces and the Theorem is proved. Thus we may assume that, if C is a disc, then ∂C bounds a disc C' in ∂M. Note that $\partial C' \cap f(B^2) = \varnothing$ and so $C' \cap f(B^2) = \varnothing$, since if $f(\partial B^2) \subset C'$ then $f|\partial B^2 \simeq 0$. Let $U_1 = V \cap \partial M$ and let U be the union of U_1 and those components of W which are discs. Now $f(\partial B^2) \subset U_1$. There is an injective map $\alpha: U \to \partial M$ such that α is the identity map on U_1 and α takes any disc component C of W to the corresponding disc C' in ∂M. Clearly $f|\partial B^2 \not\simeq 0$ in U since $f|\partial B^2 \not\simeq 0$ in ∂M. We can show $f|\partial B^2 \not\simeq 0$ in ∂V if we can show that the inclusion $U \to \partial V$ induces an injection on fundamental groups. Now ∂U consists of a finite set of scc's in ∂V. No scc in ∂U is null homotopic in ∂V. Now ∂V is a closed orientable surface and so $\pi_1(\partial V)$ is torsion-free. It follows that the inclusion map of each boundary component of ∂U in ∂V induces an injection on fundamental groups. The fact that $\pi_1(U) \to \pi_1(\partial V)$ is injective follows easily; for instance, by Van Kampen's Theorem. Let M' be the universal cover of V with covering map $\pi: M' \to V$. Let $p = i\pi: M' \to M$. Choose $f': (B^2, \partial B^2) \to (M', \partial M')$ so that $pf' = f$. We have described how the cell structure on M determines one on V. This cell decomposition lifts to a cell decomposition for M'. Let

$$\Phi' = \{\gamma' \in C_1' \mid p(\gamma') \subset \gamma \text{ for some } \gamma \in \Phi\}.$$

Clearly $c(f') \leqslant c(f)$. Suppose $c(f') = c(f)$. There is then a bijection between the set of one-cells that f' meets and the one-cells that f meets. However, we are assuming that no two distinct two-cells have more than one one-cell in common. Now if $f(B^2)$ meets a two-cell then it meets at least two of its faces. It follows that there is a bijection between the two-cells of V and the two-cells of M' that meet $f'(B^2)$. A similar argument works for the three-cells of V. It follows that the identity map $1_V: V \to V$ lifts to M'. However, this can only happen if V is simply connected, in which case ∂V is a union of two-spheres by Proposition 11.1. However, this means $f|\partial B^2 \simeq 0$ in ∂V, which is a contradiction. Hence $c(f') < c(f)$. Before we can apply our induction hypothesis we have to show that f' is minimal in the set $\mathscr{F}'$ of maps $h:(B^2, \partial B^2) \to (M', \partial M')$ satisfying the conditions:

(i) If $\gamma \in C_1'$ and $\gamma \cap h(B^2) \neq \emptyset$ then $\gamma \in \Phi'$

and

(ii) $h|\partial B^2 \not\simeq 0$ in $\partial M'$.

However, a map $h:(B^2, \partial B^2) \to (M', \partial M') \in \mathscr{F}'$ if and only if $ph \in \mathscr{F}$. For, $\pi h(\partial B^2) \subset U_1$ and we have seen that a loop in U_1 is null homotopic in ∂V if and only it is null homotopic in ∂M. But a loop l in $\partial M'$ is null homotopic if and only if $\pi(l)$ is null homotopic in ∂V. Note also that $\|ph\| = \|h\|$. Hence f' is minimal in $\mathscr{F}'$, and we can apply our induction hypothesis. Thus the patterned surface D associated with the function $k':\Phi' \to \mathbb{Z}^+$, $k'(\gamma) = |\gamma \cap f'(B^2)|$ is a compressing disc. Clearly D satisfies the conditions ($\mathscr{D}1$) and ($\mathscr{D}2$) of Section 10 for the manifold M'.

Let $\mathscr{D}'$ denote the set of all $D \in \mathscr{P}M'$ satisfying conditions ($\mathscr{D}1$) and ($\mathscr{D}2$). By Theorem 10.7 $\mathscr{D}'$ satisfies the conditions ($\mathscr{C}1$), ($\mathscr{C}2$) and ($\mathscr{C}3$) of Section 8. Note that M' is simply connected and so two-spheres in M' separate. Let $G = \pi_1(V)$. Then G acts freely on M' and $G\backslash M' \approx V$. By Theorem 8.3 $\sum_{g \in G} gD$ is equivalent to a patterned surface Q which is G-stable and every component of Q is a minimal compressing disc. However, Q is equivalent to $p^{-1}(S)$. Thus each component of S lifts to a minimal disc in M'. But the only surface covered by a disc is a disc, and so each component of S is a minimal disc. However, $\|S\| = \|f\|$, which means that S can have only one component. This completes the proof of the Theorem in the case when M is simply connected.

Suppose now that M is not simply connected. Let M' be the universal cover of M. We can now repeat the argument of the last paragraph to show that if the Theorem is true in M' then it is also true in M. ∎

The Loop Theorem 11.2 provides an algebraic criterion for a three manifold to be boundary irreducible.

11.3 Corollary. *A three manifold M is boundary irreducible if and only if the inclusion $\partial M \to M$ induces injections $\pi_1(C) \to \pi_1(M)$ for each component C of ∂M.*

12 The Sphere Theorem

In this section we will revert to assuming that $M = |K|$, where K is a simplicial complex. We consider first the case when M is simply connected. By the Hurewicz Isomorphism Theorem (see Spanier, 1966, p. 393) $\pi_2(M) \approx H_2(M)$. A closed orientable surface $S \subset M$ with specified orientation determines a unique element $[S]$ of $H_2(M)$.

12.1 Theorem. *Let M be a simply connected three-manifold, then $H_2(M)$ is generated by the set $\{[S] | S$ is an embedded two-sphere$\}$.*

Proof. First we show that $H_2(M)$ is generated by the set $\{[S] | S$ is an orientable surface$\}$. Consider the exact sequence

$$H_2(\partial M) \xrightarrow{\alpha} H_2(M) \xrightarrow{\beta} H_2(M, \partial M).$$

Let $x \in H_2(M)$. If x is in the image of α then x is a sum of elements of the form $[S]$, where S is a compact boundary component of M. Suppose then that $\beta x \neq 0$. By duality $H_2(M, \partial M) \approx H^1_f(M)$. Let ζ be a finite one-cocycle corresponding to βx. Since $H^1(M) = 0$ there is a zero-cochain z such that $\delta z = \zeta$. Define $j: S_1 K \to \mathbb{Z}^+$, $j(\gamma) = |\zeta(\gamma)| = |z(u) - z(v)|$, where u and v are the vertices of γ. Let σ be a two-simplex of K with faces γ_1, γ_2 and γ_3. Since ζ is a cocycle, γ_1, γ_2 and γ_3 can be oriented so that $\zeta(\gamma_1) + \zeta(\gamma_2) + \zeta(\gamma_3) = 0$. It is easy to see that $j(\gamma_1) + j(\gamma_2) + j(\gamma_3)$ is even, say $2m_\sigma$, and $j(\gamma_i) \leqslant m_\sigma$, $i = 1, 2, 3$. In fact it is always the case that $j(\gamma_i) = m_\sigma$ for some i. Thus by Theorems 3.2 and 8.1 there is a patterned surface W such that $j_W = j$. It is not hard to see that there is map $\psi: M \to \mathbb{R}$ such that $\psi(u) = z(u)$ if u is a vertex of K and $|\gamma|$ is mapped onto the closed interval bounded by $z(u)$ and $z(v)$, where u and v are the vertices of γ. Think of ψ as a height function. We can arrange that $W = \psi^{-1}\{n + \frac{1}{2} | n \in \mathbb{Z}\}$. Think of W as a union of the contour surfaces of the height function. Let C be a component of W. Associated with C is a one-cocycle ζ_C, where, if γ is an oriented one-simplex, $\zeta_C(\gamma) = 0$ if γ does not intersect C and $\zeta_C(\gamma) = 1$ (or -1) if γ intersects C in the direction of increasing (decreasing) height. We show that ζ is the sum of all the ζ_C's as C ranges over all the components of W. For if $\zeta_C = \delta z_C$, then it is easy to see that z differs from $\sum_{C \subseteq W} z_C$ by a constant. If we give each C an appropriate orientation then $\beta\left(x - \sum_{C \subseteq W} [C]\right) = 0$. Hence $x - \sum_{C \subseteq W} [C] \in \operatorname{Im} \alpha$. Thus x can be written as a sum of elements of the form $[S]$, where S is an orientable surface.

Also note that we can assume that either $S \subseteq \partial M$ or $S \cap \partial M = \emptyset$. In fact, we can assume that $S \cap \partial M = \emptyset$, since if $S \subseteq \partial M$ we can find a surface $S' \approx S$, such that $[S'] = [S]$ and $S' \cap \partial M = \emptyset$, by constructing a regular neighbourhood N of S and taking S' to be $\partial N \cap \mathring{M}$.

If $S \subset M$ and S is not a two-sphere then we shall show there is a disc $D \subset M$ such that $S \cap D = \partial D$ and ∂D does not bound a disc in S. Since $\partial M \cap S = \emptyset$ and $H^1(M, \mathbb{Z}_2) = 0$, S must separate M. Thus $M = M_1 \cup M_2$, where $M_1 \cap M_2 = S$. It follows from Van Kampen's Theorem that for either $i = 1$ or 2 the inclusion $S \to M_i$ cannot induce an injection on fundamental groups. Thus we can invoke the Loop Theorem to obtain a disc D as required. Carry out surgery on S along D to obtain a new surface S' (possibly not connected). We can arrange that $[S] = [S']$. By repeating this process we eventually obtain a set of two-spheres. ∎

Suppose now that M is an arbitrary three-manifold. If S is a two-sphere or projective plane in M, then there is a covering map $v: S^2 \to S$, where v is one-sheeted if S is a two-sphere and two-sheeted if S is a projective plane. Thus corresponding to S is an element $[S]$ of $\pi_2(M)$. This element is determined only up to action by elements of $\pi_1(M)$, and replacing $[S]$ by $-[S]$.

12.2 Theorem. *Let the three-manifold M be acted on by the group G so that $G \backslash K$ is finite. There is a G-set Σ of disjoint two-sided embedded two-spheres and projective planes such that $\{[S] \mid S \in \Sigma\}$ generates $\pi_2(M)$ as a $\pi_1(M)$-module.*

Proof. Let $\tilde{M}$ be the universal cover of M. There is a group $\tilde{G}$ acting on $\tilde{M}$ for which there is an exact sequence

$$1 \to \pi_1(M) \to \tilde{G} \xrightarrow{\phi} G \to 1.$$

Here if $\rho: \tilde{M} \to M$ is the covering map, then $\rho(gm) = \phi(g)\rho(m)$, $g \in \tilde{G}$, $m \in \tilde{M}$. As usual we assume that $G \backslash K = L$ is a finite complex such that $G \backslash M \approx |L|$. By Theorem 10.3 there is a $\tilde{G}$-set $\tilde{\Sigma}$ of disjoint two-spheres in $\tilde{M}$ such that $\tilde{M}[\tilde{\Sigma}]$ is irreducible. If $\tilde{S} \in \tilde{\Sigma}$ then $\rho(\tilde{S})$ is either a two-sphere or a projective plane, since these are the only surfaces covered by a two-sphere. Let $\Sigma = \{\rho(\tilde{S}) \mid \tilde{S} \in \tilde{\Sigma}\}$. The set $\{[S] \mid S \in \Sigma\}$ generates $\pi_2(M)$ as a $\pi_1(M)$-module if and only if $\{[\tilde{S}] \mid \tilde{S} \in \tilde{\Sigma}\}$ generates $\pi_2(\tilde{M})$. By Theorem 12.1, $\pi_2(\tilde{M}) = H_2(\tilde{M})$ is generated by all elements of the form $[U]$, where $U \subset \tilde{M}$ is an embedded two-sphere. Thus to prove the Theorem we need to show that, for any such U, $[U]$ is in the subgroup Q of $H_2(\tilde{M})$ generated by $\{[\tilde{S}] | \tilde{S} \in \tilde{\Sigma}\}$. This is proved by induction on the number k of components of $U \cap (\cup \{\tilde{S} | \tilde{S} \in \tilde{\Sigma}\})$. If $k = 0$ then U corresponds to a unique two-sphere U' in $\tilde{M}[\tilde{\Sigma}]$. Since $\tilde{M}[\tilde{\Sigma}]$ is irreducible $U' = \partial B$, where $B \subset \tilde{M}[\tilde{\Sigma}]$ is a three-ball. Let $\tilde{S}_i$, $i = 1, 2, \ldots, n$, be the elements of $\tilde{\Sigma}$ for which there are balls $B_i \subset B$, where B_i is bounded by one of the two-spheres obtained by cutting $\tilde{M}$ along $\tilde{S}_i$. The $\tilde{S}_i$'s can be oriented so that $[U] = \sum_{i=1}^{n} [\tilde{S}_i]$. If $k > 0$ then in some $\tilde{S}' \in \tilde{\Sigma}$ there is a disc D such that $\partial D = D \cap (\cup \{\tilde{S} | \tilde{S} \in \tilde{\Sigma}\})$. If we carry out surgery on U along D then we obtain two two-spheres U_1 and U_2 such that $[U] = [U_1] + [U_2]$ and both U_1 and U_2 intersect $U\{\tilde{S} | \tilde{S} \in \tilde{\Sigma}\}$ in fewer components than U. By our induction hypothesis $[U_1] \in Q$ and $[U_2] \in Q$. Hence $[U] \in Q$.

If any $S \in \Sigma$ is a one-sided projective plane, then we can replace S by the two-sphere S' which is the boundary of a regular neighbourhood of S. This can be done so that $gS' = (gS)'$ for every $g \in G$. ∎

In some ways the treatment of the Sphere Theorem in this section seems unsatisfactory. We would have preferred an approach along the following lines. Let M be an orientable three-manifold. Let $\mathscr{F}$ be the set of all maps $f: S^2 \to M$ which represent a nontrivial element of $\pi_2(M)$, and which are in general position with respect to the triangulation of M. Choose $f \in \mathscr{F}$ for which $\| f \|$ is minimal. Let $k: S_1 K \to \mathbb{Z}^+, k(\gamma) = | |\gamma| \cap f(S^2)|$. By Theorem 9.2, k satisfies the pattern conditions and so there is a surface S such that $j_s = k$. We would have preferred to give a detailed proof of the following theorem.

12.3 Theorem. *There is an embedding $g \in \mathscr{F}$ such that $g(S^2) = S$.*

This theorem is true. It follows from the version of the Sphere Theorem proved in Hempel (1976) where it is proved that given any $h \in \mathscr{F}$ there is an embedding $h' \in \mathscr{F}$ such that $h'(S^2) \subseteq h(S^2) \cup U$, where U is any neighbourhood of $h(\Sigma(h))$. Here $\Sigma(h)$ is the singular set of h. If we take $h = f$, where $\| f \|$ is minimal as above, then we can choose U so that $U \cap M^1 = \emptyset$. This means that $f'(S^2) \cap M^1 \subseteq f(S^2) \cap M^1$ and so by the minimality of $\| f \|$ the two sets are equal. It also follows from the minimality of $\| f \|$ that the process of simplifying $f'(S^2)$ described in Section 9 leaves $f'(S^2) \cap M^1$ unchanged, which severely limits the changes that are applied in the process. In fact we can see that there is homeomorphism $\alpha: M \to M$ fixing M^1 such that $h(f'(S^2)) = S$. Thus we can take g to be hf'.

Most of the proof of Theorem 11.2 carries over to the situation of Theorem 12.3. Unfortunately there is one place where it seems to be necessary to use a full tower construction as in the other proofs of the Sphere Theorem. The reader is invited to resolve this problem. We quit!

Notes and comments

The treatment of orbit spaces in Section 2 is based on two exercises in Armstrong (1979, p.143).

Sections 3, 5 and 6 are based on Dunwoody (1985). The proof of Theorem 3.8 was inspired by an argument of Kneser (1929).

Section 4 is based on Dunwoody and Fenn (1987).

The concept of a patterned surface developed in Section 8 is implicit in Dunwoody (1985′).

Theorems similar to Theorems 10.3, 10.5 and 10.6 were first proved by Meeks and Yau (1980, 1981) using analytic minimal surface techniques. The version of the Equivariant Sphere Theorem 10.2 given here is due to Meeks, Simon and Yau (1982). The proofs given in Section 10 are similar to those of Dunwoody (1985′). However, it should be noted that, in the proof of the Equivariant Sphere Theorem given in Dunwoody (1985′), use is made of the mistaken assertion that the intersection of two patterned surfaces cannot contain two simple closed curves which are parallel in each surface and which intersect the two-skeleton of the manifold. Edmonds (1982) was the first to show that some of the Meeks–Yau results could be obtained using nonanalytic techniques.

The Loop and Sphere Theorems were first proved by Papakyriakopoulos (1957). The proof of the Loop Theorem given in Section 11 is new, but contains some ideas from an unpublished proof of A.N. Bartholomew.

Bibliography and author index

The set at the end of each entry consists of the page numbers where the entry is quoted; other references to the author are listed after the author's name.

Armstrong, M.A.
1979. *Basic Topology*. McGraw-Hill, Maidenhead. {271}
Bartholomew, A.N. {271}
Beardon, A.F.
1983. *The Geometry of Discrete Groups*. Springer-Verlag, Berlin. {221}
Bergman, G.M.
1968. On groups acting on locally finite graphs. *Ann. of Math.* **88**, 335–40. {71}
Bieri, R.
1972. Gruppen mit Poincaré-Dualität. *Comment. Math. Helv.* **47**, 373–96. {213}
Bieri, R. and Eckmann, B.
1978. Relative homology and Poincaré duality for group pairs. *J. Pure and Applied Algebra* **13**, 277–319. {213}
Bieri, R. and Strebel, R.
1978. Almost finitely presented soluble groups. *Comment. Math. Helv.* **53**, 258–78. {213}
1980. Valuations and finitely presented metabelian groups. *Proc. London Math. Soc.* **41**, (3), 439–64. {229}
Brouwer, A.E., Cohen, A.M. and Neumaier, A.
1988. *Distance Regular Graphs*. Springer-Verlag, Berlin. {72}
Brown, KS.
1975. Homological criteria for finiteness. *Comment. Math. Helv.* **50**, 129–35. {213}
1982. *Cohomology of Groups, GTM 87*. Springer-Verlag, New York. {134}
1984. Presentations for groups acting on simply-connected complexes. *J. Pure and Applied Algebra* **32**, 1–10 {46}
Brown, M.
1960. A proof of the generalized Schoenflies theorem. *Bull. A.M.S.* **66**, 74–6. {254}
Burns, R.G.
1969. A note on free groups. *Proc. Amer. Math. Soc.* **23**, 14–17. {39}
Chiswell, I.M.
1976. The Grushko–Neumann Theorem. *Proc. London Math. Soc.* **33**, (3), 385–400. {46}
1979. The Bass–Serre Theorem revisited. *J. Pure and Applied Algebra* **15**, 117–23. {46}
Cohen, A.M. *see* Brouwer, A.E.
Cohen, D.E.
1973. Groups with free subgroups of finite index. In *Conference on Group Theory, University of Wisconsin-Parkside 1972.* Lecture Notes in Mathmatics, vol. 319, pp. 26–44. Springer-Verlag, Berlin. {100, 133}
Cohn, P.M.
1985. *Free Rings and Their Relations*, 2nd edn., London Mathematical Society Monographs 19. Academic Press, London. {28, 46}
Coldewey, H.-D. *see* Zieschang, H.
Davis, M.W.
1983. Groups generated by reflections and aspherical manifolds not covered by Euclidean space. *Ann. of Math.* **117**, 293–324. {159}
Dicks, W.
1979. Hereditary group rings. *J. London Math. Soc.* **38**, (3), 27–39. {134}

1980. *Groups, Trees and Projective Modules*. Lecture Notes in Mathematics, vol. 790. Springer, Berlin. {46, 109, 134}
1981. On splitting augmentation ideals. *Proc. Amer. Math. Soc.* **83**, 221–7. {134}
Dunwoody, M.J.
1979. Accessibility and groups of cohomological dimension one. *Proc. London Math. Soc.* **38**, (3), 193–215. {71, 100, 134}
1979′. Recognizing free factors. In *Homological Group Theory*. London Mathematical Society Lecture Notes 36, pp. 245–9. Cambridge University Press. {134}
1982. Cutting up graphs. *Combinatorica* **1**, 15–23. {72}
1985. The accessibility of finitely presented groups. *Invent. Math.* **81**, 449–57. {271}
1985′. An equivariant sphere theorem. *Bull. London Math. Soc.* **17**, 437–48. {271}
Dunwoody, M.J. and Fenn, R.A.
1987. On the finiteness of higher knot sums. *Topology* **26**, 337–43. {271}
Eckmann, B. *see also* Bieri, R.
1986. Poincaré duality groups of dimension two are surface groups. In *Combinatorial Group Theory and Topology*, pp. 35–51. Ann. of Math. Studies. Princeton University Press.
Eckmann, B. and Linnell, P.A.
1983. Poincaré duality groups of dimension two II. *Comment. Math. Helv.* **58**, 111–14. {213, 214}
Eckmann, B. and Müller, H.
1980. Poincaré duality groups of dimension two. *Comment. Math. Helv.* **55**, 510–20. {213, 214}
1982. Plane motion groups and virtual Poincaré duality of dimension two. *Invent. Math.* **69**, 293–310. {165, 178}
Edmonds, A.L.
1982. On the equivariant Dehn lemma. In *Proc. Conf. on Combinatorial Methods in Topology and Algebraic Geometry (Rochester, NY., 1982)*, pp. 141–7. Contemporary Mathematical Society, Providence, R.I. {271}
Fenn, R.A. *see* Dunwoody, M.J.
Formanek, E. {46, 213}
1973. Idempotents in Noetherian group rings. *Can. J. Math.* **15**, 366–9. {213}
Freudenthal, H.
1931. Über die Enden topologischer Räume und Gruppen. *Math. Zeit.* **33**, 692–713. {71}
1942. Neuaufbau der Endentheorie. *Ann. of Math.* **43**, 261–79. {71}
1944. Über die Enden diskreter Räume und Gruppen, *Comment. Math. Helv.* **17**, 1–38. {71}
Gerardin, P. {45}
Gersten, S.M.
1984. On fixed points of certain automorphisms of free groups. *Proc. London Math. Soc.* **48**, 72–90; Addendum, (1984), **49**, 340–2. {46}
Goldstein, R.Z. and Turner, E.C.
1986. Fixed subgroups of homomorphisms of free groups. *Bull. London Math. Soc.* **18**, 468–9. {46}
Gray, J.
1982. From the history of a simple group. *Math. Intelligencer* **4**, 59–67. {103}
Grushko, I.A.
1940. Über die Basen eines freien Produktes von Gruppen. *Mat. Sb.* **8**, 169–82. (In Russian, with German summary.) {45, 46, 77}
Haken, W.
1961. Theorie der Normal Flächen. *Acta Math.* **105**, 245–375. {248}
Hattori, A.
1965. Rank element of a projective module. *Nagoya Math. J.* **25**, 113–20. {204, 213}
Hempel, J.
1976. *3-manifolds*. Ann. of Math. Studies 86. Princeton University Press. {249, 270}
Higgins, P.J.
1966. Grushko's Theorem. *J. Algebra* **4**, 365–72. {46}
Higman, G., Neumann, B.H. and Neumann, H.
1949. Embedding theorems for groups. *J. London Math. Soc.* **24**, 247–54. {45}
Hilton, P.J. and Wylie, S.
1962. *Homology Theory*. Cambridge University Press. {216, 217, 220}
Hochschild, G.P.
1956. Relative homological algebra. *Trans. Amer. Math. Soc.* **82**, 246–69. {134}
Holt, D.F.
1981. Uncountable locally finite groups have one end. *Bull. London Math. Soc.* **13**, 557–60. {134}
Hopf, H.
1943. Enden offener Raüme und unendliche diskontinuierliche Gruppen. *Comment. Math. Helv.* **16**, 81–100. {71, 134, 213}

Howson, A.G.
1954. On the intersection of finitely generated free groups. *J. London Math. Soc.* **29**, 428–34. {39, 46}
Ihara, Y.
1966. On discrete subgroups of the two by two projective linear group over p-adic fields. *J. Math. Soc. Japan* **18**, 219–35. {46}
Johnson, F.E.A. and Wall, C.T.C.
1972. On groups satisfying Poincaré duality. *Ann. of Math.* **96**, 592–8. {213}
Kaplansky, I.
1969. *Fields and Rings*. Chicago Lecture Notes in Mathematics. University of Chicago Press. {203, 213}
Karrass, A. *see also* Magnus, W.
Karrass, A., Pietrowski, A. and Solitar, D.
1973. Finite and infinite cyclic extensions of free groups. *J. Australian Math. Soc.* **16**, 458–66. {133}
Kerckhoff, S.P.
1983. The Nielsen realization problem. *Ann. of Math.* **117**, 235–65. {166, 214}
Kneser, H.
1929. Geschlossene Flächen in dreidemensionalen Mannigfaltigkeiten. *Jahresbereicht der Deut. Math. Verein.* **38**, 248–60. {271}
Kurosh, A.G.
1937. Zum Zerlegungsproblem der Theorie der freien Produkte. *Rec. Math. Moscow* **2**, 995–1001. (In Russian, with German summary.) {35, 46, 134}
Lewin, J.
1970. On the intersection of augmentation ideals. *J. Algebra* **16**, 519–22. {134}
Linnell, P.A. {134, 213, 214} *see also* Eckmann, B.
1983. On accessibility of groups. *J. Pure and Applied Algebra* **30**, 39–46. {131}
Lyndon, R.C.
1950. Cohomology theory of groups with a single defining relation. *Ann. of Math.* **52**, 650–65. {46, 213}
Lyndon, R.C. and Schupp, P.E.
1977. *Combinatorial group theory*. Springer-Verlag, Berlin. {24}
Macpherson, H.D.
1982. Infinite distance transitive graphs of finite valency. *Combinatorica* **2**, 63–9. {72}
Magnus, W., Karrass, A. and Solitar, D.
1966. *Combinatorial Group Theory*. J. Wiley and Sons, New York.
Massey, W.S.
1967. *Algebraic Topology: An Introduction*. Harcourt, Brace and World, Inc., New York. {221}
Meeks, W.H., Simon, L. and Yau, S.-T.
1982. Embedded minimal surfaces, exotic spheres and manifolds with positive Ricci curvature. *Ann. of Math.* **116**, 621–59. {271}
Meeks, W.H. and Yau, S.-T.
1980. Topology of three-dimensional manifolds and the embedding problems in minimal surface theory. *Ann. of Math.* **112**, 441–85. {271}
1981. The equivariant Dehn's lemma and loop theorem. *Comment. Math. Helv.* **56**, 225–39. {271}
Müller, H. *see also* Eckmann, B.
1981. Decomposition theorems for group pairs. *Math. Zeit.* **176**, 223–46. {213}
Nagao, H.
1959. On GL(2, $K[x]$). *J. Poly. Osaka Univ.* **10**, 117–21. {46}
Nickolas, P.
1985. Intersections of finitely generated free groups. *Bull. Austral. Math. Soc.* **31**, 339–48. {39}
Nielsen, J.
1917. Die isomorphismen der allgemeinen, unendlichen Gruppe mit zwei Erzeugenden. *Math. Ann.* **78**, 385–97. {213}
1921. Om Regning med ikke kommutative Faktoren og dens Anvendelse i Gruppeteorien. *Math. Tidsskrift* **B**, 77–94. {37, 46}
Neumaier, A. *see* Brouwer, A.E.
Neumann, B.H. *see also* Higman, G.
1943. On the number of generators of a free product. *J. London Math. Soc.* **18**, 12–20. {45, 46, 77}
Neumann, H. *see also* Higman, G.
1948. Generalized free products with amalgamated subgroups I. *Amer. J. Math.* **70**, 590–625 {46}
1955. On the intersection of finitely generated free groups. *Publ. Math. Debrecen* **4**, 186–9; *Addendum*, (1957–58), **5**, 128. {39, 46}
Papakyriakopoulos, C.D.
1957. On Dehn's lemma and the asphericity of knots. *Ann. of Math.* **66**, 1–26. {271}
Passman, D.S.
1971. Idempotents in group rings. *Proc. Amer. Math. Soc.* **28**, 371–4. {213}
1977. *The Algebraic Structure of Group Rings*. J. Wiley and Sons, London, {213}
Pietrowski, A. *see* Karrass, A.
Piollet, D. {134}
Reidemeister, K.
1932. *Einführing in die Kombinatorische*

Topologie. Braunschweig. (Reprinted: Chelsea, New York, 1950) {46}
Richards, I.
1963. On the classification of noncompact surfaces. *Trans. Amer. Math. Soc.* **106**, 259–69. {245}
Rota, G.-C.
1986. Book reviews. *Advances in Math.* **59**, 302. {45}
Schreier, O.
1927. Die Untergruppen der freien Gruppen. *Abh. Math. Univ. Hamburg* **5**, 161–83. {37, 46}
Schupp, P.E. *see* Lyndon, R.C.
Scott, G.P. {46, 133}
1974. An embedding theorem for groups with a free subgroup of finite index. *Bull. London Math. Soc.* **6**, 304–6. {133}
Scott, G.P. and Wall, C.T.C.
1979. Topological methods in group theory. In *Homological Group Theory*, pp. 137–203. London Mathematical Society Lecture Notes 36. Cambridge University Press. {46, 134}
Serre, J.-P.
1971. Cohomologie des groupes discrets. In *Prospects in Mathematics*. Ann. of Math. Studies 70, pp. 77–169. Princeton University Press. {114, 170, 172}
1977. *Arbres, Amalgames, SL_2*. Astérisque no. 46, Société Math. de France, Paris. {45, 46, 71, 133, 134}
Simmons, G.F.
1963. *Introduction to Topology and Modern Analysis*. McGraw-Hill, New York. {124}
Simon, L. *see* Meeks, W.H.
Solitar, D. *see* Karras, A.; Magnus, W.
Spanier, E.H.
1966. *Algebraic Topology*. McGraw-Hill, New York. {268}
Specker, E.
1949. Die erste Cohomologiegruppe von Überlagerungen und Homotopieeigenschaften dreidimensionaler Mannigfaltigkeiten. *Comment. Math. Helv.* **23**, 303–33. {71, 134}
1950. Endenverbände von Räume und Gruppen. *Math. Ann.* **122**, 167–74. {71, 134}
Stallings, J.R.
1965. A topological proof of Grushko's theorem on free products. *Math. Zeit.* **90**, 1–8. {46}
1965′. Centerless groups – an algebraic formulation of Gottlieb's theorem. *Topology* **4**, 129–34. {204, 213}
1968. On torsion-free groups with infinitely many ends. *Ann. of Math.* **88**, 312–34. {71, 100, 114, 133, 134, 192}
1971. *Group Theory and Three-dimensional Manifolds*. Yale Mathematical Monographs 4. Yale University Press, New Haven. {71, 100}
Stone, M.H. {124}
Strebel, R. *see also* Bieri, R.
1977. A remark on subgroups of infinite index in Poincaré duality groups. *Comment. Math. Helv.* **52**, 317–24. {213}
Swan, R.G.
1969. Groups of cohomological dimension one. *J. Algebra* **12**, 585–610. {100, 114, 133, 134, 160}
Swarup, G.A.
1977. Relative version of a theorem of Stallings. *J. Pure and Applied Algebra* **11**, 75–82. {213}
Tits, J. {45}
Turner, E.C. *see* Goldstein, R.Z.
Vogt, E. *see* Zieschang, H.
Wagner, D.H.
1957. On free products of groups. *Trans. Amer. Math. Soc.* **84**, 352–78. {46}
Wall, C.T.C. *see also* Johnson, F.E.A.; Scott, G.P.
1967. Poincaré complexes I. *Ann. of Math.* **86**, 213–45. {134, 213}
1971. Pairs of relative cohomological dimension one. *J. Pure and Applied Algebra* **1**, 141–54 {134}
Wilkie, H.C.
1966. On non-Euclidean crystallographic groups. *Math. Zeit.* **91**, 87–102.
Wylie, S. *see* Hilton, P.J.
Yau, S.-T. *see* Meeks, W.H.
Zieschang, H.
1964. Alternierende Produkte in freien Gruppen I, II. *Abh. Math. Univ. Hamburg* **27**, 13–31; **28** (1965), 219–33. {213}
Zieschang, H., Vogt, E. and Coldewey, H.-D.
1980. *Surfaces and Planar Discontinuous Groups*. Lecture Notes in Mathematics, vol. 835. Springer-Verlag, Berlin. {102, 162, 165}

Symbol index

Subject index